ADVANCES IN LIPID RESEARCH

Volume 20

EDITORIAL ADVISORY BOARD

B. Lewis
A. V. Nichols
G. H. Rothblat

G. Schlierf
C. Sirtori
R. W. Wissler

CONTRIBUTORS

Milada Dobiášová
Hyman Engelberg
Kathryn J. Go
W. Hotz

G. K. Khuller
Gerhard M. Kostner
Bhalchandra J. Kudchodkar
J. N. Verma

Don P. Wolf

Advances in Lipid Research

Volume 20

Edited by

Rodolfo Paoletti

Institute of Pharmacology
University of Milan
Milan, Italy

David Kritchevsky

The Wistar Institute
Philadelphia, Pennsylvania

1983

ACADEMIC PRESS

A Subsidiary of Harcourt Brace Jovanovich, Publishers

New York   London
Paris   San Diego   San Francisco   São Paulo   Sydney   Tokyo   Toronto

COPYRIGHT © 1983, BY ACADEMIC PRESS, INC.
ALL RIGHTS RESERVED.
NO PART OF THIS PUBLICATION MAY BE REPRODUCED OR
TRANSMITTED IN ANY FORM OR BY ANY MEANS, ELECTRONIC
OR MECHANICAL, INCLUDING PHOTOCOPY, RECORDING, OR ANY
INFORMATION STORAGE AND RETRIEVAL SYSTEM, WITHOUT
PERMISSION IN WRITING FROM THE PUBLISHER.

ACADEMIC PRESS, INC.
111 Fifth Avenue, New York, New York 10003

United Kingdom Edition published by
ACADEMIC PRESS, INC. (LONDON) LTD.
24/28 Oval Road, London NW1 7DX

LIBRARY OF CONGRESS CATALOG CARD NUMBER: 63-22330

ISBN 0-12-024920-0

PRINTED IN THE UNITED STATES OF AMERICA

83 84 85 86   9 8 7 6 5 4 3 2 1

CONTENTS

CONTRIBUTORS ix
PREFACE xi

Apolipoproteins and Lipoproteins of Human Plasma: Significance in Health and in Disease

Gerhard M. Kostner

I. Introduction 1
II. Classification of Lipoproteins and Nomenclature 2
III. Apolipoproteins of Plasma and Lymph 5
IV. Subfractions of Intact Lipoproteins 10
V. Lipoproteins Not Commonly Found in Normal Plasma 15
VI. Role of Apolipoproteins in Lipid Metabolism 26
VII. Quantification of Apolipoproteins in the Clinical Chemical Laboratory 29
VIII. Apolipoproteins as Risk Indicators for Myocardial Infarction and Atherosclerosis 35
IX. Summary and Conclusions 39
References 41

Relationship of Cholesterol Metabolism to the Metabolism of Plasma Lipoproteins: Perspectives from Methodology

Bhalchandra J. Kudchodkar

I. Introduction 45
II. Body Cholesterol Turnover 47
III. Measurement of the True Body Cholesterol Turnover Rate by Tracer Kinetic Methods 51
IV. Total Body Cholesterol Mass and Individual Pool Sizes 58
V. Is There an Exchange of Plasma Cholesterol with Tissue Cholesterol? 80

VI. Relationship between Plasma Cholesteryl Ester Turnover and LDL ApoB Turnover	86
VII. Quantitative Relationship of Cholesterol Synthesis, Absorption, Esterification, and Catabolism with the Metabolism of Plasma Lipoproteins, Especially the Low-Density Lipoproteins	92
VIII. Significance	99
References	101

Lecithin:Cholesterol Acyltransferase and the Regulation of Endogenous Cholesterol Transport

Milada Dobiášová

I. Introduction	107
II. Reaction Pattern of LCAT	109
III. Sources, Isolation, and Properties of LCAT	117
IV. Methods of LCAT Estimation and Regulation of Its Activity	126
V. Studies on LCAT in Experimental Animals	143
VI. Clinical Studies	151
VII. Closing Remarks	183
References	185

Nicotinic Acid and Its Derivatives: A Short Survey

W. Hotz

I. Introduction	195
II. Pharmacology	196
III. Mechanism of Action	207
IV. Toxicology and Side Effects	210
V. Summary	212
References	213

Heparin and Atherosclerosis

Hyman Engelberg

I. Introduction	219
II. Heparin Actions in Systems Involved in Atherogenesis	221
III. Effects of Heparin at Different Sites	229
IV. Results of Heparin Therapy	245
References	246

Lipids of Actinomycetes

J. N. Verma and G. K. Khuller

I.	Introduction	257
II.	Lipid Composition	258
III.	Subcellular Distribution	275
IV.	Taxonomic Significance	277
V.	Influence of Environmental Factors on Lipid Composition	280
VI.	Metabolism of Lipids	288
VII.	Immunological Properties of Lipids	300
VIII.	Lipids and the Production of Antibiotics	305
IX.	Prospects	307
	References	307

The Role of Sterols in Sperm Capacitation

Kathryn J. Go and Don P. Wolf

I.	Introduction	317
II.	Methodology for the Study of Capacitation	319
III.	Sterols of Mammalian Sperm	320
IV.	Modification of Sperm Lipid Levels during Capacitation	323
V.	Effects of Sterols on Sperm Fertility	325
VI.	Factors Which May Regulate Sperm Sterol Content	326
VII.	Conclusions	328
	References	328

AUTHOR INDEX	331
SUBJECT INDEX	360
CONTENTS OF PREVIOUS VOLUMES	363

CONTRIBUTORS

Numbers in parentheses indicate the pages on which the authors' contributions begin.

MILADA DOBIÁŠOVÁ, *Institute of Nuclear Biology and Radiochemistry, Czechoslovak Academy of Sciences, 142 20 Prague 4, Czechoslovakia* (107)

HYMAN ENGELBERG, *Cedars–Sinai Medical Center, Los Angeles, California* (219)

KATHRYN J. GO,* *Department of Obstetrics and Gynecology, University of Pennsylvania, Philadelphia, Pennsylvania 19104* (317)

W. HOTZ, *Research and Development Division, Merz and Company, GmbH, 6000 Frankfurt-Am-Main, Federal Republic of Germany* (195)

G. K. KHULLER, *Department of Biochemistry, Postgraduate Institute of Medical Education and Research, Chandigarh-160012, India* (257)

GERHARD M. KOSTNER, *Institute of Medical Biochemistry, University of Graz, A-8010 Graz, Austria* (1)

BHALCHANDRA J. KUDCHODKAR, *Department of Biochemistry, Texas College of Osteopathic Medicine, North Texas State University, Denton, Texas 76203* (45)

J. N. VERMA,† *Department of Biochemistry, Postgraduate Institute of Medical Education and Research, Chandigarh-160012, India* (257)

DON P. WOLF,‡ *Department of Obstetrics and Gynecology, University of Pennsylvania, Philadelphia, Pennsylvania 19104* (317)

*Present address: Department of Physiology and Biochemistry, The Medical College of Pennsylvania, Philadelphia, Pennsylvania 19129.

†Present address: Department of Microbiology, School of Medicine, University of Pennsylvania, Philadelphia, Pennsylvania 19104.

‡Present address: Department of Obstetrics and Gynecology, University of Texas Health Science Center at Houston Medical School, Houston, Texas 77030.

PREFACE

This volume of *Advances in Lipid Research* completes the second decade of publication of our review series. We have attempted to blend reviews of established work in the lipid field with articles dealing with emerging areas of lipid research. In this volume we have collected seven essays relating to several aspects of cholesterol metabolism as well as other areas of research to which lipid metabolism is pertinent.

The first article deals with the significance of human plasma apolipoproteins and lipoproteins in health and disease and contains a review of methods of apolipoprotein quantification. The next article relates cholesterol metabolism to plasma lipoprotein metabolism. The third contribution is an extremely thorough review of lecithin:cholesterol acyltransferase (LCAT) and its role in the regulation of endogenous cholesterol transport. The next two articles are concerned with pharmaceutical agents used in treatment of hyerlipidemia and atherosclerosis. The first of these agents is nicotinic acid, which, with its derivatives, has been used for over 25 years. The other substance is heparin, which also has a long history of use in treatment of lipid abnormalities and research. The sixth article deals with the lipids of actinomycetes, an interesting and important class of microorganisms. The final essay describes the role of sterols in sperm capacitation. It describes a regulatory role for sterols that will probably be new to most lipidologists.

RODOLFO PAOLETTI
DAVID KRITCHEVSKY

Apolipoproteins and Lipoproteins of Human Plasma: Significance in Health and in Disease

GERHARD M. KOSTNER

Institute of Medical Biochemistry
University of Graz
Graz, Austria

I.	Introduction	1
II.	Classification of Lipoproteins and Nomenclature	2
	A. Lipoprotein Density Classes	3
	B. Electrophoretic Fractions	3
	C. Lipoprotein Families	4
III.	Apolipoproteins of Plasma and Lymph	5
	A. Major Apolipoproteins	5
	B. Isoforms and Genetic Variants	6
	C. Proteins Associated with Lipids and Lipoproteins	10
IV.	Subfractions of Intact Lipoproteins	10
	A. Subfractionation by Ultracentrifugation	10
	B. Subfractionation by Other Preparative Methods	12
V.	Lipoproteins Not Commonly Found in Normal Plasma	15
	A. Lipoproteins in Abetalipoproteinemia	16
	B. Lipoproteins in Obstructive Jaundice	17
	C. Lipoprotein-a	19
VI.	Role of Apolipoproteins in Lipid Metabolism	26
VII.	Quantification of Apolipoproteins in the Clinical Chemical Laboratory	29
	A. Laurell (Rocket) Electrophoresis	30
	B. Radial Immunodiffusion—The Mancini Technique	31
	C. Nephelometric Quantitation	33
	D. Radioimmunoassay, Enzyme-Linked Immunosorbent Assay, and Fluoroimmunoassay	34
	E. Standards and Control Samples	35
	F. Apolipoprotein Values of Normal Plasma	35
VIII.	Apolipoproteins as Risk Indicators for Myocardial Infarction and Atherosclerosis	35
	The Evaluation of Optimal Discriminators for Atherosclerosis by Multivariant Analysis	37
IX.	Summary and Conclusions	39
	References	41

I. Introduction

According to the definition by Alaupovic (1971), an apolipoprotein is defined as a lipid-binding protein, consisting of a single or of multiple polypeptides, with the capacity to form soluble, polydisperse lipoprotein particles; the latter are considered lipoprotein families. Among the criteria which define a certain apolipoprotein, the following seem to be of importance: (1) distinct chemical, physical, and immunochemical properties; (2) capability to bind or interact with serum lipids; (3) capacity to form discrete lipoprotein particles; and (4) integral importance for the transport or metabolism of lipids. Thus certain enzymes which have been shown to bind quite selectively to lipoproteins, as for example alkaline phosphatase to LpX, are not apolipoproteins since the interaction is not specific and the enzyme is not of importance for the general lipid metabolism. β_2-Glycoprotein I, on the other hand, is included in this article since it is believed to be of relevance for the lipolytic system, although the major part of it is present in the plasma in lipid-free form.

In 1969, when I joined the group of P. Alaupovic, we were concerned with convincing other researchers in this field of the existence of apolipoprotein C (apoC), the third immunochemically distinct entity in the lipid transport system. Today, the situation has changed dramatically. Not only has the number of known distinct units grown exponentionally, but we also have learned a great deal about the existence of isoforms, allotypes, and apolipoproteins present either in small concentrations or occurring only to an appreciable extent in diseased conditions. The nomenclature of these polypeptides, as well as that of lipoproteins in general, has often been handled in a rather arbitrary way and added to the confusion of individuals not directly acquainted with the field. One of the goals of this article, therefore, will be to strictly denominate lipoproteins, apolipoproteins, and the different classes and subclasses in terms of what they in fact represent: density classes, electrophoretic fractions, families or subfamilies, or just single polypeptides or isoforms thereof. In exceptional cases, trivial names such as "proline-rich polypeptide" might not be avoidable.

II. Classification of Lipoproteins and Nomenclature

The lipoproteins and apolipoproteins of various density classes have been treated extensively in review articles which may be consulted for specific questions (Kostner, 1975, 1981; Osborne and Brewer, 1977; Alaupovic, 1982; Assmann, 1982). Lipoproteins consist of different proportions of various lipids in addition to a number of specific polypeptides or apolipoproteins. As a consequence, physicochemical properties of intact lipoproteins vary to a considerable extent. Fractions which may be homogeneous or uniform according to one method may be mixtures of several subunits which can be demonstrated by special procedures only. This microheterogeneity is subject to change under certain metabolic, hyperlipoproteinemic, or dyslipoproteinemic conditions. It thus seems unavoidable that lipoproteins are investigated by a combination of physicochemical and immunochemical procedures.

A. Lipoprotein Density Classes

According to the concentration maxima and minima which occur along a density gradient, lipoproteins are divided into chylomicrons (CYM) with a density of <1 g/ml, very low density lipoproteins (VLDL) with $d < 1.006$ g/ml, intermediate-density lipoproteins (IDL) with $d = 1.006-1.019$ g/ml, low-density lipoproteins (LDL) with $d = 1.019-1.063$ g/ml, high-density lipoproteins (HDL) with $d = 1.063-1.21$ g/ml, and very high density lipoproteins (VHDL) with $d = 1.21-1.25$. Almost all of these density classes have been subfractionated by stepwise ultracentrifugation at increasing densities or in a density gradient giving rise to density subclasses (DeLalla and Gofman, 1954; Laggner et al., 1977b; Patsch et al., 1978). As will be shown later, there exists no single density class which might be considered as immunochemically homogeneous, and distinct entities from any given density fraction can be separated. The most homogeneous fraction in this respect is LDL, which under normal fasting conditions may contain more than 90% of lipoprotein B (LpB).

B. Electrophoretic Fractions

There is a great variety of methods in current use which separate lipoproteins according to charge. Electrophoretic methods on the one hand may be subdivided, whether or not the supporting media cause sieving effects of lipoproteins. On the other hand, isoelectric focusing (IEF) of intact lipoproteins proved to be the most powerful procedure for exhibiting microheterogeneities (Kostner et al., 1969). The most common electrophoretic procedures using paper, cellulose acetate, or agarose gels separate the lipoproteins into α-lipoproteins, β-lipoproteins, and pre-β-lipoproteins. By the use of refined methodology it has been impossible to split all these electrophoretic fractions into three to five bands (Papadopoulos, 1978). Depending on their position in the gel or on the paper strip, names such as pre-β_1-lipoprotein, α_2-lipoprotein, fast- or slow-β-lipoprotein, inter-β-lipoprotein, and many more have been created by numerous authors. Since the electrophoretic migration of subfamilies depends so much on individual systems, I consider electrophoresis reliable only for defining or quantitating major fractions, unless separations are performed by specialists in one and the same laboratory.

Lipoprotein density, as well as electrophoretic, classes are composed of a number of subfractions which have been demonstrated by various procedures. Since apolipoproteins are not covalently bound to lipids and secondary complex formation within the bloodstream may be a common feature of all hydrophobic material, any separation procedure applied may cause artifact formation and may fail to mirror the true *in vivo* situation. However, there are methods which are considered "safe," which possibly disrupt only weak linkages. Among these, ultracentrifugation, precipitation and electrophoresis, and column chromatography in the absence of detergents and dissociating agents have been successfully applied for the purification of intact lipoprotein entities, so-called lipoprotein families.

C. LIPOPROTEIN FAMILIES

Lipoprotein families are characterized by their protein moiety. They behave as homogeneous fractions, metabolically as well as physicochemically. At a density of approximately 1.020 g/ml, some of the lipoprotein families are complexed by neutral lipids, predominantly triglycerides, and fall apart during the action of lipoprotein lipase. Today we know almost 10 distinct lipoprotein families, most of them being found in the HDL class. Table I lists most of these lipoprotein families and relates them to density fractions, on the one hand, and to electrophoretic classes on the other.

The most abundant family in human serum is LpB, the major fraction of LDL. Since more than 90% of LDL consists of LpB, which migrates as β-lipoprotein, the terms LDL, β-lipoprotein, and LpB have frequently been used interchangeably. This might be quite safe under normal conditions but leads to tremendous errors in diseased states. Lipoprotein A is the major lipoprotein of HDL. Lipoprotein C exists in HDL in a free form (Kostner and Alaupovic, 1972) but is complexed with other lipoproteins in VLDL and chylomicrons. Lipoproteins D, E, F, G, and H are considered

Table I
LIPOPROTEIN FAMILIES OF HUMAN PLASMA

Lipoprotein family	Major density class	Electrophoretic mobility
LpA	HDL	α
LpB	LDL (VLDL)	β (pre-β)
LpC	HDL (VLDL)	α (pre-β)
LpD	HDL	α
LpE	HDL (VLDL)	α_2 (pre-β)
LpF, LpG	HDL	α
Lp(a)	HDL_1	pre-β_1

to be minor lipoprotein families as far as concentration is concerned, but they may be of great importance for lipoprotein metabolism.

III. Apolipoproteins of Plasma and Lymph

A. MAJOR APOLIPOPROTEINS

According to the concept of Alaupovic (1982), lipoprotein families consist in their protein moiety of one or several related polypeptides which, as it turned out, probably after they had been detected and denominated, behave very similarly metabolically. At the beginning, apolipoproteins were named simply according to their occurrence in lipoprotein families (Table II). In the case of LpA it was possible to separate nonidentical subunits, which were called apoA-I, apoA-II, and possibly apoA-III, after delipidation (Kostner and Alaupovic, 1971; Kostner, 1974b). The protein moiety of LpB consists of two, probably identical subunits (apoB). There is still some dispute as to whether the protein moiety of LpB is homogeneous. Apolipoprotein B has a molecular weight of 550,000 and is manufactured primarily in the liver. The apoB of intestinal origin has approximately half that molecular weight and has been called B-48, in contrast to the liver apoB which was designated B-100. Information is lacking on how B-100 and B-48 interrelate.

Together with apoA and apoB, the apolipoproteins of the C family are considered as the major apolipoproteins of plasma and lymph of fasting

Table II
APOLIPOPROTEINS OF HUMAN SERUM

Apolipoprotein	Synonym	Molecular weight $\times 10^{-3}$	Major density class
AI		28.5	HDL
AII		17.5	(HDL_2), HDL_3
AIII	D	22	HDL, VHDL
B		550	LDL
CI		7	Chylomicron, VLDL (HDL)
CII		8.5	Chylomicron, VLDL (HDL)
CIII		8.5	Chylomicron, VLDL (HDL)
D		22	HDL_3, VHDL
E		36	VLDL, IDL, HDL_c
F		30	HDL
G		72	VHDL, $d = 1.21$ g/ml bottom
(a)	Sinking pre-β	65	HDL_1

Table III

APODIPOPROTEINS FORMING NO SEPARATE LIPOPROTEIN FAMILY
AND PROTEINS ASSOCIATED WITH LIPIDS AND LIPOPROTEINS

Apolipoprotein	Synonym	Molecular weight $\times 10^{-3}$	Major density class
AIV		46	Chylomicron, $d = 1.21$ g/ml bottom
H	β_2-Glycoprotein I	54	Chylomicron, $d = 1.21$ g/ml bottom
PRP	Proline-rich protein	74	Chylomicron, $d = 1.21$ g/ml bottom
S (SAA)	Serum amyloid antigen	10–15	HDL (VLDL)
D_2		7	HDL
E–AII complex		46	HDL_c

man. Apolipoprotein C is a mixture of three immunochemically distinct units which are of great importance for the action of lipoprotein lipases (LPL). The apolipoproteins which were detected later because of their much lower serum concentration are apoD, -E, -F, -G, and -H. They are listed in Table II together with the major apolipoproteins. Among these proteins I have listed apo-a, the characteristic antigenic determinant of lipoprotein-a [Lp(a)]. Lipoprotein-a was detected by Berg (1963) and was considered at that time as a genetic variant of β-lipoprotein or, better, LpB. Today we know that the a-peptide is inherited as a quantitative genetic trait and can be demonstrated in the plasma of almost every individual. Apolipoprotein-a has the highest molecular weight of all the apolipoproteins detected so far. Table III lists another class of proteins associated with lipids. They may be of equal importance for lipid metabolism or transport, yet separate lipoprotein families of them have not been defined until now.

B. Isoforms and Genetic Variants

Apolipoproteins purified to homogeneity by electrophoretic criteria have been shown in the past to be heterogeneous by isoelectric focusing. Since in many cases pooled human plasma was used as starting material, there was some confusion as to whether the separated bands represented isoforms comparable to isoenzymes, or whether genetic variants might have been separated. As it turned out, both possibilities do occur. In a general fashion they might be called isoelectric subspecies. With respect to this it must be mentioned that even minute structural alterations or aggregations which occur during purification cause a tremendous heterogeneity in isoelectric focusing experiments. This can be caused by carbamylation in the presence of urea, interaction of NH_2 groups with aldehydes (which are created by oxidation of unsaturated lipids), deamidation, or proteolysis. Thus whenever such isoforms are to be demonstrated, a most gentle treatment of the fractions is mandatory. The most prominent isoelectric subspecies as well as genetic variants of plasma apolipoproteins are listed in Table IV.

1. Subspecies of ApoA

According to Zannis et al. (1980), six different isoelectric subspecies of apoA can be demonstrated by a combination of sodium dodecyl sulfate–polyacrylamide gel electrophoresis (SDS–PAGE) and IEF. Some of them are biosynthesized primarily in the intestine, and others are synthesized in the liver. Apolipoprotein AI_4 is the most abundant in fasting plasma, followed by apoA-I_5. The other isoforms occur in rather small amounts only.

In addition to these isoforms, a variety of genetic variants have been described in the literature. One of them might be the apoA-I of Tangier patients exhibiting a polypeptide with altered lipid-binding properties (Assmann, 1979). Another example demonstrating that structural alterations of apolipoproteins may lead to dyslipoproteinemias is the "Milano" variant described by Franceschini et al. (1980). This apoA-I, in contrast to

Table IV
ISOELECTRIC SUBSPECIES AND GENETIC VARIANTS
OF HUMAN PLASMA APOLIPOPROTEINS

Apolipoprotein	Isoprotein (partially deriving from different organs)	Genetic variants (allotypes)
AI	AI_1-AI_6	Tangier, Milano, Marburg, Giessen
B	B-100, B-74, B-48, B-26	Ag: a_1, c, d, g, h, i, t, x, y, z
CI	CI_1, CI_2	
CII		CII^0, CII^1
CIII	$CIII_0-CIII_3$	
E	E_1-E_4 (sialylated forms)	E^2, E^3, E^4
AIV		AIV^1, AIV^2
H	H_1-H_4	

normal apoA-I, contains one molecule of $\frac{1}{2}$ Cys per monomer unit, giving rise to the formation of dimers linked by disulfide bridges. Due to these modifications, lipoprotein metabolism is altered and serum apoA-I concentrations are drastically reduced. For the sake of completeness it should be remembered that subspecies of apoA-I have been separated preparatively by DEAE column chromatography (Shore and Shore, 1968; Kostner, 1981). Whether these subfractions correspond to the isoelectric species remains to be determined.

ApoA-II usually gives only one band upon IEF if freshly isolated. There is, on the other hand, an appreciable portion of apoA-II monomer present in HDL, comprising some 10% of total apoA-II. Factors influencing this apoA-II monomer:dimer ratio are unknown. Apolipoprotein AII monomer has also been found to be linked with apoE by disulfide bridges (Weisgraber *et al.*, 1978). This "bastard apolipoprotein" occurs primarily in a subclass of HDL called HDL_C .

2. Subspecies of ApoB

There have been numerous reports in the literature that apoB might consist of nonidentical subunits of various molecular weights. Since most of these reports are more or less contradictory, I will restrict myself here to a description of the more generally accepted apoB subspecies. One major puzzle in studying the distribution of this peptide is the fact that during separation and purification, proteolysis is a general feature. This might be caused by proteolytic enzymes of endogenous or exogenous origin, or even by chemical cleavage of the protein.

Because of the high molecular weight and instability after delipidation of apoB, IEF is not generally applicable. There have been a number of genetic variants described which were detected by xenoantibodies of multiply transfused individuals (for a review see Kostner, 1976). These allotypes of apoB are poorly characterized and there is no information on the nature of the structural variations. According to Bütler *et al.* (1974), five closely linked loci with condominant alleles give rise to the formation of 10 so-called "Ag factors" which were called $Ag-a_1$, -c, -d, -g, -h, -i, -t, -x, -y, and -z. The significance of these Ag factors for possible causes of dyslipoproteinemias or atherogenesis is completely unknown.

Isoforms of apoB have been best characterized by Kane *et al.* (1980). According to their results, the majority of apoB found in fasting plasma consists of B-100. B-100 is of hepatic origin and thus found in VLDL, LDL, and also Lp(a) (Kostner *et al.*, unpublished). Two complementary fragments of this protein, called B-74 and B-26 on the basis of their molecular weights relative to B-100, are believed to be also formed in the liver.

Another protein found as an integral fraction of apochylomicron was called B-48 because its molecular weight was only 48% of that of B-100. B-48 is biosynthesized in the intestine and may accumulate in the plasma if chylomicron or chylomicron-remnant removal is impaired.

3. Subspecies of ApoC

Apolipoprotein CI separates into two major fractions by IEF. The nature of this heterogeneity has not been investigated so far. Apolipoprotein CII of normals is homogeneous upon IEF. There has, however, been a report in the literature of a genetic variant of apoC-II which was detected by IEF (Havel et al., 1979). The presence of this variant causes dyslipoproteinemia. Apolipoprotein CIII occurs in normal plasma with different amounts of neuraminic acid. According to the number of neuraminic acid molecules per apoC-III unit, we distinguish between apoC-III$_0$, apoC-III$_1$, apoC-III$_2$, and apoC-III$_3$. Since this sugar strongly influences the charge of the polypeptide, apoC-III subspecies can easily be separated by simple electrophoresis in polyacrylamide gels. In a mixture with other apolipoproteins there is some overlap, primarily of apoC-III with apoC-II, and thus IEF should be applied for clearcut separations. These isoelectric subspecies of apoC-III are isoproteins and not genetic variants (allotypes). Their physiological significance remains to be determined.

4. Isoelectric Subspecies of ApoE

The microheterogeneity of apoE has first been demonstrated by Utermann (1975) by IEF. Because apoEs are sialylated to a variable degree, influencing their isoelectric points, these subspecies have been designated as isoproteins. The apoE subspecies of much greater importance, however, are the genetic variants. According to one report (Zannis et al., 1982), three genetic variants, E2, E3, and E4, are known. Differences between these allotypes have been demonstrated in the primary structure, where one or two molecules of ½ Cys are substituted for Arg. In heterozygotes two genetic variants coexist which may be sialylated to a variable extent, giving rise to a sometimes complex IEF pattern. Treatment of the sample with neuraminidase may help to clearly type a given individual.

5. Subspecies of Other Apolipoproteins

There were additional reports in the literature of isoelectric heterogeneity of apoA-IV, which is of genetic origin (Assmann, 1982). Apolipoprotein H (β_2-glycoprotein I), on the other hand, exhibits at least four isoproteins upon IEF (Polz et al., 1981). The existence of subspecies of other apolipoproteins is currently under investigation.

C. Proteins Associated with Lipids and Lipoproteins

There are reports in the literature of "apolipoproteins" which accumulate mainly in diseased conditions. One class of these is certainly the so-called serum amyloid antigen (SAA or apoS) proteins which were detected in HDL of glucose-infused patients (Malmendier et al., 1979). Serum amyloid antigen proteins are a mixture of 5-10 polypeptides, with molecular weights ranging from 10,000 to 15,000, which are immunochemically related to amyloid. Because of their hydrophobic nature they accumulate primarily in HDL upon entry into the bloodstream. Since a significance for lipid metabolism has not been demonstrated so far, SAA should probably not be considered as apolipoproteins according to the criteria given earlier. Another polypeptide of low molecular weight has been described by Lim et al. (1976) and designated as D_2-peptide. It has a molecular weight of 7000 and its concentration is increased in the serum of abetalipoproteinemic individuals. The true nature of this apolipoprotein remains to be determined before it can be integrated into the current classification system. The term "D_2" refers to its elution from DEAE columns at position 2, and thus "D_2" has nothing to do with "apoD" (or apoA-III) according to the ABC nomenclature.

IV. Subfractions of Intact Lipoproteins

A. Subfractionation by Ultracentrifugation

Lipoproteins must be considered as a continuous spectrum of lipid:protein associates with hydrated densities ranging from less than 1 to approximately 1.25 g/ml. In fasting normal human plasma, there are peak concentrations of individual classes at distinct densities. Under these conditions a stepwise flotation at increasing densities may lead to homogeneous fractions in a wider sense. Postprandially, under dyslipoproteinemic conditions, and in plasma of species other than man, the lipoprotein density profile may change completely, making separation in a density gradient necessary. Very low density lipoproteins have been separated into five subfractions by stepwise flotation (Gustafson et al., 1965), LDL into three to five subfractions (Lee and Alaupovic, 1970), and HDL into three major fractions (Anderson et al., 1977).

Isopycnic separation of lipoproteins from whole serum in a density gradient has been described among others by Redgrave et al. (1975) and by Chapman et al. (1981). This method has also been successfully applied for subfractionation of isolated LDL (Lee and Downs, 1982; Kraus and Burke, 1982), of HDL (Cheung and Albers, 1979; Tall et al., 1982), and of all major lipoprotein density classes (Lindgren et al., 1972). An interesting variant of this method has been published by Chung et al. (1980); a separation of all major density classes is achieved within 1–2.5 hours in a density gradient using a vertical rotor.

Subfractionation of Plasma Lipoproteins by Rate Zonal Ultracentrifugation

This method was first applied for the separation of lipoproteins by Wilcox and Heimberg (1968). We have modified this procedure in such a way that baseline separation of HDL_2 from HDL_3 is achieved by using whole serum as starting material (Patsch et al., 1974). This method proved also to be of value for subfractionation of isolated VLDL (Patsch et al., 1978), LDL (Patsch et al., 1975), and HDL (Laggner et al., 1977b). In order to achieve an optimal separation of VLDL, IDL, and LDL, on the one hand, and of HDL_2 from HDL_3, on the other, the zonal centrifugation method was designed as a two-step procedure. In the first step, VLDL together with LDL are separated from HDL_2 and HDL_3; in the second run a subfractionation of lipoproteins with $d < 1.063$ g/ml is achieved. We have used in our experiments primarily Beckman equipment and zonal rotors Ti-14, Ti-15, and Z-60 Ti, although ultracentrifuges and rotors from other companies probably work equally well.

The separation of the main lipoprotein density fractions by this method is, briefly, as follows: A nonlinear gradient ranging from 1.0 to 1.4 g/ml is pumped into the Ti-14 rotor. Normolipemic serum (10–15 ml), adjusted to $d = 1.40$ g/ml by the addition of solid NaBr, is mixed with 10 ml of NaBr solution ($d = 1.4$ g/ml) and injected from the bottom into the Ti-14 rotor at a speed of 3500 rpm. The rotor is then accelerated to 41,000 rpm (approximately $120,000g$). Separation proceeds for 24 hours at 15°C. The rotor content is now displaced by pumping $d = 1.4$ g/ml NaBr solution into the rotor, and the effluent is monitored photometrically at 280 nm. In the second step the combined VLDL + LDL fraction is separated in a linear gradient from $d = 1.00$–1.30 g/ml at 42,000 rpm and 15°C for 140 minutes.

Due to the relatively high gradient volume, lipoproteins elute in their "purest" form, making any washing unnecessary. Because of this and the relatively short exposure time to high salt concentrations and high g forces, we believe that the peptide distribution of these fractions comes close to that observed in native plasma. The relative content of individual peptides in zonal rotor eluates is shown in Fig. 1. The particular advantage of rate zonal ultracentrifugation lies in the fact that lipoprotein separation becomes immediately evident from densitometric evaluation of the eluate, allowing one to cut any number of fractions at any peak maximum, minimum, or shoulder. By calibrating the system, zonal rotors can replace the analytical ultracentrifuge for many types of problems. Plasma samples ranging from as small as 0.5–1 ml to as large as up to 50 ml have been successfully separated. Since animal plasma as well as plasma from dyslipoproteinemic individuals may contain grossly altered lipoprotein patterns with respect to density distribution, rate zonal ultracentrifugation was also successfully applied for the abnormal lipoprotein in Type III individuals (Patsch et al., 1975) and for LpX (Patsch et al., 1977), in addition to many others.

B. Subfractionation by Other Preparative Methods

Because of the complexity of human plasma lipoprotein distribution, one single separation method will never be capable in resolving all the different subspecies or families. Naturally, the higher the amount and kind of separation steps, the greater the danger of creating artifacts. Nevertheless, there are numerous reports in the literature in which methods such as isotachophoresis (Bittolo Bon et al., 1981), gradient gel electrophoresis (Blanche et al., 1981), immune-specific adsorbers (Kostner and Holasek, 1970), lectins (Suenram et al., 1979), and many others have been applied to resolve the whole lipoprotein spectrum. One of the most powerful methods for separating intact lipoproteins, column chromatography over hydroxyapatite, was first used in our laboratories (Kostner and Alaupovic, 1972; Kostner and Holasek, 1977).

The dominant principles in column chromatographic separation techniques include steric exclusion, ion exchange, and adsorption. All three proved to be useful for the separation of plasma lipoproteins. They have been applied for special purposes to further subfractionate lipoprotein density fractions obtained by ultracentrifugation.

1. Steric-Exclusion Chromatography

In principle, this method can be applied for the separation of lipoproteins from whole serum. Since, however, some lipoproteins coelute with serum proteins, purified density fractions should be applied. One example of separating lipoproteins by column chromatography over BioGel A-5m is given in Fig. 2. In the first step, phosphotungstate precipitation of

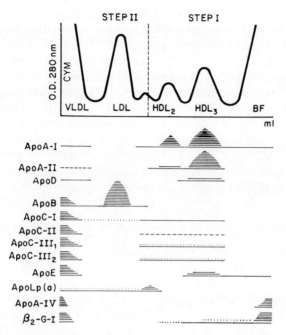

FIG. 1. Apolipoprotein distribution in lipoprotein density fractions obtained by rate zonal ultracentrifugation. The lipoproteins were separated in a Ti-14 zonal rotor in a nonlinear gradient by a two-step procedure, and the apolipoproteins were quantitated immunochemically. CYM, chylomicrons; BF, bottom fraction. [Adapted from Kostner (1975).]

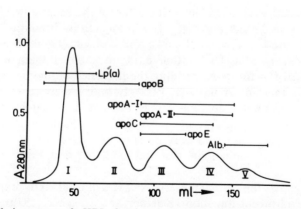

FIG. 2. Elution pattern of a HDL_1 fraction obtained by sodium phosphotungstate precipitation and ultracentrifugation over a BioGel A-5m column. Alb, albumin. [From Kostner and Laggner (1979).]

high-molecular-weight lipoproteins was performed according to Kostner and Alaupovic (1972). The precipitate was solubilized by dialysis against Tris-HCl buffer and ultracentrifuged at $d = 1.06$ g/ml in NaCl. Under these conditions, VLDL + LDL float, whereas some coprecipitated "HDL" sediment, in addition to Lp(a) and small amounts of LpB with hydrated density characteristics of HDL (Kostner, 1972). All these lipoproteins can be resolved into five peaks over BioGel A-5m. This is not only the simplest way to obtain large amounts of purified Lp(a), but it is also the simplest way to accumulate minor lipoproteins from the HDL_1 class which are rich in LpC and LpE. Another example of steric-exclusion chromatography is the separation of intact, freshly isolated or aged, whole HDL over Sephacryl S-300 (Kostner and Laggner, 1979).

2. Ion-Exchange Column Chromatography

This method gives no baseline separation of HDL subfractions and an overlap of some six peaks if total HDL free of apoB is applied to DEAE–Sephacel. From the pattern of peaks 1–4 in PAGE it becomes evident that the major fraction of HDL represents LpA. In addition there were subfractions containing apoA + apoC-II, and fractions with apoA + apoC-III polypeptides. Whether these classes in fact exist in native plasma or are artifacts created during separation remains to be determined. In this respect it is of interest to note that similar subfractions have been obtained by a different procedure, one using hydroxyapatite.

3. Hydroxyapatite Column Chromatography

Hydroxyapatite (HA) is a powerful agent for the separation of proteins, nucleic acids, and other polymers. It was applied for lipoprotein subfractionation for the first time by Roelcke and Weicker (1969). These authors succeeded in separating LDL into five fractions, one of them being Lp(a). We have used HA for demonstrating the presence of lipoprotein families of HDL (Kostner and Alaupovic, 1972). In later experiments, HDL_2 and HDL_3 were separated individually by HA column chromatography, giving rise to five and six subfractions, respectively (Kostner and Holasek, 1977). This separation could only be achieved if self-prepared HA was used, because the commercial material exhibited flow rates which were too slow.

Figure 3 shows the elution pattern from HA of total HDL prepurified by preparative ultracentrifugation (Kostner and Laggner, 1979). Although the chemical and physicochemical properties of all eight subfractions were significantly different, they were all within the limits characteristic

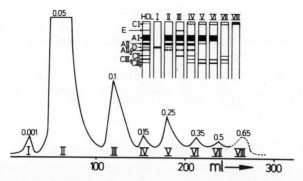

FIG. 3. Hydroxyapatite column chromatography of total HDL purified by ultracentrifugation (numbers above peaks, molarities of the elution buffer). The insert shows the pattern in 10% polyacrylamide gels containing 8 M urea (Roman numerals, number of fractions). [From Kostner and Laggner (1979).]

for HDL. From the PAGE pattern in urea-containing gels it becomes evident that four lipoprotein families could be obtained in a pure form: LpD (peak I), LpA (peak II), LpC (peak VII), and LpB from HDL (peak VIII). Peak III contained in addition to LpA some apoE and apoC-II, and peaks IV, V, and VI contained a mixture of apoA and apoC peptides in various proportions. If HDL are eluted from HA columns not stepwise but rather with a sodium phosphate gradient, we could divide the eluate into many more fractions differing in their protein moiety, as judged from the PAGE pattern. Whether all these forms truly exist *in vivo* and are of physiological significance remains to be investigated. It should, however, be remembered here that the circulating lipoproteins represent a mixture of particles of hepatic and intestinal origin which undergo a constant degradation by lipolytic enzymes and transferases. In addition, exchange and net transfer of all kinds of lipids and apolipoproteins is a common feature. All of these processes necessarily lead to a complex mixture of lipoproteins. I consider column chromatography on HA to be one of the most powerful methods to separate and thus demonstrate such subfamilies. Obviously before using such methods the actions of enzymes, especially of lecithin : cholesterol acyltransferase (LCAT), need to be blocked. We have achieved this by adding 5 mM sodium iodoacetate to the starting plasma.

V. Lipoproteins Not Commonly Found in Normal Plasma

Under any circumstances when lipid metabolism is disturbed, an alteration of the lipoprotein profile becomes apparent. Such disturbances may be caused genetically or may be of a secondary nature, that is, caused by various diseases. It would be beyond the scope of this article even to list in tabulary form all of the described primarily and secondary dyslipoproteinemias and explore the changes in the lipoprotein profiles brought about by them. Thus I will restrict myself here to the description of "abnormal" lipoproteins which have been investigated in my laboratory only.

A. Lipoproteins in Abetalipoproteinemia

Abetalipoproteinemia (ABL) is a rare inborn error of metabolism. In this disease apoB-containing lipoproteins are lacking in the plasma and chyle. Abetalipoproteinemic individuals can neither absorb nor secrete triglycerides (TG) from the liver or intestine, and thus their plasma triglyceride values are virtually zero (for a review see Kostner, 1975). We have investigated three affected siblings from heterozygous parents which were homozygous for this disease (Kostner et al., 1976b). The following lipoprotein abnormalities were observed. A lack of VLDL and LDL, and diminished concentrations of HDL. Most of the HDL consisted of the HDL_2 subspecies. Analytical isoelectric focusing of these sera revealed the presence of four lipoprotein bands, as opposed to the more than eight bands observed in normal plasma (Fig. 4). The apolipoproteins B-100, B-48, and $CIII_1$ were completely absent. Although the mother of the patients was strongly Lp(a) positive, all three children lacked completely a lipoprotein containing the a-protein, or the a-protein in its lipid-free form. From this observation it must be concluded that the metabolism of apo-a is strongly related to that of LpB, and that the occurrence of apo-a in the plasma requires the presence of LpB. We have infused into these individuals intravenously an artificial triglyceride emulsion (Intralipid) and followed the redistribution of apolipoproteins between HDL and VLDL-LDL within 24 hours (Kostner et al., 1976b). There was a remarkable shift of all apoA and apoC proteins to the $d < 1.063$ g/ml fraction. After 24 hours the starting situation was reached again with only trace amounts of apoA-I in the $d < 1.063$ g/ml fraction. The most interesting observation, however, was that apoC-III_1, not present in the starting plasma, occurred in the triglyceride-rich fraction, but not in HDL during triglyceride clearance. Whether this apoC-III_1 resulted from apoC-III_2, due to the action of a neuraminidase, or was newly biosynthesized can not be answered at this time and remains an interesting question. During the clearance of triglyceride we also observed a threefold increase of LCAT activity as measured by the Stokke and Norum procedure (1971).

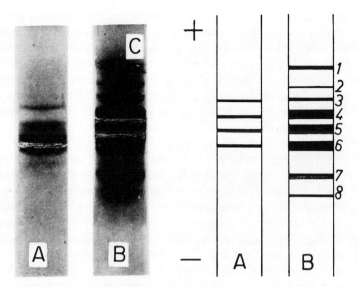

FIG. 4. Isoelectric focusing pattern of (A) intact lipoproteins from plasma of a patient with abetalipoproteinemia, and (B) plasma of a normal control. Lipoproteins were prestained with Sudan black. Bands 1–3 are LpC subclasses, 4–6 are LpA subclasses, and 7 and 8 are LpB subclasses. [From Kostner et al. (1974a).]

B. Lipoproteins in Obstructive Jaundice

Most of the work on abnormal lipoproteins occurring in the plasma of jaundiced individuals was published by Seidel and his collaborators (for a review see Seidel, 1977). He called the abnormal lipoprotein of LDL LpX. Under conditions of obstructive liver disease, however, there are gross changes of all serum lipoprotein density classes. We have separated, for example, the lipoproteins of LDL by HA column chromatography and demonstrated the presence of at least three abnormal fractions: LpX, LpY, and LpB_c (Kostner et al., 1976a). Lipoprotein X contained all apoC proteins in almost equal proportions, LpY contained in addition to apoC large amounts of apoB and was rich in triglycerides as compared to normal LpB. Lipoprotein B_c (cholestatic LpB) differed from normal LpB by the presence of unusually high amounts of apoC proteins. In LpX not all apoproteins were equally exposed to the surface, as revealed by immunochemical methods. Intact LpX gave an immunochemical reaction only with antibodies to apoC-I and apoD. After delipidation, a positive reaction was also obtained with antibodies against apoC-II, apoC-III, apoA-II, albumin, and immunoglobulins G and A.

Lipoprotein X was structurally completely different from spherical lipoproteins of normal human serum. By negative-stain electron microscopy, flattened disks were observed which formed stacked rouleaux, caused by the presence of phosphotungstate (Fig. 5). In subsequent experiments, small-angle X-ray scattering studies were performed with homogeneous LpX preparations (Laggner *et al.*, 1977a). The electron-density profile across the bilayer plane of LpX, as obtained from the scattering curves, proved that a lipid bilayer is the underlying structural principle. In addition it was revealed that apolipoproteins are partly bound within the polar head group arrays and partly occluded in soluble form in the interior of the vesicle. The thickness of the lipid bilayer was found to be 5.1 nm and the diameter of the whole vesicle was found to be greater than 30 nm. These results were confirmed by nuclear magnetic resonance measurements (Hauser *et al.*, 1977). Here it was found that the packing of the bilayer was tighter and the segmental motion of both the polar head groups and the fatty acid chains were significantly reduced as compared to pure phosphatidylcholine bilayers. We also demonstrated an asymmetric distribution of individual phospholipids between the two layers. Sphingomyelin and lysophosphatidylcholine were preferentially

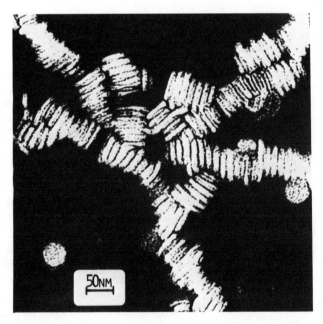

FIG. 5. Negative-stain electron microscopy of LpX, the abnormal lipoprotein of patients suffering from cholestasis. The flattened disks are artifacts created by the staining technique.

located on the inner layer, while phosphatidylcholine was present on the outer layer, thus resembling the asymmetry of other membranes.

The higher the content of LpX in a given serum, the lower, usually, the apoC content in the HDL. Under certain conditions an LpE-containing lipoprotein was found in HDL_2 which sometimes comprised up to 50% of the total HDL mass.

C. LIPOPROTEIN-a

Lipoprotein-a was for a long time believed to represent a genetic variant of LpB (for a review see Kostner, 1976). Harvie and Schultz (1970) finally demonstrated by a radioimmunoassay technique that Lp(a) can be found in almost every human. We have investigated this lipoprotein with respect to structure, metabolism, and epidemiology; Lp(a) can be demonstrated in the serum by electrophoresis and by immunochemical techniques (Fig. 6). Some of the chemical and physicochemical characteristics of Lp(a) are summarized in Table V and compared with those of LpB. Because of a twofold higher particles mass, Lp(a) migrates slower in PAGE and appears as a distinct band in 3.75% gels. Lipoprotein-a, on the other hand, exhibits a greater sialic acid content compared to LpB and thus migrates faster in cellulose acetate and agarose gels. According to its position in electropherograms, Lp(a) is also called by some authors "pre-β_1-lipoprotein."

Table V
COMPOSITION AND PROPERTIES OF Lp(a) AND LpB

Component	Lp(a)	LpB	Parameter	Lp(a)	LpB
Lipoprotein	(% by weight)		Hydrated density	1.085 g/ml	1.033 g/ml
Protein	26	22			
Carbohydrates	8	2			
Lipids	66	76	$F_{1.20}$	−24 S	−50 S
Lipids	(% by weight)		Diameter	25 nm	21 nm
Phospholipids	30	31			
Free cholesterol	17	15			
Cholesteryl ester	48	46	MW × 10^{-6}	5.5	2.4
Triglycerides	5	8			
Sugars	(μg/mg protein)		Electrophoretic mobility	pre-β_1	β
Hexose	150	56			
Hexose–NH_2	85	30			
Sialic acid	66	11	Isoelectric point	4.5	5.5

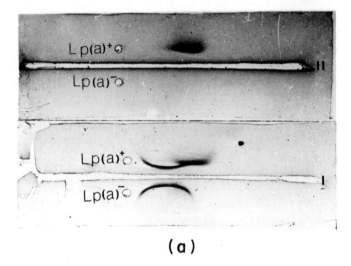

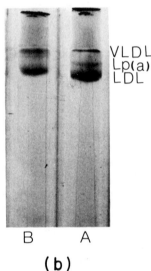

Fig. 6. Immunoelectrophoresis and polyacrylamide gel electrophoresis of (a) Lp(a)-positive and (b) Lp(a)-negative plasma. Trough I contains antibodies against Lp(a) and LpB, and trough II contains antibodies against Lp(a) only. The polyacrylamide gels are 3.5%, and the lipoproteins were prestained with Sudan black.

Unlike all other apoB-containing lipoproteins, Lp(a) sediments at $d = 1.063$ g/ml and thus must be considered as a HDL. In this fraction Lp(a) is the only lipoprotein with a molecular weight exceeding 10^6. With precipitating agents, on the other hand, such as heparin–Mn^{2+}, dextran sulfate–Mg^{2+}, or sodium phosphotungstate–Mg^{2+}, Lp(a) is removed together with LDL and VLDL. This explains in part the discrepancies observed in quantitating HDL cholesterol first by ultracentrifugation and second by precipitation methods. The purification of Lp(a) is outlined in Section IV,B.

1. Polypeptide Content of Lp(a)

The nature of the protein moiety of Lp(a) was for a long time a matter of dispute. In addition to apoB and the a-polypeptide, albumin, apoC, and apoA-III (apoD) have been demonstrated by various investigators. If Lp(a) is isolated from fresh plasma in the quickest possible way as outlined above, it only contains apoB (approximately 90%) and apo-a (10%). Lipoprotein-a is very unstable and precipitates spontaneously upon storage, a process which is partly reversible by adding neutral salts. During aggregation, Lp(a) may adsorb apolipoproteins and serum proteins unspecifically, which may explain the findings mentioned previously.

The a-protein, synonymously called "a-peptide" in this article, has the highest molecular weight of all the apolipoproteins found so far in human plasma. After delipidation of intact Lp(a), there is only one band visible in 3% SDS–PAGE, representing a complex of apoB and a-protein. Upon treatment with reducing agents, this band splits into the two fractions with apparent molecular weights of 550,000 for apoB and more than 650,000 for apo-a. Lipoprotein-a not only self-associates spontaneously, but it also adsorbs all kinds of serum proteins and enzymes. During isolation of Lp(a), proteases frequently are copurified, leading to a degradation of the a-peptide. This gives rise to the appearance of multiple bands in PAGE, giving the composition of this apolipoprotein a complex appearance (Jürgens *et al.*, 1977).

2. Metabolism of Lp(a)

Because apoB is the major protein in Lp(a), one may assume that its metabolism proceeds via routes of other apoB-containing lipoproteins. We have performed a series of experiments in man aimed at elucidating the relation of Lp(a) metabolism to LpB (Krempler *et al.*, 1978, 1979, 1980). In one experiment, autologous, ^{125}I-labeled VLDL was injected intravenously into strongly or weakly Lp(a)-positive donors and the specific radioactivity time course was followed for 5 days. Figure 7 shows the result of a representative experiment. The radioactivity decay curve of VLDL crossed the activity curve of LDL before the latter reached its maximum, confirming the precursor–product relationship of these two

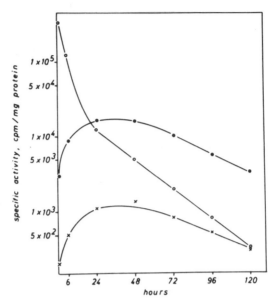

FIG. 7. Die-away curve of the specific radioactivity of apoB in VLDL (○), LDL (●), and Lp(a) (×). ^{125}I-Labeled autologous VLDL was injected intravenously and the specific radioactivity was determined after separation of lipoproteins and apolipoproteins. [From Krempler et al. (1979).]

lipoproteins. There were only trace amounts of activity in the Lp(a) fraction (logarithmic scale!) which was due to contamination, and no precursor–product relationship could be demonstrated in any patient. Other individuals who continued fasting for a prolonged time did not show any reduction of their plasma Lp(a) levels. From these experiments, together with other observations to be mentioned later, we concluded that Lp(a) does not have a TG-rich lipoprotein as a precursor, but rather is biosynthesized in the liver and secreted most probably as such, that is, in the form that is found in circulation.

In patients with liver cirrhosis or obstructive jaundice, the plasma Lp(a) levels are drastically reduced, and many of those individuals exhibit Lp(a) levels of virtually zero (Marth et al., 1982). Heavy alcohol drinkers also show a significant reduction of plasma Lp(a) levels, even at a time when all other clinical tests for liver disease are still within normal limits. Alcohol seems to be the only agent known thus far that is capable of reducing Lp(a) serum concentrations. In our studies we also found that prolonged fasting, cholesterol- and lipid-poor diet, and lipid-lowering drugs such as cholestyramin or clofibrate analogs have no influence on plasma Lp(a)

Table VI
METABOLIC PARAMETERS OF Lp(a) AND LpB
OF NORMOLIPEMIC VOLUNTEERS

Parameter	Lp(a)[a]	LpB[a]
$t_{1/2}$ (days)	3.9 ± 0.8	3.6 ± 0.6
Fractional catabolic rate (day^{-1})	0.26 ± 0.06	0.38 ± 0.08
Synthesis rate (mg/kg/day)	4.6 ± 3.6	60 ± 9
Intravascular pool (%)	76 ± 8	64 ± 9

[a] Values represent mean ± standard deviation of 14 individuals studied.

concentrations, even if VLDL and/or LDL are drastically reduced (Vessby et al., 1982). The metabolic parameters of Lp(a) were studied by injecting radiolabeled Lp(a) intravenously into some 20 volunteers. In previous experiments it was ascertained that more than 95% of the radiolabel resided in the protein moiety, and upon mixing with serum there was no redistribution of label between other lipoprotein classes. On the basis of this, a peptide exchange could be excluded. Table VI lists the obtained metabolic parameters of Lp(a) in comparison to those of LpB (LDL). In summary, it was found that their half-lives are similar, though that of Lp(a) is somewhat longer than that of LDL. The percentage intravascular pool of Lp(a) is greater than that of LDL. The serum concentration of Lp(a) highly significantly correlated with the rate of synthesis but not at all with the fractional catabolic rate.

3. Binding of Lp(a) to Specific Cell Surface Receptors

Since receptor physiology has gained access to lipoprotein laboratories, there has been an overwhelming number of papers dealing with the specific binding of lipoproteins to surface receptors in general. Apolipoprotein B was the first specifically bound lipoprotein to be detected (Brown et al., 1976) and is still under investigation. Since Lp(a) contains apoB as a major protein it can be anticipated that it behaves similarly if not identically to LpB (or LDL) in this respect. On the other hand, it has been observed that desialylated LDL exhibit a higher affinity to the LpB receptor. Since Lp(a) contains a sixfold higher concentration of sialic acid as compared to LDL, quantitative differences may be expected. There is, however, the possibility that the specific a-antigen may either affect the binding constant or even be recognized by separate receptors of different

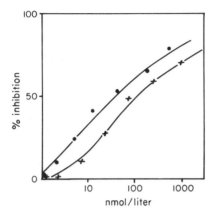

FIG. 8. Inhibition of binding of ^{125}I-labeled Lp(a) to cultured human fibroblasts by the addition of excess Lp(a) (●) and LDL (×) in increasing concentrations.

specificity. In order to study this possibility, cultured human fibroblasts from apparently healthy male and female individuals have been incubated with LDL, Lp(a), and apoB-free Lp(a) (Krempler et al., 1983). Figure 8 shows a characteristic displacement curve in which a constant amount of ^{125}I-labeled LDL was incubated with cultured human fibroblasts in the presence of increasing amounts of Lp(a), the vice versa. It was found that Lp(a) could replace LpB and that LpB did replace Lp(a) to a similar degree. The mean binding constant of Lp(a) was 1×10^{-8}, and that of LpB was 7×10^{-9}, some 30% lower. Purified apo-a peptide was than recombined with LDL lipids and incubated with fibroblasts. In these cases there was no specific binding observed at all. Fibroblasts from one familial homozygous hypercholesterolemic individual exhibited binding neither of LpB nor of Lp(a). In another set of experiments we could also demonstrate that Lp(a) is not only bound to fibroblasts, but is internalized into them as well; furthermore, Lp(a) reduced the activity of HMG-CoA reductase to a similar degree as did LDL.

From these experiments it can be concluded that the *in vivo* metabolism of Lp(a) proceeds via similar routes as described for LpB. Lipoprotein-a is partly catabolized after cell surface binding via specific receptors. This binding is mediated through the apoB protein which is present on Lp(a). The a-antigen possesses no recognition sites in the cell system studied, that is, fibroblasts from several normal and one familial homozygous patient. From the lower binding constant one may expect that the catabolic rate *in vivo* of Lp(a) might be reduced compared to LpB. This in fact has been confirmed by injecting ^{125}I-labeled Lp(a) and ^{131}I-labeled LpB simultaneously into several individuals (Krempler et al., 1983).

4. The Significance of Lp(a) for Atherosclerosis and Myocardial Infarction

Since the detection of Lp(a) in 1963, there have been intensive investigations into its function and significance for lipid metabolism. Unfortunately, all of these attempts have failed, and it must be admitted that the role of Lp(a) in lipid metabolism remains obscure. On the other hand, Dahlen et al. (1972) reported that individuals with angina pectoris and coronary artery diseases exhibited an extra pre-β_1 electrophoretic band much more frequently than controls. In these and subsequent experiments, the extra pre-β band was equated with Lp(a), and the authors came to the conclusion that Lp(a) must be a risk factor for atherosclerosis and coronary artery diseases. This band, however, was not quantifiable, and in addition there was the danger that the pre-β_1 band might be confused with a double pre-β band, which in fact is a VLDL variant.

We have quantitatively measured Lp(a) by the Laurell technique, using antibodies specific for the a-antigen, in the plasma of 76 post-myocardial infarction (MI) patients aged between 40 and 60 years, and in over 100 control subjects matched for age, sex, and social status. Patients and controls were divided into normolipemics and hyperlipemics typed according to Fredrickson (Kostner et al., 1981). From Table VII it can be seen that Lp(a) levels in fact were significantly higher in MI survivors than in controls, independent of the Fredrickson type. Lipoprotein-a levels were then correlated with concentrations of apoA-I, apoA-II, and apoB, as well as with serum lipids and lipoprotein lipids of these individuals, but there was no significant correlation found. From these studies we concluded that Lp(a) in fact does represent a risk factor for myocardial infarction, which is independent of other well-recognized risk and antirisk factors. There seemed to be a threshold level for Lp(a), which was in the range of 25–30 mg/dl, below which a MI risk could not be proven statistically. In another trial in collaboration with D. Seidel (unpublished data) we measured Lp(a) concentrations in more than 200 individuals with angiographically proven atherosclerosis and compared them with individuals with no atherosclerosis detectable by angiography. Also in those experiments we found a 2- to 2.5-fold higher relative risk for individuals with Lp(a) levels exceeding 25 mg/dl.

These experiments indicate that quantitation of Lp(a) may help to reveal risk factors for atherosclerosis and myocardial infarction in early life. This conclusion can be drawn on the basis of other studies indicating that Lp(a) levels stay constant for a long period of time independently of lifestyle.

Table VII
PLASMA Lp(a) VALUES OF PATIENTS SUFFERING FROM MYOCARDIAL INFARCTION (MI) AND OF CONTROL INDIVIDUALS GROUPED ACCORDING TO FREDRICKSON TYPES

Subject	Total number	Lp(a): <25 mg/dl	% of total 25–50 mg/dl	>50 mg/dl
Normolipemics				
Controls	64	70.2	20.1	9.7
MI patients	43	51.4	23.4	25.2
Type IIa				
Controls	14	66.9	24.8	8.3
MI patients	16	52.4	27.4	20.2
Type IIb				
Controls	10	67.2	25.2	7.6
MI patients	9	53.3	26.9	19.8
Type IV				
Controls	27	56.8	27.2	16.0
MI patients	14	60.7	30.4	8.9
Totals				
Controls	115	64.3	25.9	9.8
MI patients	82	53.7	27.3	19.0

VI. Role of Apolipoproteins in Lipid Metabolism

It is beyond the scope of this article to review in detail the relationship of individual apolipoproteins to lipid and lipoprotein metabolism. These aspects have been reviewed (Osborne and Brewer, 1977; Havel, 1980; Kostner, 1982b). Here I will only try to give an overview of general features of this subject in order to make the following section more easily understandable. Since lipids are insoluble in aqueous solutions, they must be solubilized by amphiphilic substances such as phospholipids and proteins. In fact, the existence of lipids without any protein in plasma has never been observed, and even fatty acids are bound specifically to albumin. There are several functions which apolipoproteins must fulfill : (1) solubilization and transport of lipids; (2) direction of lipids to certain organs; (3) mediation of the secretion of lipids from cells of origin; (4) making lipids accessible to enzymes; (5) regulating enzyme activities of lipids in the circulation; and (6) exchange or transfer of individual lipids from one fraction to another. Although we have direct and indirect proof that different apolipoproteins fulfill these tasks, a lot of work still needs to be done in this area, and many theories are either hypothetical or unverifiable *in vivo*.

With this in mind, the role of apolipoproteins can be summarized as follows (Table VIII). There are three classes of lipoproteins in the plasma: the triglyceride-rich lipoproteins, the cholesterol- and cholesterol ester-rich lipoproteins, and the phospholipid protein-rich lipoproteins. As outlined earlier, almost all known apolipoproteins are found in VLDL and HDL, and LDL also contain in addition to apoB some apoD, -E and -C polypeptides. There are, however, quantitative differences in concentration and distribution among all the lipoprotein classes. From models of inborn errors of metabolism, for example, abetalipoproteinemia, Tangier disease, apoC-II and apoE deficiency, LCAT and LPL deficiency, and some others, we concluded how apolipoproteins may function under normal conditions. It is, however, quite surprising how well the patients affected by most of these diseases can handle the absence of one or another apolipoprotein or enzyme—and many of these diseases have been detected only late in life or during a routine screening or checkup. This may be an indication that many apolipoproteins function similarly or can replace one another to a certain degree.

From ABL we learned the apoB is absolutely necessary for the absorption of dietary fat, especially triglyceride and possibly cholesterol. The serum levels of affected individuals are virtually zero. Thus apoB must also be necessary for the secretion of triglycerides from the liver. This model also indicates that apoA-containing lipoproteins of HDL are secreted probably without any triglyceride, possibly from the liver. From the

Table VIII
FUNCTION OF APOLIPOPROTEINS

Apolipoprotein	Function
AI	LCAT activation; reverse cholesterol transport
AII	LCAT inhibition (?); activation of liver lipase (?)
AIII	Involved in LCAT reaction
B	Lipid absorption; specific binding to cell receptors
CI	LCAT reaction
CII	Activation of lipoprotein lipase (LPL)
CIII	Modulation of the LPL reaction
D	See AIII
E	Remnant catabolism in the liver; binding to the B–E receptor; specific binding to the E receptor
H (β_2-glycoprotein I)	Triglyceride metabolism
F, G, (a), and others	Unknown

pioneering work of Brown *et al.* (1976), on the other hand, we know that cell surfaces possess specific receptors for apoB which help to supply cells of endocrine glands, for instance, with cholesterol. Since apoB receptors have been observed on almost all cells of every tissue, apoB might be responsible for the catabolism of the "remnant" LpB under normal conditions. But there must exist other routes as well for cholesteryl ester-rich lipoproteins, since familial homozygous hypercholesterolemic individuals completely lack this receptor, and there is no evidence of impaired endocrine function. Low-density lipoproteins are also catabolized there, although with a reduced fractional rate. Together with triglycerides and cholesterol, lipid-soluble vitamins are absorbed from the diet with the aid of apoB. Since this process is impaired in ABL, patients suffer from avitaminosis, and this seems to be the major or only defect in this disease.

The A apoproteins are complexed primarily to phospholipid-rich HDL. ApoA-I and apoA-II are integral components of chylomicrons (Kostner and Holasek, 1972), but are quickly lost from this fraction and transferred to HDL during lipolysis. Nascent HDL from the liver, on the other hand, are rich in apoA-I, apoA-II, and probably phospholipids, and are possibly free of triglycerides and cholesteryl esters. Apolipoprotein E can also be released from the liver in a complex with apoA-I, and in some animals, such as rats and dogs, this seems to be the major lipoprotein class which is biosynthesized. High-density lipoproteins and apoA have several functions: acceptance of material released during lipolysis from triglyceride-rich lipoproteins, especially phospholipids, free cholesterol, and apoC; this transforms the HDL or subfractions thereof into an optimal substrate for LCAT, and it has been shown that LCAT needs some cofactors, primarily apoA-I and possibly apoD (Kostner, 1974a), to exert maximal activity.

Apolipoprotein D may also function only as a carrier of LCAT (Fielding and Fielding, 1982) or as an acceptor of the product formed during enzymatic action that is, cholesteryl ester. The cholesteryl ester formed does not reside in HDL but rather is transported by specific carriers to VLDL, LDL, and other HDL subfractions (Barter *et al.*, 1982), giving rise to cholesteryl ester-rich LDL and possibly to the formation of HDL_2. The latter lipoprotein is not observed in species devoid of the cholesteryl ester transport protein. In Tangier disease as well as in ABL, LCAT functions normally. In the first case, an LCAT–LpD complex may transfer the cholesteryl ester directly to VLDL or LDL, and in ABL, where an acceptor for cholesteryl ester in the $d < 1.063$ g/ml fraction is missing, the accumulation of cholesteryl esters leads to the formation of HDL_2, which is the only HDL observed in this disease.

From apoC-II deficiency we know that LPL needs this peptide as a cofactor for clearing chylomicrons. Apolipoprotein CIII certainly influences this lipolytic process and is believed to inhibit LPL. The C apoproteins released during lipolysis are transfered to HDL but can recycle once triglyceride-rich particles enter the bloodstream again. These particles must not necessarily contain apoB since it has been shown that artificial triglyceride emulsions avidly take up all three apoC peptides (Kostner *et al.,* 1976b).

Apolipoprotein E has been shown not only to bind stronger to the B receptor than LDL does, but also to bind specifically to the E receptor of the liver, which is not recognized by apoB (Mahley *et al.,* 1981). Apolipoprotein E thus probably has a similar function in animals, where apoE is the major lipoprotein class, as apoB is in humans. In humans, apoE is thought to be involved in the removal of chylomicron remnants, which are taken up within minutes from the liver. In type III hyperlipoproteinemia, as well as in ApoE deficiency, chylomicron remnants containing mainly B-48 accumulate in the plasma.

None of the other apolipoproteins listed in Table VIII have been characterized with respect to their function in lipid metabolism. The fact that their plasma concentrations are in most cases small by no means implies an inferior significance. The low concentration may only indicate increased turnover. Of the three apolipoproteins which preferentially bind to chylomicrons, but after lipolysis are found essentially lipid free in the $d = 1.21$ g/ml bottom [ApoA-IV, proline-rich protein, and apoH (β_2-glycoprotein I)], we have done some experiments with the latter peptide (Polz and Kostner 1979a,b; Polz *et al.,* 1981). Apolipoprotein H binds avidly to Intralipid with an association constant of approximately 10^8. Nakaya *et al.* (1980) have shown that it mediates the activity of LPL *in vitro*. In our experiments performed in rats *in vivo,* apoH accelerated the clearance of triglycerides if the removal system was oversaturated (Wurm *et al.,* 1982).

VII. Quantification of Apolipoproteins in the Clinical Chemical Laboratory

Because of the fundamental role of apolipoproteins in the physiology and pathophysiology of lipid transport and metabolism, the measurement of individual apolipoproteins should make possible a better characterization at the cellular level of defects present in lipid disorders than lipid measurements. At any rate, one may obtain additional information and round up clinical pictures. There are many aspects which must be considered in quantifying apolipoproteins which cannot be treated adequately here and should be investigated in more specialized reviews (Avogaro *et al.*, 1979; Kostner, 1982a; Schwandt, 1982). There are many precautions which must be considered if accurate apolipoprotein measurements are to be performed. One of the major obstacles lies in the fact that parts of the antigenic determinants are hidden, because polypeptide domains merge into the lipid phase and are inaccessible to the antibody. Since apolipoproteins are distributed among various lipoprotein classes, and size and lipid content varies considerably from one fraction to another, the kind and space of determinants accessible to the antibody are variable depending on the individual lipid concentrations of the plasma. Therefore it is absolutely essential that lipids are removed before immunoquantitation, or at least that apolipoproteins are put into a uniform state. The second major problem arises from the fact that apolipoproteins are hard to isolate in pure form, are easily modified chemically during the isolation process or during storage, are not soluble in monomeric form in aqueous buffers in the absence of dissociating agents, and are heterogeneous, exhibiting multiple bands upon IEF (see Section III,B). This has led to tremendous confusion in the past as far as mean values of normal and diseased plasma are concerned. Another drawback has arisen because of the different specificities and avidities of the antibodies used. Until recently, commercial batches were hardly available and thus comparisons from laboratory to laboratory were not possible. We have put some effort into these problems and tried not only to quantitate as many apolipoproteins as possible, but also to apply most of the immunochemical methods currently known.

Immunochemical Methods for Apolipoprotein Quantification

Almost every immunochemical method described has been applied in this field. No general recommendation can be given as to which particular procedure should be used. This will depend on the number of samples to be assayed per day, the ease of producing pure standards and monspecific antibodies, and the equipment available in a given laboratory. Except for the enzymes LPL and LCAT, all of the apolipoproteins described so far occur at concentrations exceeding 1 mg/dl, and thus concentration is not a limiting criterion for the selection of a method.

A. LAURELL (ROCKET) ELECTROPHORESIS

This method was first used in our laboratory for apolipoprotein quantification (Kostner *et al.*, 1974a,b). In subsequent experiments it was applied by various research groups to quantify all apolipoproteins of the A, B, D, D, E, H, and (a) families. The method is advantageous for measuring large particles, e.g., VLDL apoB, LpB, or Lp(a), because radial immunodiffusion (RID) requires 3–5 days before equilibrium is reached. Otherwise we prefer the latter method because rocket electrophoresis causes problems in standardization. We usually run our rockets overnight for 12–16 hours at a field strength of approximately 1 V/cm measured in the agarose gel. Antibodies should be monospecific and of high affinity, otherwise tailing of the peak occurs. Each plate must contain at least three standards of different concentrations per 10 samples. Evaluation of the peaks can be done by measuring areas or heights, depending on the particular antigen and antibody. The processing of samples is essentially identical to that described for RID.

B. Radial Immunodiffusion—The Mancini Technique

I consider this technique as the method of choice if some 10 to 50 samples have to be assayed in one set. There are no problems with standardization; they usually are stable for months and samples can be frozen at $-20°C$ or below and kept for 2–4 months before being assayed. Once samples are thawed, they must be measured immediately, and freezing and thawing should be avoided. As in the case of Laurell electrophoresis, the admixture of intensifiers to the agarose gel, for example, Dextran T-70 or polyethylene glycol at concentrations between 2 and 4%, may greatly improve the sensitivity and accuracy of the method. The avidity of the antibody usually is not critical, but it should be absolutely monospecific. Antigens on large molecules, as for example apoB on LDL or Lp(a), need a diffusion time of a minimum of 3 days. If diffusion proceeds at 37°C instead of at room temperature, the time can be reduced to 24 hours.

The processing of samples before the assay needs special attention and no general rule applicable for all apolipoproteins can be recommended. Different treatments of samples for individual apolipoprotein quantitations are listed in Table IX.

1. RID of ApoA Peptides

For the exposure of antigenic determinants, a variety of delipidating and dissociating agents have been applied. We prefer the use of 2% Triton X-100, which either is incorporated into the agarose gel or is used as a diluent. In this way, apoA-I, apoA-II, and apoA-III (apoD) expose all their antigenic determinants and may be measured accurately. Other nonionic detergents may be used as well, but these need evaluation of

Table IX
PRETREATMENT OF LIPOPROTEINS FOR THE DETERMINATION
OF INDIVIDUAL APOLIPOPROTEINS
BY VARIOUS IMMUNOCHEMICAL TECHNIQUES

Apolipoprotein	Treatment
AI, AII, AIII (D)	1–2% Triton X-100 in dilution buffer or directly in the agarose gel; for nephelometry also 0.5% Tween 20
B	No agents for RID and Laurell electrophoresis; for other methods, treatment with lipase
CI, CII, CIII	Triton X-100 (0.5–1%) or mixing 1:1 with tetramethylurea followed by dilution with 4 M urea; in LpX, delipidation with diisopropyl ether/n-butanol, (6/4) followed by dilution with 4 M urea
E	Dilution with 2 M guanidine hydrochloride; for radioimmunoassay also 0.5% sodium dodecyl sulfate
H	No agents; in the presence of chylomicrons, 1:1 dilution with tetramethylurea followed by dilution with normal buffers
a [Lp(a)]	In all assays as intact particle

correct concentrations at first. The diffusion time is 24 hours, and staining is not necessary. The circles which form must be symmetrical and should have sharp edges.

2. RID of ApoB Peptides

Apolipoprotein B is the major protein in VLDL, LDL, and Lp(a). Because of striking differences in size, it would be desirable to delipidate these lipoproteins, rendering them uniform. Unfortunately, lipid-free apoB is very unstable and precipitates or aggregates after such treatment. With three to four antibody batches from different animal species we found it unnecessary to delipidate the serum for apoB quantitation provided the agarose gel had a concentration of <1% and the diffusion time was adequately long. Shorter times are sufficient if diffusion is performed at 37°C or if plasma samples are pretreated with lipase (see later). We do not recommend the use of detergents for apoB quantitation by either method.

3. RID of ApoC, ApoE, and Other Apolipoproteins

Apolipoprotein C readily dissociates from VLDL as well as from HDL; they are freely exchangeable and do not need strong dissociating agents. Samples can be applied either after dilution with 4 M urea buffers or after mixing 1:1 with tetramethylurea or with 0.5–1% Triton X-100 buffers. Triton X-100 may also be directly admixed to the agarose. If, on the other hand, apoC of LpX has to be measured, much stronger delipidating agents are required. We use an organic solvent mixture of diisopropyl ether/*n*-butanol (6/4) followed by dilution of the sample with 4 *M* urea (Kostner *et al.*, 1979).

Apolipoprotein E quantitation, in our opinion, needs the most careful selection of diluent, otherwise comparable results cannot be obtained. This is because antigenic determinants of apoE are partly hidden in $d <$ 1.063 g/ml lipoproteins, and in HDL apoE is complexed with apoA and phospholipids. We have tried all anionic, cationic, and neutral detergents available, lipase treatment, urea, and others, and have come to the conclusion that best results can be obtained if the samples are diluted 1:1 with 2 *M* guanidine HCl followed by further dilution, depending on the system, with 4 *M* urea. This processing proved to be optimal for Laurell electrophoresis, RID, and nephelometry, and should also be applicable for radioimmunoassay (RIA) and enzyme-linked immunosorbent assay (ELISA) techniques.

No other apolipoproteins have been quantitated except for apoH. With apoH, whether samples were delipidated or not made no difference in the results, unless large amounts of chylomicrons were present. In this case, samples were mixed 1:1 with tetramethylurea and further diluted appropriately with normal buffers (Polz and Kostner, 1979b). Lipoprotein-a is the only lipoprotein which can be assayed by all methods without any delipidating or dissociating agent, since the specific antigen (a-peptide) occurs only on one homogeneous fraction.

C. Nephelometric Quantitation

There are two principles of nephelometry in use: endpoint nephelometry and rate nephelometry. Since light scattering is strongly influenced by the size of immune complexes or the morphology of the precipitate, it is necessary to render the antigens into a homogeneous form. This might not be so critical in endpoint measurements since lipoproteins are relatively small compared to the immune precipitate. In rate nephelometry, however, this is essential because the speed of precipitate formation is a measure of antigen concentration. These kinetics strongly depend on the diffusion constant of the antigen, which is a result of size, morphology, and charge. We also have quantified nephelometrically apolipoproteins AI, AII, D, (a), and E, and have found that the same detergents or diluents which are listed in Table IX for RID can be applied.

For apoA-I and apoA-II, Triton X-100 may be replaced by 0.5% Tween 20, and for Lp(a) no detergent should be used.

Nephelometry of ApoB

There are some reports in the literature in which detergents such as Thesit, Triton X-100, or Apovax at different concentrations have been recommended for nephelometry of apoB. We have tried them all and have come to the conclusion that none of them can be recommended (Da Col and Kostner, 1983). Up to a plasma triglyceride concentration of less than 350 mg/dl, 0.2 g/liter of Thesit or Apovax may be used, but this works only reasonably well with the endpoint procedure. At higher triglyceride concentrations and in rate nephelometry we only obtained accurate results after lipase treatment of the samples. As a final assay procedure we can recommend the following steps. Mix 100 μl of serum with 10 units of Lipase–TG (Calbiochem) in 10 μl of buffer and incubate for 6–18 hours at 37°C. This enzyme hydrolyses all the triglycerides as well as the bulk of the phospholipids, up to a serum concentration of 2 g/dl. The apoB-containing lipoproteins [VLDL, LDL, and Lp(a)] are rendered into a homogeneous fraction, and antigenic determinants are completely exposed. The samples are diluted 1 : 30 with buffer and mixed 1 : 1 with antibody solutions of appropriate concentrations. The latter solution contains 4–5% polyethylene glycol to accelerate the antigen–antibody reaction. Since endpoint measurements are somewhat more sensitive, sample dilutions of 1 : 50 or 1 : 100 are required. With the equipment from Beckman we received an excellent linearity from 50 to 200 mg/dl of apoB. Rate nephelometry works fast and reliably, and some 20 assays per hour can easily be performed. There is no need for a special purified apoB standard. The lipase treatment mentioned previously proved also to be optimal for apoB measurements by other immunochemical methods, for example, RID, Laurell, and others, and thus may be generally recommended.

D. Radioimmunoassay, Enzyme-Linked Immunosorbent Assay, and Fluoroimmunoassay

These techniques were used in the past for almost every apolipoprotein with some success. They can be recommended only if large sample numbers have to be processed routinely. Otherwise, the tedious preparation of purified, labeled antigens, or the pretreatment of antibodies with radioactive tracers or enzymes, does not pay off. In principal, the same detergents or buffers may be used as mentioned previously for individual apolipoproteins.

E. STANDARDS AND CONTROL SAMPLES

As mentioned earlier, apolipoproteins are sometimes hard to purify and are not stable for long periods of time. The use of secondary standards must thus be recommended. Such secondary standards in our laboratory are normolipemic complete human sera which are stored at 4°C, −20°C, or sometimes in lyophilized form. They are standardized by themselves by running fresh and highly purified apolipoproteins—or, in the case of LpB and Lp(a), pure intact lipoproteins—in the appropriate system at increasing concentrations, together with various dilutions of the secondary standard. Care must be taken that the antigens are not aggregated and contain a representative mixture of isoforms if polymorphism is a feature of the apolipoprotein. The concentration of the antigen in the primary standard may be obtained by a Lowry assay, amino acid analysis, or gravimetry. Control samples usually are handled in the same way as secondary standards, and there should be control sera available with high, medium, and low concentrations.

F. APOLIPOPROTEIN VALUES OF NORMAL PLASMA

The most recent compilation of published plasma apolipoprotein values appeared in the monograph by Assmann (1982). In these tables, variations of 100% and more are common features. This probably reflects different methodologies, antibody specificities, sample preparations, and determinations of standard concentrations. We have measured in our laboratory almost all the apolipoproteins by various immunochemical techniques and have derived the mean values shown in Table X. The current commercial preparations of specific antisera and standards, and the nomination of an international standard committee, probably will help to clarify matters in this area within the near future.

VIII. Apolipoproteins as Risk Indicators for Myocardial Infarction and Atherosclerosis

The measurement of lipids, lipoprotein lipids, and lipoprotein concentrations has become a general practice not only in lipid research clinics but also in routine clinical laboratories. It seems to be generally accepted that

Table X
AVERAGE APOLIPOPROTEIN VALUES IN PLASMA OF HEALTHY MALE AND FEMALE NORMOLIPEMIC INDIVIDUALS[a]

Apolipoprotein	Plasma concentrations
AI	
Male	115 ± 10
Female	122 ± 10
AII	40 ± 5
AIII (D)	8 ± 2
B	80 ± 10
CI	6 ± 2
CII	4.5 ± 1
CIII	11 ± 2
E	7 ± 2
H	20 ± 5
Lp(a) (intact)	1–100 (trimodal distribution)

[a] These values were determined in our laboratory by various immunochemical methods and are used as "normal" values.

disturbances of lipid metabolism are frequently accompanied by atherosclerosis and premature myocardial infarction. In the Fredrickson area the typing system by normal electrophoresis contributed greatly to the definition of the genetic disorders or secondary forms of dys- and hyperlipoproteinemias which are mostly connected with cardiovascular diseases. Later it was found that HDL may play a significant protective role against atherosclerosis, and the following questions arose: (1) how should HDL be measured, and (2) since HDL is heterogeneous, what part of it should be measured? After it had been realized that many more apolipoproteins existed than was originally thought, and in addition that individual apolipoproteins may be distributed over the whole spectrum of lipoprotein classes, it became hard to foresee which particular fraction, if any, might be used as the best discriminator for cardiovascular disease. Since there are only three classes of lipids, four to five main classes of lipoproteins, but some 20 individual apolipoproteins which govern the whole intravascular lipid metabolism, the chances are high that apolipoprotein quantitation can greatly improve our diagnostic methods.

In order to correctly elaborate which specific lipoprotein or apolipoproteins might be independent risk indicators for atherosclerosis, a prospective study of many years' duration would need to be conducted in which a great number of assays of several thousand people would be performed.

During such studies, however, new developments in lipoprotein research and the detection of new apolipoproteins could spoil the whole project. We thus have simplified matters and measured retrospectively apolipoproteins and lipids in patients affected by atherosclerosis, myocardial infarction, and stroke. The results of this study do not provide a final answer to this problem, but certainly can be used as a basis to design large prospective trials.

The Evaluation of Optimal Discriminators for Atherosclerosis by Multivariant Analysis

Several of such studies are currently in progress in my laboratory, and some have already been published (Pilger *et al.*, 1983). With multiple regression discriminant analysis, which has been especially adapted to this problem (Pfeiffer and Dutter, 1981), it is possible to correlate different parameters. Those being of least significance may be sorted out, while the others are connected by a mathematical term: $y = a_1B_1 + a_2B_2 + a_3B_3$, etc., where a stands for serum concentrations of individual parameters, and B is a factor which has to be calculated from different models. In the study of patients with peripheral vascular diseases we determined some 20 parameters, including apolipoproteins AI, AII, and B, total apoA, LDL apoB, triglyceride, LDL-C, HDL-C, VLDL triglyceride, and ratios thereof. From the 2^{20} models we sorted out by an economic selection strategy those which either gave the best segregation or were the most practicable. Cutoff points of the calculated y values were set according to a minimal rate of error (sum of false-positive and false-negative results). Table XI shows some of the models and the resulting error rates. Considering all 20 parameters, which obviously is rather unfeasible in practice, gave an error rate of less than 10%. The mathematical treatment of values for apoA-I, apoA-II, apoB, LDL-C, HDL-C, and triglyceride made it possible to correctly classify patients and controls in 87% of the cases. By measuring only apoA-I, apoA-II, and apoB we received an error rate of 21%, which was far below the error rate obtained by measuring only LDL-C and HDL-C (34%). Table XI also lists some of the error rates observed in univariate analysis. By including some other well-recognized risk factors, such as cigarette smoking, systolic and diastolic blood pressure, hyperuricemia, and diabetes, we reached a minimum error rate of some 4% in combination with apolipoprotein measurements (Fig. 9).

In an ongoing study of patients suffering from myocardial infarction, finally, where we also measured Lp(a), the following independent discriminators have been evaluated: ApoA-I, apoA-II, apoB, Lp(a), total

Table XI
EVALUATION OF THE ATHEROSCLEROSIS RISK
BY MEASURING SERUM APOLIPOPROTEIN AND LIPID PARAMETERS[a]

Disease	Measured parameters	Error rate[b] (%)
MI[c]	HDL-C/LDL-C	35
	ApoA-I/apoB	26
	ApoA-I, apoA-II, apoB, Lp(a), total cholesterol, triglyceride, HDL-C, LDL-C	5.3
Stroke	ApoA-I	10.2
	ApoA-II	13.3
	HDL–lecithin	16–20
	ApoA-I, apoA-II, HDL-C, triglyceride, blood pressure	2.5
PVD[d]	HDL-C/LDL-C	34
	ApoA-I, apoA-II, apoB	21
	ApoA-I, apoA-II, apoB triglyceride, HDL-C, LDL-C, number of cigarettes, blood pressure, insulin, uric acid	4.4

[a] The error rate was obtained by univariate or multivariate evaluation of the results. Cutoff points were set at minima (see Fig. 9). The data are from several retrospective studies. The results on MI and stroke are preliminary.
[b] Sum of false-positive and false-negative results.
[c] MI, myocardial infarction.
[d] PVD, peripheral vascular diseases.

cholesterol, triglyceride, HDL-C, and HDL phospholipid. With this model, an error rate of some 5% may be obtained.

In another study, to be published soon (G. Kostner *et al.*, in preparation), we measured the lipid and lipoprotein parameters of 72 patients suffering from stroke in conjunction with those of more than 60 control individuals matched for age and sex. Here we also observed the "stroke paradox," infrequently mentioned in the literature: that triglyceride and LDL-C might be higher in controls when compared with stroke patients. In addition to these unexpected findings we also demonstrated that total apoB is lower in these patients than in the controls, reaching significance at a p level of 5%. By univariate analysis apoA-I was by far the best discriminator, yielding an error rate of 10%. The error rate using only apoA-II was 13% and was 16–20% using HDL–lecithin. Thus, here also, apolipoproteins seem to be superior to lipid or lipoprotein lipids. In the multivariate analysis we calculated several models with some 10 variables, resulting in a complete segregation of patients from controls. The most practicable model was that with the four parameters apoA-I, apoA-II,

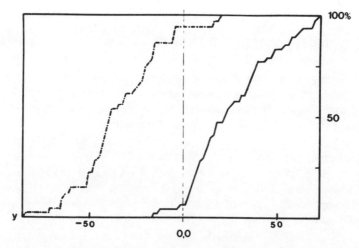

FIG. 9. Cumulative frequency distribution (in %) of the discriminant function y of patients suffering from peripheral vascular disease (dotted line) and of matched controls (solid line). y was calculated from several parameters, including apolipoproteins, lipids, and other risk factors, using the formula given in the text.

HDL-C, and triglyceride. In combination with the values for blood pressure it gave an error rate of 2.5% (Table IX). For the moment it may appear that this number of parameters is too big for screening risk factors in a large population. In this respect it must be remembered that other diseases, for example, liver diseases or thrombosis, also require the application of a number of tests. With the aid of multichannel autoanalyzers and the technology currently available, all that may be needed in the future is a few drops of specific antibodies to demonstrate quite accurately the presence of atherosclerotic diseases, making the many invasive techniques unnecessary. We certainly have not yet arrived at a point where further research in this area should cease, and I consider such studies as the basis of further developments and as evidence that lipids and especially lipoproteins are related to atherogenesis to a greater extent than is admitted by many investigators.

IX. Summary and Conclusions

When DeLalla and Gofman (1954) presented their work "Ultracentrifugal Analysis of Serum Lipoproteins" more than 25 years ago, we were thinking about lipoproteins in terms of density fractions. In the 1970s the electrophoresis concept was pushed by Fredrickson and his colleagues (Fredrickson *et al.*, 1967). There is no doubt that both these lipoprotein research centers have fertilized entire investigations in this field and still have a tremendous impact on our current knowledge. It was, however, not until 1966, when Gustafson, Alaupovic, and Furmann first described the presence of a third lipoprotein family, LpC, that researchers in this area became aware of the dominant role of apolipoproteins in the transport and metabolism of plasma lipids. Lipoprotein density fractions and electrophoretic classes in the mean time have not lost their importance; they still exist and the application of methods yielding those fractions is still going on in lipoprotein laboratories. Yet we need to recognize that the whole lipid transport system is far more complex than was believed some 10 or 20 years ago. Lipoprotein density fractions consist of varying numbers of families; some of them comigrate upon electrophoresis, and the protein moiety of them is mostly composed of nonidentical polypeptides.

There are a number of inborn errors of metabolism, for example, ABL, Tangier disease, and enzyme defects, which have taught us a lot about the functions and interplay of the complex apolipoprotein system. In dyslipoproteinemia, abnormal lipoproteins occur in the plasma and apolipoproteins, which are hardly recognized in normal fasting plasma, suddenly become prominent. There still exist, however, apolipoproteins and lipoproteins, one of which certainly is Lp(a), whose function and biological significance remains completely unknown.

The structure and the molecular arrangement of lipids and apolipoproteins within a lipoprotein particle has been the subject of intensive investigations, and almost every physicochemical method available has been applied to reveal the morphology of individual lipoproteins in closest detail. Lipoproteins and apolipoproteins have often also served as model substances for cell membranes.

After the purification of individual apolipoproteins succeeded in many laboratories and specific antibodies were available, clinical chemists and epidemiologists became interested in this area of research. Apolipoprotein quantification currently is most prominent for the prediction of atherosclerotic risk in preventive medicine. Here we certainly are still at the beginning, and the application of multivariate analysis in large prospective studies may answer the question of whether apolipoprotein analysis can in fact replace lipid measurements, and of which combination of parameters may be optimal for segregating healthy individuals from diseased ones. An investigator in this field must be aware of the complexity of this area of research and observe the system as a whole. We for our part have tried to increase our understanding by pursuing different lines of inquiry in lipoprotein research, as have other lipid research centers.

Acknowledgments

The work cited in this article was supported by the Österreichische Fonds zur Förderung der wissenschaftlichen Forschung, Project No. 3734, 4478, and 4118. The technical assistance of E. Schön and G. Grillhofer is appreciated.

References

Alaupovic, P. (1971). *Atherosclerosis* **13**, 141.
Alaupovic, P. (1982). *Ric. Clin. Lab.* **XII**, 3.
Anderson, D. W., Nichols, A. V., Forte, T. M., and Lindgren, F. T. (1977). *Biochim. Biophys. Acta* **493**, 55.
Assmann, G. (1979). *Atheroscler. Rev.* **6**, 1–28.
Assmann, G. (1982). *In* "Lipidstoffwechsel und Atherosclerose" (G. Assmann, ed.), pp. 14–53. Schattauer, Stuttgart.
Avogaro, P., Bittolo Bon, G., Cazzolato, G., and Quinci, G. B. (1979). *Lancet* **I**, 901.
Barter, P. J., Hopkins, G. J., Gorjatschko, L., and Jones, M. E. (1982). *Atherosclerosis* **44**, 1982.
Berg, K. (1963). *Acta Pathol. Microbiol. Scand.* **59**, 369.
Bittolo Bon, G., Cazzolato, G., and Avogaro, P. (1981). *J. Lipid Res.* **22**, 998.
Blanche, P. J., Gong, E. L., Forte, T. M., and Nichols, A. V. (1981). *Biochim. Biophys. Acta* **665**, 408.
Brown, M. S., Ho, Y. K., and Goldstein, J. L. (1976). *Ann. N.Y. Acad. Sci.* **275**, 244.
Bütler, R., Brunner, E., and Morganti, G. (1974). *Vox Sang.* **26**, 485.
Chapman, M. J., Goldstein, S., Lagrange, D., and Laplaud, P. M. (1981). *J. Lipid Res.* **22**, 339.
Cheung, M. C., and Albers, J. J. (1979). *J. Lipid Res.* **20**, 200.
Chung, B. H., Wilkinson, T., Geer, J. C., and Segrest, J. P. (1980). *J. Lipid Res.* **21**, 284.
Da Col, P., and Kostner, G. M. (1983). *Clin. Chem.* **29**, 1045.
Dahlen, G., Erison, C., Furberg, C., Lundkvist, L., and Scärdsudd, K. (1972). *Acta Med. Scand. Suppl.* **531**, 1.
DeLalla, O. F., and Gofman, J. W. (1954). *Methods Biochim. Anal.* **1**, 459–478.
Fielding, C. J., and Fielding, P. E. (1982). *Med. Clin. North Am.* **66**, 363.
Franceschini, G., Sirtori, C. R., Capurso, A., Weisgraber, K. H., and Mahley, R. W. (1980). *J. Clin. Invest.* **66**, 892.
Fredrickson, D. S., Levy, R. I., and Lees, R. S. (1967). *New Engl. J. Med.* **276**, 32, 94, 148, 215, 273.
Gustafson, A., Alaupovic, P., and Furmann, R. H. (1965). *Biochemistry* **4**, 596.
Gustafson, A., Alaupovic, P., and Furmann, R. H. (1966). *Biochemistry* **5**, 632.
Harvie, N. R., and Schultz, J. S. (1970). *Proc. Natl. Acad. Sci. U.S.A.* **66**, 99.
Hauser, H., Kostner, G. M., Müller, M., and Skrabal, P. (1977). *Biochim. Biophys. Acta* **489**, 246.
Havel, R. J. (1980). *Ann. N.Y. Acad. Sci.* **384**, 16.
Havel, R. J., Kotite, L., and Kane, J. P. (1979). *Biochem. Med.* **21**, 121.
Jürgens, G., Marth, E., Kostner, G. M., and Holasek, A. (1977). *Artery* **3**, 13.
Kane, J. P., Hardmann, D. M., and Paulus, H. E. (1980). *Biochemistry* **77**, 2465.
Kostner, G. M. (1972). *Biochem. J.* **130**, 913.
Kostner, G. M. (1974a). *Scand. J. Clin. Lab. Invest.* **33** (Suppl. 137), 19.
Kostner, G. M. (1974b). *Biochim. Biophys. Acta* **336**, 383.

Kostner, G. M. (1975). *In* "Lipid Absorption: Biochemical and Clinical Aspects" (K. Rommel, ed.), pp. 203–239. MTP Press, Heidelberg.
Kostner, G. M. (1976). *In* "Low Density Lipoproteins" (C. E. Day, ed.), pp. 229–269. Plenum, New York.
Kostner, G. M. (1981). *In* "High Density Lipoproteins" (C. E. Day, ed.), pp. 1–42. Dekker, New York.
Kostner, G. M. (1982a). *Ric. Lab.* **XII,** 155.
Kostner, G. M. (1982b). *Ber. Österr. Ges. Klin. Chem.* **5,** 86.
Kostner, G. M., and Alaupovic, P. (1971). *FEBS Lett.* **15,** 320.
Kostner, G. M., and Alaupovic, P. (1972). *Biochemistry* **11,** 3419.
Kostner, G. M., and Holasek, A. (1970). *Lipids* **5,** 501.
Kostner, G. M., and Holasek, A. (1972). *Biochemistry* **11,** 1217.
Kostner, G. M., and Holasek, A. (1977). *Biochim. Biophys. Acta* **488,** 417.
Kostner, G. M., and Laggner, P. (1979). *Rep. HDL Methodol. Workshop* (NIH Publ. No. 79-1661, pp. 343–355.).
Kostner, G. M., Albert, W., and Holasek, A. (1969). *Hoppe Seylers Z. Physiol. Chem.* **354,** 1347.
Kostner, G. M., Holasek, A., Bohlmann, H. G., and Thide, H. (1974a). *Clin. Sci. Mol. Med.* **46,** 457.
Kostner, G. M., Patsch, J., Sailer, S., Braunsteiner, H., and Holasek, A. (1974b). *Eur. J. Biochem.* **45,** 611.
Kostner, G. M., Laggner, P., Prexl, H. H., and Holasek, A. (1976a). *Biochem. J.* **157,** 401.
Kostner, G. M., Bohlmann, H. G., and Holasek, A. (1976b). *J. Mol. Med.* **1,** 311.
Kostner, G. M., Avogaro, P., Bittolo Bon, G., and Cazzolato, G. (1979). *Clin. Chem.* **25,** 939.
Kostner, G. M., Avogaro, P., Cazzolato, G., Marth, E., Bittolo Bon, G., and Quinci, G. B. (1981). *Atherosclerosis* **38,** 51.
Krauss, R. M., and Burke, D. J. (1982). *J. Lipid Res.* **23,** 97.
Krempler, F., Kostner, G. M., Bolzano, K., and Sandhofer, F. (1978). *Atherosclerosis* **30,** 57.
Krempler, F., Kostner, G. M., Bolzano, K., and Sandhofer, F. (1979). *Biochim. Biophys. Acta* **575,** 63.
Krempler, F., Kostner, G. M., Bolzano, K., and Sandhofer, F. (1980). *J. Clin. Invest.* **65,** 1483.
Krempler, F., Kostner, G. M., Roscher, A., and Sandhofer, F. (1983). *J. Clin. Invest.* **71,** 1431.
Laggner, P., Glatter, O., Müller, K., Kratky, O., Kostner, G., and Holasek, A. (1977a). *Eur. J. Biochem.* **77,** 165.
Laggner, P., Stabinger, H., and Kostner, G. M. (1977b). *Prep. Biochem.* **7,** 33.
Lee, D. M., and Alaupovic, P. (1970). *Biochemistry* **9,** 2244.
Lee, D. M., and Downs, D. (1982). *J. Lipid Res.* **23,** 14.
Lim, C. T., Chung, J., Keyden, H. J., and Scanu, A. M. (1976). *Biochim. Biophys. Acta* **420,** 332.
Lindgren, F. T., Jensen, L. C., Wills, R. D., and Stevens, G. R. (1972). *Lipids* **7,** 194.
Mahley, R. W., Hui, D. Y., and Innerarity, T. L. (1981). *J. Clin. Invest.* **68,** 1197.
Malmemdier, C. L., Christophe, J., and Ameryckx, J. P. (1979). *Clin. Chim. Acta* **99,** 167.
Marth, E., Cazzolato, G., Bittolo Bon, G., Avogaro, P., and Kostner, G. M. (1982). *Ann. Nutr. Metab.* **26,** 56.
Nakaya, Y., Schaefer, E. J., and Brewer, H. B. (1980). *Biochem. Biophys. Res. Commun.* **95,** 1168.
Osborne, J. C., and Brewer, H. B. (1977). *Adv. Protein Chem.* **31,** 235–337.
Papadopoulos, N. M. (1978). *Clin. Chem.* **24,** 277.

Patsch, J. R., Sailer, S., Kostner, G. M., Sandhofer, F., Holasek, A., and Braunsteiner, H. (1974). *J. Lipid Res.* **15**, 356.
Patsch, J. R., Sailer, S., and Braunsteiner, H. (1975). *Eur. J. Clin. Invest.* **5**, 45.
Patsch, J. R., Aune, K. C., Gotto, A. M., and Morrisett, J. D. (1977). *J. Biol. Chem.* **252**, 2113.
Patsch, W., Patsch, J. R., Kostner, G. M., Sailer, S., and Braunsteiner, H. (1978). *J. Biol. Chem.* **253**, 4911.
Pfeiffer, K. P., and Dutter, R. (1981). *EDV Med. Biol.* **4**, 115.
Pilger, E., Pristautz, H., Pfeiffer, K. P., and Kostner, G. M. (1983). *Artherosclerosis* **3**, 57.
Polz, E., and Kostner, G. M. (1979a). *FEBS Lett.* **102**, 183.
Polz, E., and Kostner, G. M. (1979b). *Biochem. Biophys. Res. Commun.* **90**, 1305.
Polz, E., Wurm, H., and Kostner, G. M. (1981). *Artery* **9**, 305.
Redgrave, T. G., Roberts, C. K., and West, C. E. (1975). *Anal. Biochem.* **65**, 42.
Roelcke, D., and Weicker, H. (1969). *Z. Clin. Chem.* **7**, 467.
Schwandt, P. (1982). *Klin. Wochenschr.* **60**, 637.
Seidel, D. (1977). *Klin. Wochenschr.* **55**, 611.
Shore, B., and Shore, V. (1968). *Biochemistry* **7**, 2773.
Stokke, K. T., and Norum, K. T. (1971). *Scand. J. Clin. Lab. Invest.* **27**, 21.
Suenram, A., McConnathy, W. J., and Alaupovic, P. (1979). *Lipids* **14**, 505.
Tall, A. R., Blum, C. B., Forester, G. P., and Nelson, C. A. (1982). *J. Biol. Chem.* **257**, 198.
Utermann, G. (1975). *Hoppe Seylers Z. Physiol. Chem.* **356**, 1113.
Vessby, B., Kostner, G. M., Lithell, H., and Thomis, J. (1982). *Atherosclerosis* **44**, 61.
Weisgraber, K. H., Bersot, T. P., and Mahley, R. W. (1978). *Biochem. Biophys. Res. Commun.* **85**, 287.
Wilcox, H. G., and Heimberg, M. (1968). *Biochim. Biophys. Acta* **15**, 424.
Wurm, H., Beubler, E., Polz, E., and Kostner, G. M. (1982). *Metabolism* **31**, 484.
Zannis, V. I., Breslow, J. L., and Katz, A. J. (1980). *J. Biol. Chem.* **255**, 8612.
Zannis, V. I., Breslow, J. L., and Utermann, G. (1982). *J. Lipid Res.* **23**, 911.

Relationship of Cholesterol Metabolism to the Metabolism of Plasma Lipoproteins: Perspectives from Methodology

BHALCHANDRA J. KUDCHODKAR

Department of Biochemistry
Texas College of Osteopathic Medicine
North Texas State University
Denton, Texas

I.	Introduction	45
II.	Body Cholesterol Turnover	47
III.	Measurement of the True Body Cholesterol Turnover Rate by Tracer Kinetic Methods	51
	A. Compartmental Analysis	51
	B. Isotopic Steady-State Method	57
	C. Input–Output Analysis	58
IV.	Total Body Cholesterol Mass and Individual Pool Sizes	58
	A. Total Body Cholesterol Mass	59
	B. What Are the Exchangeable Pools?	69
V.	Is There an Exchange of Plasma Cholesterol with Tissue Cholesterol?	80
VI.	Relationship between Plasma Cholesteryl Ester Turnover and LDL ApoB Turnover	86
VII.	Quantitative Relationship of Cholesterol Synthesis, Absorption, Esterification, and Catabolism with the Metabolism of Plasma Lipoproteins, Especially the Low-Density Lipoproteins	92
VIII.	Significance	99
	References	101

I. Introduction

Four approaches to the study of body cholesterol metabolism are currently available: (1) the chemical and/or isotopic balance methods; (2) the tracer kinetic methods (compartmental three-pool model and input–output analysis); (3) isotopic steady-state kinetics; and (4) combined sterol balance with tracer kinetics (Grundy and Ahrens, 1969; Goodman *et al.*, 1973; Samuel and Lieberman, 1973; Morris *et al.*, 1957; Samuel *et al.*, 1978). All these methods are assumed to measure daily cholesterol turnover (biosynthesized + dietary absorbed). In combination with others that measure absorption (Grundy and Ahrens, 1966; Borgstrom, 1969; Sodhi *et al.*, 1974; Zilversmit, 1972; Quintao *et al.*, 1971), they can be used to calculate the daily synthesis and absorption of both the exogenous and the endogenous cholesterol secreted in the gastrointestinal (GI) tract. The above parameters can also be measured by a combination of the intestinal perfusion technique with the sterol balance method (Grundy and Metzger, 1972; Mok *et al.*, 1979). In addition, the tracer kinetic methods singly or in combination with the sterol balance method can be used for the calculation of the total body mass of cholesterol as well as of the sizes of the rapidly, slowly, and very slowly exchanging pools (Goodman *et al.*, 1973; Samuel and Lieberman, 1973; Samuel *et al.*, 1978). These calculations, however, are based on a number of assumptions and require the monitoring of the plasma cholesterol specific activity curve for a period of at least 12 (combined sterol balance and tracer kinetic analysis) or 45 weeks (tracer kinetic analysis alone).

The theoretical basis and calculations underlying these various methods have been reviewed (Grundy and Ahrens, 1969; Sodhi *et al.*, 1979). The purpose of this article is to present data (obtained from the literature) which show that (1) the true body cholesterol turnover (synthesis in the absence of dietary cholesterol) may not be just equal to the amount of cholesterol excreted in the feces, but may be equal to the sum of that cholesterol excreted in the feces and that reabsorbed from the gastrointestinal tract; (2) the tracer kinetic methods measure not only the cholesterol synthesis + dietary absorbed, but also the endogenously absorbed cholesterol; (3) the total body mass of cholesterol (as well as the three projected pools of body cholesterol) can be calculated rapidly and accurately using specific activity data obtained over a period of 5–6 weeks. Alternatively, it can be calculated using the isotopic sterol balance method alone or in combination with tracer kinetics; (4) both the synthesized and the absorbed (exogenous and/or endogenous) cholesterol removed by the liver is quantitatively transported first to the plasma before being secreted (as such or after it is converted to bile acids) into the bile; and (5) equilibration of plasma cholesterol specific activity with tissue cholesterol may not involve exchange, but may involve the receptor and nonreceptor pathways of cholesterol uptake. Based on reanalyses of published data, an overall model for cholesterol metabolism is proposed which integrates cholesterol metabolism (synthesis, absorption, esterification, and catabolism) and the metabolism of plasma lipoproteins.

II. Body Cholesterol Turnover

At present, body cholesterol turnover is defined as the cholesterol which is biosynthesized plus that which is absorbed from the diet. Therefore, *in the absence of cholesterol in the diet,* the total cholesterol turnover equals endogenous synthesis (Grundy and Ahrens, 1969). In general, the rate of cholesterol synthesis in any organ appears to parallel the activity of tissue cell turnover and/or lipoprotein synthesis (Sodhi and Kudchodkar, 1974; Sodhi, 1975). Although nearly all tissues synthesize cholesterol at varying rates, the liver is the main organ responsible for its catabolism. The liver secretes cholesterol, as such and after its conversion to bile acids, into the bile. A part of the cholesterol secreted into the intestinal lumen via the bile is reabsorbed (enterohepatic cycle) (Sodhi, 1975; Grundy, 1978) and transported to the liver in cholesteryl ester (CE)-rich chylomicron remnants (Nestel *et al.,* 1963; Goodman, 1965). The endogenously absorbed cholesterol delivered to the liver activates the feedback inhibition of cholesterol synthesis in the liver (Grundy *et al.,* 1969; Grundy, 1978). Since some of the cholesterol is converted to bile acids, some is converted to hormones, and some is excreted as such, the cholesterol lost from the body is regained by new synthesis, which is equal to the amount lost from the body (Grundy and Ahrens, 1969). Thus, in a steady state, the daily need of the cholesterol by the body is met *both by de novo synthesis and by reabsorption* of cholesterol (endogenously synthesized) secreted in the intestinal lumen, mainly via bile. The true rate of body cholesterol turnover should therefore be equal to the sum of the endogenously absorbed and the excreted cholesterol, and not just to the excreted cholesterol alone. It will be equal to excretion under those conditions in which there is no reabsorption (complete interruption of the enterohepatic cycle). If the diet contains cholesterol, the true rate of body cholesterol turnover will equal the total absorbed cholesterol (dietary + endogenous) plus the biosynthesized cholesterol. ("True body cholesterol turnover" is used in order to distinguish it from "body cholesterol turnover," which at present is believed to equal biosynthesized + dietary absorbed.) Since, in a steady state, synthesis equals catabolism, the daily true body cholesterol turnover (synthesis + absorbed) will be equal to the amount of cholesterol catabolized daily. The liver, which is the primary organ for the catabolism of cholesterol, excretes cholesterol, as such and after its conversion to bile acids, into the bile (some cholesterol catabolism also occurs through the exfoliation of cells over skin surface and intestines and in hormone-producing organs). Therefore, in a steady state, cholesterol and newly synthesized bile acids excreted into the bile will be equal to the biosynthesized + absorbed (exogenous and/or endogenous) cholesterol. These types of projections for the daily true body cholesterol turnover (Table I) suggest that the daily catabolism of cholesterol in normolipemic persons of normal weight is between 1400 and 1500 mg (18–21 mg/kg), out of which about 1000–1200 mg (14–16 mg/kg) is excreted in the intestinal (GI) lumen as such and about 300–400 mg (4–5 mg/kg) after conversion to bile acids. Of the cholesterol excreted in the GI lumen, about 500–700 mg (7–9 mg/kg) is reabsorbed and the remaining 400–500 mg (6–7 mg/kg) is excreted in the feces along with bile acids. The cholesterol excreted in the feces (fecal neutral and acidic steroids of cholesterol origin) is made up by the new synthesis. The daily new synthesis of cholesterol is about 700–900 mg (10–12 mg/kg).

As stated earlier, in a steady state, assuming no absorption, true body cholesterol turnover (synthesis + absorbed) will be equal to total body cholesterol synthesis. Reexamination of sterol balance (synthesis) data available in subjects in whom the absorption of cholesterol could be considered negligible suggests that total body cholesterol synthesis is equal to synthesis plus absorbed (Table II). Thus in adult subjects with abetalipoproteinemia (Illingworth et al., 1980) or malabsorption (Vuoristo et al., 1980), total body cholesterol synthesis is nearly equal to the values for synthesis plus absorbed measured in adult normal subjects. Similarly, the indirectly calculated true body cholesterol turnover values in homozygous receptor-negative and -defective patients of Bilheimer et al. (1979) are similar to the values for total body cholesterol synthesis reported by Deckelbaum et al. (1977) in two similar types of patients which had, however, complete biliary diversion. These values in young patients also seem to be similar to the total body synthesis values obtained in young subjects with abetalipoproteinemia (Illingworth et al., 1980). Data available from Ahrens' laboratory (Pertsemlidis et al., 1973a) also show that total body cholesterol synthesis in two dogs (E and F) with complete biliary diversion is similar to the calculated synthesis plus absorbed in the same dogs. The just-mentioned data, therefore, strongly support the suggestion that the true body cholesterol turnover is equal to synthesis plus total absorbed and not just to synthesis + dietary absorbed cholesterol (Table II). Although the true body cholesterol turnover was similar in conditions with an intact and an interrupted enterohepatic cycle, there were marked differences in the ratio of biliary cholesterol to bile acids (newly synthesized). The ratio was lower in adult subjects ($\simeq$ 3.0) compared to young subjects ($\simeq$ 10.0), and was markedly reduced ($\simeq$ 0.12) upon biliary diversion. In dogs with an intact enterohepatic cycle the ratio was ~14.0, and was reduced to ~0.25 upon biliary diversion. Interruption of cholesterol absorption alone had no effect on the ratio.

Table I
TRUE BODY CHOLESTEROL TURNOVER RATES DERIVED BY A COMBINATION OF THE INTESTINAL PERFUSION TECHNIQUE WITH THE MEASUREMENT OF FECAL BILE ACIDS OR BY A COMBINATION OF THE STEROL BALANCE TECHNIQUE WITH THE METHODS MEASURING ABSORPTION[a]

	Parameters of cholesterol metabolism[b] (mg/day)						
	Catabolism (secretion in GI tract)					Anabolism	
Subjects	BC + FBA	BC	FBA[c]	FENS[c]	Synthesis (FENS + FBA)	Absorbed (BC − FENS)	Total (synthesis + absorbed)
Normolipemic[d]	1445 ± 281	1062 ± 292	383 ± 152	402 ± 115	725 ± 213	720 ± 370	1445 ± 281
Normolipemic[e]	1415 ± 332	1048 ± 324	373 ± 90	504 ± 190	870 ± 204	544 ± 136	1415 ± 332
Type IV + V[d]	1918 ± 642	1464 ± 458	454 ± 262	760 ± 441	1081 ± 572	837 ± 210	1918 ± 642
Type IV + V[e]	2043 ± 949	1471 ± 680	572 ± 372	822 ± 331	1119 ± 582	924 ± 434	2043 ± 949
Type IV[e]	1890 ± 576	1402 ± 428	488 ± 138	746 ± 278	1000 ± 371	890 ± 365	1890 ± 576

[a] In a steady state, biliary cholesterol + newly synthesized bile acids (catabolism) = synthesis + absorbed (anabolism) = true body cholesterol turnover.
[b] Abbreviations: BC, biliary cholesterol (includes cholesterol secreted into the GI tract from other sources); FBA, fecal bile acids (equals bile acid synthesis); FENS, fecal endogenous neutral sterols.
[c] These are excretion values.
[d] Data from Mok et al. (1979) not included in pooled analyses. Secretion measured by the intestinal perfusion technique.
[e] Data pooled from the literature. Secretion estimated: fecal endogenous neutral sterols/(1 − fraction absorbed).

Table II

DATA SHOWING THAT VALUES OBTAINED FOR CHOLESTEROL SYNTHESIS PLUS TOTAL ABSORBED (TRUE BODY CHOLESTEROL TURNOVER) ARE SIMILAR TO THE VALUES OBTAINED FOR SYNTHESIS (EXCRETION) ALONE WHEN CHOLESTEROL ABSORPTION IS BLOCKED

		True body cholesterol turnover (mg/kg/day)		
Species and type of subject	Number of subjects	Synthesis + total absorbed	Synthesis (no absorption)	Reference
Humans				
Normolipemic	11	19.2 ± 3 (14–23)[a]		Pooled data
Severe malabsorption	24		26.8 ± 16 (9–90)	Vuoristo et al. (1980)
	21		23.2 ± 6[b] (15–36)	
Hypercholesterolemic				
Receptor negative or defective	6	34.5 ± 10[c] (29–55)		Bilheimer et al. (1979)
21 years, 96 kg	1	19.0		Deckelbaum et al. (1977)
Biliary diversion	2		68.4; 31.0	
Abetalipoproteinemic				Illingworth et al. (1980)
5–6 years, 15 kg			43.2	
5–6 years, 18 kg			26.5	
23–26 years, 53 kg			19.3	
Dogs				
Dog E				Pertsemlidis et al. (1973a)
Preoperative		91.1[d]		
Biliary diversion			87.5[e] (91.0)[f]	
Dog F				
Preoperative		81.5[d]		
Biliary diversion			86.8[e]	

[a] Numbers in parentheses denote the range of values.
[b] After removing three subjects, one with very low (number 20) and two with very high (numbers 2 and 26) cholesterol synthesis values.
[c] Calculated assuming 60% absorption, the mean absorption value on the very low cholesterol diet.
[d] Calculated assuming 90% absorption, the mean absorption of dogs C and D on the low cholesterol diet.
[e] Mean of all studies.
[f] Mean of the first three studies.

III. Measurement of the True Body Cholesterol Turnover Rate by Tracer Kinetic Methods

Over the past 30 years simple techniques utilizing isotopes have been developed to measure body cholesterol turnover and the total body miscible pool based on the disappearance of labeled cholesterol from plasma, which proceeds at a series of decreasing exponential rates for periods of as long as 50–60 weeks (Morris *et al.*, 1957; Chobanian *et al.*, 1962; Goodman and Noble, 1968; Samuel and Lieberman, 1973; Goodman *et al.*, 1973). At present, it is believed that the production rate of cholesterol calculated by the various tracer kinetic methods (two- and three-pool models, input–output analysis) gives values for cholesterol synthesis + dietary absorbed, since these values compare favorably to those measured by more direct sterol balance methods (Grundy and Ahrens, 1969; Sodhi and Kudchodkar, 1973a; Samuel *et al.*, 1978). In order to obtain this parameter, it is necessary to analyze the curves over a period of 45 weeks or longer. Evidence presented later suggests that all of the previously mentioned tracer kinetic methods, along with the earlier one-pool model (Chobanian *et al.*, 1962) and the isotopic steady-state method (Morris *et al.*, 1957; Wilson, 1964), could be used for the calculation of true body cholesterol turnover, i.e., cholesterol synthesis + total absorbed (endogenous plus dietary).

A. Compartmental Analysis

One of the parameters of cholesterol metabolism which is obtained by the kinetics of plasma cholesterol specific activity assuming a two- or a three-pool model is the amount of cholesterol transported to pool B from pool A (R_{AB}) (Nestel *et al.*, 1969; Goodman *et al.*, 1973). It is calculated as the product of the mass of cholesterol in pool A (M_A) and the rate of its transfer to pool B (k_{AB}), that is, $R_{AB} = M_A \times k_{AB}$. The value for R_{AB} is generally greater than the production rate PR_A, which is considered to be equal to synthesis or synthesis + dietary absorbed. Our analyses presented in the following suggest that calculated values for R_{AB} are similar to the values for cholesterol synthesis plus total absorbed (endogenous plus dietary) as measured by a combination of the sterol balance method with the methods measuring cholesterol absorption. Thus the values for synthesis plus total absorbed (mg/kg/day) versus the values for R_{AB} (mg/kg/day) in various groups of subjects were as follows: normolipemic (N), 19.2 ($N = 11$) versus 20.2 ($N = 21$); hypercholesterolemic (C), 16.4 ($N = 9$) verus 17.8 ($N = 15$); hypercholesterolemic subjects with xanthomatosis (C_{xan}), 15.4 ($N = 18$) versus 16.5 ($N = 7$); hypertriglyceridemic (T), 24.4

(N = 12) versus 22.0 (N = 8); hypercholesterolemic and hypertriglyceridemic (CT), 25.3 (N = 16) versus 25.6 (N = 9); and hypercholesterolemic and hypertriglyceridemic with xanthomatosis (CT_{xan}), 26.0 (N = 10) versus 25.4 (N = 5), where N equals the number of subjects. These data were obtained by separately pooling the data on cholesterol balance studies and studies on kinetic analyses of the plasma cholesterol specific activity curve by the two-pool model in different types of subjects studied in different laboratories (for references, see Sodhi et al., 1980). For the purpose of these analyses, concentrations of 250 mg and 150 mg/100 ml of plasma were arbitrarily assumed to be the upper limits of normal for plasma cholesterol and triglycerides, respectively. The subjects in groups C and CT were further separated into C_{xan} and CT_{xan} based on the presence or absence of xanthomatosis.

Since this kind of comparison is fraught with difficulties of interpretation, data obtained in the same subject by the simultaneous use of the two methods were reanalyzed (Sodhi and Kudchodkar, 1973b; Carter et al., 1979; Kekki et al., 1977). The data presented in Table III show that synthesis + total absorbed values are nearly similar to the R_{AB} values calculated in the same subject. There was a high degree of correlation ($r = 0.92$,

Table III
COMPARISON IN THE SAME SUBJECT OF VALUES FOR SYNTHESIS PLUS TOTAL ABSORBED MEASURED BY THE COMBINED STEROL BALANCE METHOD WITH THE VALUES FOR THE AMOUNT OF CHOLESTEROL TRANSPORTED TO POOL B (R_{AB}) CALCULATED BY TRACER KINETICS (TWO-POOL MODEL)

Type of subject[a]	Number of subjects	Synthesis (S) + absorbed (A) (mg/day)	R_{AB} ($M_A \times k_{AB}$) (mg/day)	% Difference $\left(\dfrac{R_{AB} - (S + A)}{(S + A)} \times 100\right)$
N[b]	3	1414	1464	+3.5
C[b]	5	879	1068	+21.5
T[b]	3	1691	1784	+5.5
CT[b]	4	2092	2123	+1.5
C_{xan}[c]	Di	574	574	0
C_{xan}[c]	De	584	478	−18
Mixed[d]	7	930[e]	950	+2

[a] Abbreviations: N, normolipemic; C, hypercholesterolemic; T, hypertriglyceridemic; CT, hypercholesterolemic + hypertriglyceridemic; xan, subject with xanthomatosis.
[b] Mean of subjects studied by Sodhi and Kudchodkar (1973).
[c] Subjects studied by Carter et al. (1979).
[d] Subjects studied by Kekki et al. (1977).
[e] Calculated assuming 40% absorption.

$p < 0.001$, $N = 17$) between the values obtained by the two different approaches. Further evidence in support of the above statement comes from the reanalysis of data in monkeys and rabbits in whom data on cholesterol metabolism was obtained in the same laboratory by the simultaneous use of the two methods. Thus, the R_{AB} values obtained by the two-pool (African green monkeys, studied by Parks *et al.*, 1977) or the three-pool model (rhesus monkeys, studied by Eggen, 1976) were similar to the values for endogenously absorbed + excreted endogenous cholesterol (synthesis + dietary absorbed) as measured by the sterol balance technique. This was also true of rabbits (Table IV) studied in two different laboratories (Massaro and Zilversmit, 1977; Huff and Carroll, 1980).

The exceptions to the just-mentioned suggestion seem to be the rat and the squirrel monkey. In these two species the rates of excretion of cholesterol from pool A (k_A) were greater than the rates of transfer of cholesterol from pool A to pool B (k_{AB}). As shown in Table V, in these two species, synthesis + absorbed as measured by the modified sterol balance method, or the turnover values calculated by the isotopic steady-state method, equaled the values calculated as the production rate ($PR_A = M_A \times K_A$), and not the values calculated as the amount of cholesterol transferred to pool B from pool A ($R_{AB} = M_A \times k_{AB}$) as in other species (Table IV).

That the turnover values obtained by tracer kinetics include absorbed cholesterol is suggested from the data presented in Table VI. When the dietary absorbed cholesterol values (as measured by Borgstrom's fecal analysis method) were added to the R_{AB} values obtained during the cholesterol-free diet period, the sum of the two values was nearly similar to the R_{AB} values calculated during the cholesterol-feeding period. The little data that are available in animals (Eggen, 1976) and in humans (Smith *et al.*, 1976) suggest that values calculated as cholesterol transported to pool C (or pool three) may represent values for cholesterol absorption alone.

In a mixed group of subjects, values for "miscible-pool" cholesterol turnover obtained by Chobanian *et al.* (1962) using the one-pool model, were 23.7 ± 6.8 mg/kg/day. These values compare favorably with the 21 ± 5 mg/kg/day measured by the sterol balance technique for synthesis plus total absorbed. In 10 subjects (11 studies) examined by Grundy and Ahrens (1969), the mean value for miscible-pool cholesterol turnover calculated by the one-pool model was 1449 mg/day (1318 mg/day if subject number 4 is removed from the analysis). The mean value for R_{AB} (calculated using cpm injected, C_A, C_B, and $t_{1/2}$ of A and B) was 1376 mg/day (1353 mg/day, $N = 10$). These analyses suggest that the miscible-pool cholesterol turnover value calculated by the one-pool model may also

Table IV

Comparison of Values for Synthesis + Total Absorbed Obtained by the Combined Sterol Balance Method with Those Calculated by Tracer Kinetics Using Compartmental Analysis (R_{AB}), Isotopic Steady State (ISS), and Input–Output Analysis[a]

Species	Number of animals	Diet[b]	True body cholesterol turnover (mg/day)				Reference
			Synthesis + total absorbed[c]	R_{AB}	ISS	Input–output	
Monkey							
Africans green[d]	9	HC	108	115	—	—	Parks et al. (1977)
Rhesus[e]							
Low responder	5	Basal	85	89	—	—	Eggen (1976)
		HC	188	193	—	—	
High responder	5	Basal	95	101	—	—	Eggen (1976)
		HC	226	214	—	—	
Rhesus	8	HC	225	—	230	—	Eggen (1974)

Baboon	8	HC	434	—	493	—	Eggen (1974)
Rabbit	6	Basal	128	126[f]	—	—	Massaro and Zilversmit (1977)
Rabbit	6	Basal	133	132	—	—	Huff and Carroll (1980)
Dog							Pertsemlidis et al. (1973a)
C		HC	1621	—	—	1630	
D		HC	1674	—	1670[g]	1730	
F		Basal	1328[h]	—	—	1140 ⎫	
			1135			⎭	
Bile diverted			1084	—	—	1320 ⎫	
			1891[i]			⎭	

[a] The data were obtained in the same animal in the same laboratory using the two methods simultaneously.
[b] HC, high cholesterol.
[c] Synthesis + total absorbed = cholesterol secreted in GI lumen + fecal bile acids.
[d] The percentage absorption value used for the calculation was from the same species studied in the same laboratory, but from a different study (St. Clair et al., 1981).
[e] Values of R_{AB} represent values calculated as Q22 by the three-pool model. The body weight is assumed to be 6 kg.
[f] In this study the rate of transfer of cholesterol from pool A to B is denoted as "kba."
[g] Calculated using Fig. 4, day 496 (Pertsemlidis et al., 1973a).
[h] Cholesterol + bile acids excreted in urine (via bile) (brackets denote studies 1, 2, and 4).
[i] Mean value for cholesterol + bile acids excreted in urine + feces.

Table V
DATA SHOWING THAT IN RATS AND IN SQUIRREL MONKEYS THE TRUE BODY CHOLESTEROL TURNOVER MEASURED BY THE MODIFIED STEROL BALANCE METHOD OR BY THE ISOTOPIC STEADY-STATE METHOD EQUALS PR_A AND NOT R_{AB} (ISOTOPE KINETICS, TWO POOL) AS IN OTHER ANIMAL SPECIES

	Sterol balance (mg/kg/day)		Isotope kinetics (mg/kg/day)		
			Two-pool method		
Species	Present (synthesis)	Modified (synthesis + absorbed)	PR_A	R_{AB}	Steady-state turnover
Rat[a]	33 ± 9	58 ± 8	57 ± 10	35 ± 6	53[b]
Squirrel monkey[c]	52	92	93	55	92

[a] Mean values for six groups of rats fed low cholesterol semisynthetic diets with different fats (calculated from Feldman et al., 1978a,b).

[b] Mean value for a group of rats fed diet CEL (calculated from Mathe and Chevallier, 1979).

[c] Values based on sterol balance (Eggen, 1974) and isotope kinetic (Lofland et al., 1970; Raymond et al., 1976) data.

Table VI
DATA IN RABBITS SHOWING THAT R_{AB} VALUES MAY INCLUDE ABSORBED CHOLESTEROL VALUES[a]

Strain of rabbit	Number	Diet	Dietary cholesterol absorbed[b] (mg/day)	R_{AB} $(M_A \times kba)$[c] (mg/day)	R_{AB} control + dietary absorbed (mg/day)
Becken					
Females	2	Control	—	123	—
Females	2	Atherogenic	76	216	199
Becken					
Males	2	Control	—	140	—
Males	2	Atherogenic	82	219	222
Jackson					
Females	2	Control	—	116	—
Females	2	Atherogenic	75	182	191

[a] Analysis of data from Massaro and Zilversmit (1977).

[b] Measured by the fecal analysis method of Borgstrom (1969).

[c] The rate of transfer of cholesterol from pool A to pool B is denoted as "kba" in this study.

represent the value for synthesis + total absorbed and not just synthesis or synthesis + dietary absorbed.

An intriguing aspect of the production rate values obtained by the two-pool model was that the values were more nearly equal to the values calculated for the secretion of cholesterol in the GI lumen (except in the rat and squirrel monkey). In the eleven studies carried out by Grundy and Ahrens (1969), the mean values obtained by the one-pool model differed from the values for k_{PR} (1017 mg/day) by an average of 40% (28%, N = 10). When the fecal bile acid values (k_{FBA}; newly synthesized bile acids secreted into bile) available in each subject were added to the k_{PR} values, the sum of two (PR_A + FBA = 1281 mg/day, N = 10) differed from the values obtained by the one-pool model by only 2%. In 16 subjects studied by Sodhi and Kudchodkar (1973b), the PR_A values (1089 ± 476 mg/day) were similar to the values for cholesterol secreted in the GI lumen (1054 ± 538 mg/day) calculated in the same subjects. This was also true of animal studies. In low- as well as in high-responding (LR and HR, respectively) rhesus monkeys studied by Eggen (1976), the values (in mg/kg/day) for cholesterol secreted in the intestinal lumen on the basal diet were 10.3 and 10.4, while those for the production rate were 11.2 and 10.2, and those for excretion were 8.4 and 7.6, respectively. On the atherogenic diet, the values for secretion were 24.0 and 26.4, those for PR_A were 23.2 and 29.4, and those for excretion were 20.1 and 23.5, respectively.

B. Isotopic Steady-State Method

Calculations of cholesterol turnover values by the isotopic steady-state method from the data available in baboons and rhesus monkeys studied by Eggen (1974), and in dog D studied by Pertsemlidis *et al.* (1973a), show that these values are nearly identical to the values for cholesterol + newly synthesized bile acids excreted in the GI lumen (Table IV). In the four subjects studied by Grundy and Ahrens (1969), the values for the daily cholesterol turnover calculated by the isotopic steady-state method (1258 mg using absorption values obtained by method II) tended to be similar to the true body cholesterol turnover rate calculated either by the one-pool method (1330 mg) or as R_{AB} by the two-pool method (1356 mg). Similarly, in rats as well as in squirrel monkeys values for turnover calculated by the isotopic steady-state method were nearly equal to the values for synthesis + absorbed (Table V). These results suggest that the isotopic steady-state method also measures the true body cholesterol turnover rate, that is, cholesterol synthesis + total absorbed and not just cholesterol synthesis + dietary absorbed.

C. Input–Output Analysis

Analysis of the data on cholesterol turnover obtained by input–output analysis suggests that input values calculated by the analysis of the plasma cholesterol specific activity curve *for a period of 5–10 weeks* may also represent values for cholesterol synthesis + total absorbed and not synthesis + dietary absorbed as is presently believed. Thus in a mixed group of subjects, the mean value for daily cholesterol input was 19.8 mg/kg (Samuel *et al.*, 1978), a value similar to the 18–21 mg/kg obtained for cholesterol synthesis + total absorbed by the combined sterol balance technique. Further evidence in support of this is obtained by reanalysis of the data available in the dogs studied by Pertsemlidis *et al.* (1973a). Only when the calculated endogenously absorbed cholesterol values were added to the sterol balance values (synthesis plus dietary absorbed) were the turnover values obtained by the two methods were identical (Table IV).

Thus, the fact that concordant results are obtained by the various tracer kinetic analyses of plasma cholesterol specific activity curves and the combination of sterol balance with different techniques of measuring absorption, methods which are so different in approach, suggests that synthesis + total absorbed (endogenous + dietary) or synthesis (when total absorption is zero) values obtained by these various methods probably approximate the true body cholesterol turnover (synthesis in the absence of dietary cholesterol) rate. This belief is further strengthened by our reanalysis of data on total body cholesterol mass.

IV. Total Body Cholesterol Mass and Individual Pool Sizes

Tracer kinetic methods permit calculation of the total body mass of cholesterol (M_T). At present, in order to obtain this parameter it is necessary to analyze the plasma cholesterol specific activity curve obtained over a period of at least 45 weeks (Goodman *et al.*, 1973; Samuel and Lieberman, 1973), which is a major drawback of these methods. To circumvent this problem, Samuel *et al.* (1978) used the daily cholesterol turnover values measured by the sterol balance method (synthesis + dietary absorbed) with their input–output analysis and obtained a minimum value for M_T by analyzing the curve for a period of only 10–12 weeks. For obvious reasons calculated values for M_T cannot be validated in man but can be validated in animals. The results of such comparisons, however, are inconclusive: while Lofland *et al.* (1970) found that in squirrel monkeys (on a cholesterol-free diet) chemically measured values were similar to those calculated by the input-output analysis or by the two-pool method ($M_A + M_B = M_T$), Wilson in baboons (1970) and Hough and Basset in rabbits (1975) found that the M_T values calculated by the two-pool method were considerably smaller than chemical estimates. To account for the differences Wilson (1970) introduced pool C, which was assumed to exchange very slowly with plasma cholesterol. Pertsemilidis et al. (1973b) have compared the chemically obtained values for M_T in dogs with those obtained by input-output analysis. Only in one out of six was the chemically measured value identical to the calculated value. In the remaining five the M_T was underestimated by 30-50%. Findings in guinea pigs (Green et al., 1976) were also similar. As stated earlier, the two-pool model was modified to a three-pool model (Goodman et al., 1973). Values for M_T ($M_A + M_B + M_C$) obtained by this method in one study have been shown to be nearly the same as those measured chemically (Raymond et al., 1976).

The M_T by the isotopic dilution principle is calculated as injected dose of radioactivity/specific activity of plasma cholesterol at zero time (time at which the label is infused) extrapolated from the linear portion of the curve. Daily body cholesterol turnover is then calculated as $M_T \times \beta$, where β = slope of the linear portion of the curve. Since turnover = $M_T \times \beta$, M_T = turnover/β (Zilversmit, 1960; Chobanian et al., 1962). Evidence presented later suggests that M_T can most accurately be calculated using the above formula. The values for turnover (synthesis + total absorbed) could be obtained either by combined sterol balance or by tracer kinetics (R_{AB} or input rate). The advantage of this new approach is that M_T can be obtained from the analysis of the curve obtained over a period of 5-6 weeks (or as soon as the plasma cholesterol specific activity begins to decline linearly).

A. TOTAL BODY CHOLESTEROL MASS

Recalculation of M_T using the formula R_{AB}/β (where $\beta = 0.693/t_{1/2}$ days of the second exponential in the two-pool model) showed that the value (mean ± standard deviation) for M_T in five cholesterol-fed baboons studied by Wilson was 1384 ± 331 mg/kg compared to 1339 ± 207 mg/kg obtained chemically (Wilson, 1970). Mean values for synthesis + total absorbed as well as β values could be computed from the isotopic sterol balance data in eight cholesterol-fed baboons studied by Eggen (1974). Values of M_T calculated using the formula (synthesis + total absorbed)/β were 1360 mg/kg. The M_T value on a low cholesterol diet was 1267 mg/kg ($N = 3$) in baboons studied by Wilson (1972) and 1268 mg/kg in baboons studied by Eggn [endogenous cholesterol secretion and absorption was

Table VII

Data In Squirrel Monkeys Showing That Whole Body Cholesterol Mass Calculated Using the Formula (Turnover/β) Is Similar to That Obtained by Chemical Analysis of the Carcass[a]

	Body cholesterol mass (mg)	
Dietary group	Chemical	Turnover/β[b]
Lard	1402	1463
	(1269–1664)[c]	
Lard + cholesterol	2583	2594
	(1585–3203)	
Safflower + cholesterol	2890 (2524)[d]	2513
	(1308–6172)	
Coconut oil + cholesterol	2304	2297
	(1507–3922)	
Butter + cholesterol	3037	3135
	(1827–4328)	
Corn oil + cholesterol		
Hyporesponder		2513
Hyperresponder		2717

[a] Analysis of data from Lofland et al. (1970).

[b] Turnover in squirrel monkeys = synthesis + absorbed (endogenous + exogenous) or production rate (PR_A). $\beta = 0.693/t_{1/2}$ days of the second exponential in a 5–6 week study. β values used for calculation are from other studies ($\beta = 0.063$ on a cholesterol-free diet and 0.037 on a cholesterol-containing diet).

[c] Numbers in parentheses denote the range of values.

[d] After removing one monkey with the highest body cholesterol mass.

calculated using the percentage cholesterol absorption value from baboons on a similar diet studied by Mott et al. (1980)]. In rabbits on a control diet, chemically measured values for M_T were 1001 ± 95 mg/kg (Hough and Basset, 1975). Those obtained by calculations were 992 mg/kg from one study (Huff and Carroll, 1980) and 1001 mg/kg from another (Massaro and Zilversmit, 1977) (β values for these calculations were obtained roughly from the specific activity time curves published in the two references). In squirrel monkeys fed high cholesterol with different fats (Lofland et al., 1970), calculated values (synthesis + absorbed or PR_A/β) were also nearly equal to the chemically measured values (Table VII). The same was also true for rats.

That this remarkable similarity of M_T values obtained by the two approaches with those measured chemically is not accidental is suggested by the analysis of available data in dogs studied by Pertsemlidis et al. (1973a,b). Complete sets of data required for these calculations were available in three out of six dogs. In each of these dogs, the M_T obtained by our modification was virtually identical to that obtained chemically (Table VIII). Values of M_T were also calculated in control dogs A and B in whom endogenously absorbed cholesterol was calculated using percentage absorption from dog C on a low cholesterol diet. In dog E it was calculated using the β value from the similarly treated dog F (Study 4). In dog A the calculated values were grossly overestimated. In dogs B and E the difference between the two methods was less than 2 g (Table VIII).

Raymond et al. (1976) have shown that in squirrel monkeys the M_T values calculated by using the three-pool model ($M_A + M_{B,min} + M_{C,min}$) were close to those measured chemically in the same animals. Values of M_T calculated by the three-pool model in different groups of subjects are available from Goodman's laboratory (Smith et al., 1976; Goodman et al., 1980). Comparison of those values with those calculated by our modification show (Table IX) that the recalculated values agree closely with the M_T values obtained by the addition of mass of cholesterol in pool A (M_A),

Table VIII

IDENTITY OF WHOLE BODY CHOLESTEROL MASS OBTAINED BY THE FORMULA (SYNTHESIS + TOTAL ABSORBED)/β WITH THAT OBTAINED CHEMICALLY IN THE SAME ANIMAL[a]

Dog	Study	Synthesis + total absorbed[b] (mg/day)	β (day^{-1})	Body cholesterol mass (g)		
				(Synthesis + total absorbed)/β	Chemical	Samuel's method
C	5	1604	0.0396	40.5	39.6	39.8
D	4	1052	0.0315	33.4	33.8	24.6
	7	1674	0.0495	33.8	33.8	29.8
F	1–4	1891[c]	0.0587	32.2	32.4	19.8
			0.0603	31.4	32.4	21.2
E	1–4	2824[c]	0.0603[d]	46.8	45.5	—
B	2	1192[e]	0.0315	37.8	38.8	25.9
A	2	1807[e]	0.0198	91.3	52.5	38.2

[a] Analysis of data from Pertsemlidis et al. (1973a,b).

[b] Total absorbed = exogenous + endogenous. Absorption of endogenous cholesterol was calculated as described in the text.

[c] In these dogs with complete bile diversion the values represent mean values for total endogenous excretion.

[d] β Value from the similarly treated dog F.

[e] Calculated using percentage absorption from dog C, Study 2.

Table IX

COMPARISON OF VALUES OBTAINED FOR WHOLE BODY CHOLESTEROL MASS (M_T) CALCULATED BY THE THREE-POOL MODEL WITH THOSE OBTAINED BY THE MODIFIED METHOD USING R_{AB} VALUES OR THE SYNTHESIS + TOTAL ABSORBED VALUES[a]

Type of subject	Three-pool model (mg/kg)		Modified method[b] (mg/kg)	
	$M_1 + M_{2,\min} + M_{3,\min}$	$M_1 + M_{2,\max} + M_{3,\min}$	R_{AB}/β	(Synthesis + absorbed)/β
N	982	1062	1082	1022
C	1362	1446	1402	1238
T	1005	1109	1102	1203
CT	1152	1255	1278	1270

[a] Analysis of data pooled from literature.
[b] Subjects with xanthomatosis have been removed from these groups.

maximum mass in pool B ($M_{B,\max}$), and minimum mass in pool C ($M_{C,\min}$) ($M_T = M_A + M_{B,\max} + M_{C,\min}$).

Our analysis also suggested that mass of body cholesterol calculated using values for excretion (synthesis) alone nearly equals $M_A + M_B$, while that calculated using reabsorbed equals M_C (Fig. 1). This supports the earlier suggestion that cholesterol calculated as that transported to pool C in a three-pool model may represent total absorbed cholesterol. Interestingly, in baboons studied by Wilson (1970) it was found that when carcass radioactivity originating from dietary labeled cholesterol was divided by dietary cholesterol specific activity, the mass values obtained resembled closely the values of cholesterol in pool C [$MT_{chem} - (M_A + M_B)$], so that when these newly calculated values were added to the calculated "exchangeable" cholesterol values ($M_A + M_B$), the sum was identical to the chemically measured M_T in each baboon (Table X).

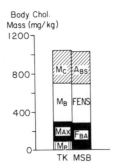

FIG. 1. Total body cholesterol mass as determined by the modified tracer kinetic (TK) and the modified sterol balance (MSB) methods. The values for the mass of body cholesterol calculated using the values for cholesterol excretion (FENS + FBA, i.e., fecal endogenous neutral sterols + fecal bile acids) equaled the values for $M_A + M_B$, and those calculated using the values for reabsorbed cholesterol (A_{BS}) nearly equalled the values for the mass of cholesterol in pool C (M_C).

Table X
BODY CHOLESTEROL MASS BY THE ISOTOPIC STEADY-STATE METHOD[a]

Baboon	Dietary [³H]cholesterol (mg/day)		Total body radioactivity[b] (dpm)	Total body cholesterol mass (g)			
				Calculated[c]			
	Fed specific activity	Plasma specific activity from fed		Fed specific activity	Plasma specific activity	Total (Fed + plasma specific activity)	Chemical
1	3169	2472	42.2 × 10⁶	13.316	17.071	30.4	30.8
2	3239	2527	48.0 × 10⁶	14.819	18.994	33.8	33.4
3	5603	2970	59.8 × 10⁶	10.642	20.134	30.8	31.2
4	5634	3355	61.9 × 10⁶	10.986	18.450	29.4	28.4
5	5147	3340	67.7 × 10⁶	13.153	20.269	33.1	29.5

[a] Analysis of data from Wilson (1970).
[b] Total body cholesterol radioactivity originating from dietary [³H]cholesterol.
[c] Total body cholesterol radioactivity originating from dietary [³H]cholesterol/fed or plasma-specific activity.

Our analysis also suggested that when body pools of cholesterol are grossly expanded on a cholesterol diet or when there is an accelerated synthesis of cholesterol (e.g., ileal bypass, biliary diversion, bile acid-binding resins, gross obesity), the true body cholesterol turnover calculated by the tracer kinetic methods (R_{AB} or input rate) is not similar to that obtained by the combined sterol balance method (synthesis plus total absorbed). The values obtained by the tracer kinetic methods are lower (the reason for this is given later), and consequently the calculated M_T values are lower than those calculated by the combined sterol balance method. The latter values are, however, correct, since the calculated values for M_T in dogs with biliary diversion (E and F) are identical to those measured chemically (Table VIII).

The failure of the R_{AB} or of the input values calculated by tracer kinetics to give values identical to those measured by combined sterol balance in the previously mentioned conditions may be due to the fact that the tracer kinetic analysis assumes that all synthesized (and absorbed) cholesterol passes through the plasma at least once. Any newly synthesized (or absorbed) cholesterol that bypasses the plasma, such as that synthesized in the liver, that absorbed and removed by the liver that passes directly, as such or after its conversion to bile acids, via bile into the feces, or that stored in the liver, would not be reflected in the kinetic calculations. However, that very portion would be included in the measurement of turnover by the combined sterol balance method. Similarly, cholesterol synthesized or absorbed by the intestine and excreted directly into the feces would not be reflected in the kinetic calculations (Sodhi and Kudchodkar, 1973; Samuel *et al.*, 1978).

This concept is best illustrated by the comparison of data on synthesis plus total absorbed measured by combined sterol balance with data calculated by input–output analysis in dogs (Table XI). Whereas in dogs C and D the calculated cholesterol input rates are identical to the values for synthesis + total absorbed, in the bile-diverted dog (F) the input rate (1140–1320 mg) is not similar to synthesis measured by sterol balance (1600–2430 mg). This is expected since the cholesterol synthesized by the intestines or that transported to the intestines by blood lipoproteins will be excreted directly in the feces as a result of its not being absorbed. In dog F, the cholesterol excreted in the feces (synthesized and/or removed but not absorbed and hence not transported to plasma) was 451–534 mg. Removal of this from total synthesized (sterol balance) gives values similar to those obtained by input–output analysis, suggesting that input–output analysis (and other tracer kinetic methods) measures only the cholesterol entering the plasma. The nonidentity of the values calculated by tracer kinetics with those measured by combined sterol balance in

Table XI

DATA ILLUSTRATING THE CONCEPT THAT THE TRACER KINETIC METHOD MEASURES ONLY THE CHOLESTEROL (SYNTHESIZED + ABSORBED) ENTERING PLASMA[a]

Dog	Cholesterol input (mg/day)		Body cholesterol mass (g)			Tracer kinetics corrected	
	Sterol balance	Tracer kinetics	Chemical	Sterol balance	Tracer kinetics	Input[b] (mg/day)	Mass (g)
C	1604	1630	39.6	40.5	41.1	—	—
D	1674	1730	33.8	33.8	34.9	—	—
F[c]	1891[d]	1320	32.4	32.2	22.5	1887	32.1

[a] Analysis of data from Pertsemlidis et al. (1973a,b).
[b] Turnover value by tracer kinetics + fecal excretion by sterol balance. Cholesterol excreted in feces was not absorbed due to the lack of bile, and therefore was not transported from the liver to the plasma. Note that the amount excreted in urine (1328 mg/day) as measured by sterol balance is nearly the same as that calculated by tracer kinetics (1320 mg/day).
[c] Bile diverted in urine.
[d] Urinary (= biliary) excretion (1328 mg/day) + fecal excretion (563 mg/day) = 1891 mg/day.

gross obesity, ileal bypass, or bile acid-binding resin feeding may be due to the fact that all the cholesterol synthesized by the intestines in these conditions may not be absorbed and thus may not be transported into the plasma. This cholesterol, however, will be excreted in the feces and thus accounted for by the sterol balance method. A marked increase in intestinal cholesterol synthesis under these conditions has been shown (Grundy et al., 1971; Grundy, 1978).

In both the hypo- (LR) and hyperresponding (HR) rhesus monkeys studied by Eggen (1976) with tracer kinetics and with sterol balance, the M_T values calculated by the two approaches were nearly identical during the basal period but differed by 1000 mg/kg during cholesterol feeding. The values obtained by the combined sterol balance method were higher and tended to be similar to those measured chemically in other experiments (Bullock et al., 1975). Fecal endogenous cholesterol excretion data (sterol balance) in these monkeys suggested that the dietary absorbed cholesterol was retained in the body (8.6 mg/kg/day in LR, 11.5 mg/kg/day in HR). When M_T values calculated by the tracer kinetic method were recalculated using the correct input rate (R_{AB} + dietary retained), they were similar to those obtained by the combined sterol balance method (Fig. 2).

Data available in guinea pigs (Green et al., 1976) further support the

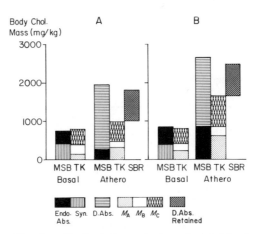

FIG. 2. Relationship of total body cholesterol mass calculated by the modified isotopic sterol balance method with that calculated by the modified tracer kinetic (two-pool model) method. (A) Hyporesponders, (B) hyperresponders. On a high cholesterol (Athero) diet the total body cholesterol mass (M_T) calculated by the modified tracer kinetic (TK) method is lower than that calculated by the modified sterol balance (MSB) method. However, when the mass of cholesterol contributed by the dietary cholesterol retained in the body (D. Abs.) was calculated (SBR) and added to the mass obtained by TK, it nearly equalled the M_T calculated by MSB. The data indicate that values for input calculated by TK may fail to reflect cholesterol not transported to the plasma. Endo. Abs., endogenously absorbed; syn, synthesis. [Based on the analysis of data from Eggen (1976).]

suggestion that absorbed cholesterol stored in the liver also is not reflected in the input values calculated by tracer kinetics. In these guinea pigs, the input of endogenous cholesterol on a control diet was 25 mg/day. The calculated input upon cholesterol feeding was 37 mg/day. The chemically measured total body cholesterol mass in these cholesterol-fed guinea pigs was 1790 mg, and was considerably higher than the value of 1606 mg calculated using the input of 37 mg/day obtained by the input–output analysis. If the chemically measured cholesterol stored in the body (4 mg/day; 3.6 mg in the liver and most of the remaining in the intestines) was added to the input (37 mg + 4 mg = 41 mg) and the body cholesterol mass was recalculated [the β value used for the calculation was calculated as $0.693/t_{mean}$ (days); t_{mean} by input–output analysis represents mean transit time or turnover time, which is $1.44 \times t_{1/2}$ days; the identity of the chemically measured value with that calculated using β suggests that t_{mean} in this short-term experiment may represent $t_{1/2}$ days and not the transit time], it yielded nearly the same value (1775 mg) as the chemical mass (1790 mg). Thus, the failure of the tracer kinetic methods to account for the cholesterol stored in the liver and intestines seems to be the reason for the underestimation of the input rate and, therefore, the M_T values on high cholesterol diet. On such diets, the values obtained by modified combined sterol balance are valid since they are similar to those measured chemically. These observations also suggest that cholesterol stored in the body (including the liver) may not equilibrate or exchange significantly with the cholesterol in the plasma compartment.

Interestingly, in these cholesterol-fed guinea pigs the corrected input (37 mg + 4 mg = 41 mg) was identical to the sum of the input during control period (25 mg) and input through dietary cholesterol absorption (16 mg) as measured by Borgstrom's (1969) method (25 mg + 16 mg = 41 mg). Since the total body cholesterol mass calculated using the above input value is similar to the chemically measured mass, these data indicate that in these guinea pigs there was no compensatory decrease in whole body cholesterol biosynthesis upon cholesterol feeding.

The identity of the values for the true body cholesterol turnover (synthesis + total absorbed or luminal cholesterol + fecal bile acids) measured by the combination of sterol balance and the methods measuring cholesterol absorption with the values calculated by the various tracer kinetic methods, the methods which measure cholesterol entering the plasma, suggests that under normal conditions both the absorbed (exogenous and/or endogenous) and the biosynthesized cholesterol is transported first to the plasma before being excreted, as such or after its conversion to bile acids, into the bile. Cholesterol not transported to the plasma is mostly stored in the liver. Similarly, the identity of chemically measured values for the total body cholesterol mass with those calculated by the kinetic analysis of plasma cholesterol specific activity over a period of 4–6 weeks, and the combination of isotopic sterol balance and absorption methods, validates the methods and the concept underlying these methods, that is, the true body cholesterol turnover rate includes endogenously absorbed plus endogenous excretion and not just excretion of cholesterol alone. Since in the absence of dietary cholesterol all the cholesterol in the body has to originate from biosynthesis, it is suggested that the true body cholesterol turnover values and not the values for excretion alone represent the values for total body cholesterol synthesis.

While the source of endogenously reabsorbed cholesterol is not known at present, the following observations suggest that it is likely to be extrahepatic. Both *in vivo* and *in vitro* studies suggest that hepatic biosynthesis (as calculated by incorporation of labeled precursors in cholesterol or by measurement of HMG-CoA reductase) is markedly reduced upon cholesterol feeding (Tomkins *et al.*, 1953; Dietschy and Siperstein, 1967), as well as upon the feeding of a cholesterol-free semisynthetic diet (Carroll, 1971; Reiser *et al.*, 1977; McNamara *et al.*, 1982). Under these conditions, synthesis of cholesterol as measured by the sterol balance method (total endogenous excretion-dietary absorbed) is also markedly decreased, suggesting that the excretion of endogenous cholesterol in the feces may represent the cholesterol synthesized in the liver. Furthermore, Dietschy and his co-workers (Nervi *et al.*, 1975; Anderson *et al.*, 1979) have shown that the hepatic synthesis of cholesterol is negatively correlated with the cholesteryl ester content of the liver. Our analysis of data (Mathe and Chevallier, 1976, 1979; Mathe *et al.*, 1977; Kellog, 1974) on cholesterol synthesis as measured by the sterol balance method in rats fed a variety of diets showed a significant negative correlation (Fig. 3; $r = -0.89$, $N = 20$ groups) with the amount of cholesteryl esters in the liver, especially when it was expressed as the percentage of total cholesterol in the liver.

Further support for this concept comes from the comparison of values for the true body cholesterol turnover obtained by the various modified approaches with the total body cholesterol synthesis value measured *in vivo* using ^3H_2O in Dietschy's laboratory (Jeske and Dietschy, 1980; Turley *et al.*, 1983). The data presented in Table XII show that in rats, the values for total body cholesterol synthesis (mg/day) measured by these various approaches are nearly the same. It is also interesting to note that while the values for the endogenous cholesterol excreted out of the body (total body cholesterol synthesis, as is presently believed) are vaguely similar to the values for synthesis of cholesterol by the liver only, the values for the reabsorbed cholesterol (calculated by the sterol balance method) are closely similar to the values for cholesterol synthesis by the remaining tissues of the carcass (total body minus liver).

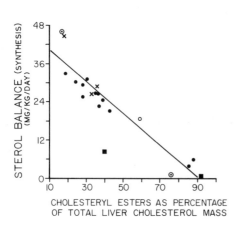

FIG. 3. Relationship of cholesterol synthesis (sterol balance) with the mass of cholesteryl esters in the liver. The biosynthesis of cholesterol (excretion − intake) was negatively correlated ($r = -0.89$, $N = 20$) with the mass of cholesteryl esters (as percentage of total liver cholesterol) in the livers of rats fed different cholesterol-free and cholesterol-containing diets. The data suggest that cholesterol synthesis as measured by sterol balance may primarily reflect the cholesterol synthesis activity of the liver. [Based on data from Mathe and Chevallier (1976, 1979), Mathe *et al.* (1977), and Kellogg (1974).]

Interestingly, when the flux of cholesterol from liver to plasma (or turnover of liver cholesterol) was calculated by the isotope kinetic method [(mass of cholesterol in liver $\times$ α)/2, where α is the rate constant ($0.693/t_{1/2}$ days of exponential A] describing the biexponential specific radioactivity time curve of plasma cholesterol; in this calculation it is assumed that α represents the rate constant for the input of cholesterol (nonradioactive) from liver to plasma as well as the removal of cholesterol transported to the plasma from extrahepatic tissues], the value obtained for the liver cholesterol turnover was nearly the same as that measured as synthesis by the conventional sterol balance method (Table XIII). This, however, was not true in conditions where there was an accumulation of cholesterol in the liver (e.g., feeding cholesterol or cholesterol-free casein-containing diets). Under these conditions, the flux of cholesterol from liver to plasma calculated as the product of liver cholesterol mass (control period) and α (experimental period) divided by 2 was similar to the synthesis of cholesterol as measured by the sterol balance method during the experimental period (Table XIII). These data support the suggestion made earlier that the cholesterol stored in the liver may not exchange or equilibrate with the cholesterol in the plasma compartment.

Thus these exhaustive analyses suggest that the presently established methods have the potential for in-depth studies of cholesterol metabolism. However, they need to be rigorously reevaluated using this modified approach. These analyses suggest that under normal conditions, all the cholesterol in the body is in rapid equilibrium (5–6 weeks as opposed to 50–60 weeks) with plasma cholesterol. The concept of rapidly, slowly, and very slowly exchangeable pools, therefore, may not be valid. Similarly, these analyses raise questions about the exchangeability of cholesterol between the plasma, liver, and other body pools of cholesterol. As shown in the next section, our analysis suggests that the equilibration of plasma cholesterol with the body tissues may not involve the exchange of cholesterol as is presently believed (Goodman and Noble, 1968; Grundy and Ahrens, 1969), but may involve the receptor and nonreceptor pathways discovered by Brown and Goldstein (1976).

B. WHAT ARE THE EXCHANGEABLE POOLS?

Upon *in vivo* labeling of plasma cholesterol, the labeled cholesterol disappears from the plasma in a series of declining exponential rates for periods of as long as 50–60 weeks. These exponentials are believed to be due to the differential exchange of plasma cholesterol with tissue cholesterol (Grundy and Ahrens, 1969; Goodman *et al.*, 1973). Based on this, the tissues in the body have been grouped into three pools (Goodman *et al.*,

Table XII

DATA IN RATS SHOWING THAT THE VALUES FOR TOTAL BODY CHOLESTEROL SYNTHESIS MEASURED *in Vivo* BY USING ^3H_2O ARE SIMILAR TO THE TRUE BODY CHOLESTEROL TURNOVER VALUES AS MEASURED BY THE VARIOUS MODIFIED APPROACHES[a]

Body weight (g)	Diet	Daily synthesis of cholesterol					
		Whole body (mg)	Whole body (mg/kg)	Liver (mg)	Liver (mg/kg)	Extrahepatic (mg)	Extrahepatic (mg/kg)
		Synthesis		^3H_2O incorporation			
				Synthesis		Synthesis	
172	Chow	23.9	139.0	12.1	70.1	11.8	68.9
203	Chow (middark)	14.0	69.0	6.3	31.2	7.7	37.8
320	Chow	22.1	69.0	14.6	45.6	7.5	23.4
178	Chow + cholesterol	10.5	59.1	0.54	3.0	10.0	56.2
154	Fast (48 hours)	6.8	43.9	1.06	6.9	5.7	37.1
279	Semisynthetic (−EFA)[b]	11.8	42.2	3.7	13.2	8.1	29.0

		Modified sterol balance			
		Synthesis + endogenously absorbed	Synthesis		Endogenously absorbed
335	Chow	27.1 80.9	15.4	45.9	11.8 35.5
465	Chow	31.4 89.0	26.6	57.2	14.8 31.8
330	Chow	21.4 65.0	14.8	45.0	6.6 20.0
470	Chow + cholesterol	16.5 35.1	0.47	1.0	16.0 34.0
310	Semisynthetic	17.7 57.0	9.9	32.0	7.8 25.0
491	Semisynthetic + cholesterol	10.6 21.6	2.0	4.1	8.6 17.5
458	Semisynthetic + hypothyroid	11.7 25.6	3.9	8.5	7.8 17.0
		Isotopic steady state			
		Endogenous turnover			
171	Semisynthetic + cholesterol	11.0 64.3			
466	Semisynthetic + cholesterol	22.2 47.6			
		Isotope kinetics, two-pool			
		Production rate (PR$_A$)			
240	Chow	22.5 93.7			
377	Semisynthetic	15.4 40.9			
310	Semisynthetic	17.7 57.0			

[a] Based on data from Jeske and Dietschy (1980), Turley et al. (1983), Miettinen et al. (1981), Mathe and Chevallier (1979), Wilson (1964), Lei and Lei (1981), and Feldman et al. (1978a,b).
[b] EFA, essential fatty acids.

Table XIII

DATA SHOWING THAT THE VALUES FOR THE FLUX OF CHOLESTEROL FROM LIVER TO PLASMA AS CALCULATED BY THE KINETICS OF PLASMA CHOLESTEROL SPECIFIC ACTIVITY ARE SIMILAR TO THE VALUES FOR CHOLESTEROL SYNTHESIS (EXCRETION) AS MEASURED BY THE STEROL BALANCE METHOD[a]

Species	Diet	Mass of cholesterol in the liver (chemical) (mg)	k_{FT} of pool A (α)[b] (day^{-1})	Flux of cholesterol from liver to plasma[c] (mg/day)	Sterol balance (mg/day)	
					Synthesis[d]	Turnover
Rabbit	Stock	264	0.578	76	76	75
	HC[e]	2450	0.231	282	29	107
	Stock[f]	264	0.231	30	29	—
	Stock	240	0.578	69	74	74
	Casein	343	0.231	40	29	29
	Stock[f]	240	0.231	28	29	—
Rat	Stock	44	0.693	15	15	19
	Casein	48	0.347	8	9	13
	Hypothyroid	48	0.198	5	4	10

Squirrel monkey	LC	132	0.462	31	33	33
Baboon	LC	1700	0.231	196	196	196
Dog						
A (Study 2)	LC	1950	0.231	225	258	278
B (Study 2)	LC	1234	0.257	159	225	239
C (Study 5)	HC	1808	0.217	196	92	1419
	LC[g]	1234	0.217	133	127[h]	—
D (Study 4)	HC	1505	0.210	158	183	573
	LC[g]	1234	0.210	130	131[i]	—

[a] Based on the analysis of data from Massaro and Zilversmit (1977), Huff and Carroll (1980), Hough and Basset (1975), Mathe and Chevalier (1976, 1979), Feldman et al. (1979a,b), Raymond et al. (1976), Bullock et al. (1975), Lehner et al. (1972), Eggen (1974), Wilson (1970, 1972), Pertsemlidis et al. (1973a,b).

[b] k_{FT}, fractional turnover rate; α is the rate constant $(0.693/t_{1/2}$ days of exponential A) describing the biexponential specific radioactivity time curve of plasma cholesterol.

[c] Calculated as [mass of cholesterol in liver (chemical) $\times \alpha$]/2, assuming that α represents the rate constant for the input of cholesterol (nonradioactive) from the liver to the plasma, as well as the rate constant for the removal of the same cholesterol transported to the plasma from the peripheral tissues.

[d] Synthesis = total endogenous cholesterol excreted in the feces (turnover) − exogenous absorbed cholesterol.

[e] HC, high cholesterol; LC, low cholesterol.

[f] Flux of cholesterol from the liver to the plasma calculated as [liver cholesterol mass (control) $\times \alpha$ (experimental period)]/2.

[g] Liver cholesterol mass of dog B.

[h] Mean of Studies 3 and 5.

[i] Mean of Studies 3 and 4.

1973). While cholesterol in the tissues of pool A is believed to exchange rapidly, that in the tissues of pools B and C is believed to exchange slowly and very slowly, respectively, with plasma cholesterol (Goodman and Noble, 1968; Wilson, 1970; Samuel and Lieberman, 1973). The mechanism responsible for the exchange is not known, but is believed to be a physical process involving a collision between lipoprotein unesterified cholesterol (UC) and tissue UC (Gurd, 1960; Bell, 1976). The process does not involve any *net* transfer of cholesterol and is not affected by metabolic events, e.g., the time required for the equilibration of specific activity among different plasma lipoproteins or between plasma lipoprotein and red blood cell cholesterol is the same under a variety of conditions affecting cholesterol and/or lipoprotein metabolism (Sodhi, 1975; Bell, 1976).

The assumption that the rapid exponential portion of plasma cholesterol specific activity is due to exchange can be questioned on the following grounds:

1. Besides the intravascular (plasma) pool, lipoproteins are known to be present in the extravascular pool (tissue fluids). Studies on lipoprotein–apoprotein turnover have shown that the intra- and extravascular (IV and EV, respectively) pools are in rapid equilibrium (Julien *et al.*, 1981). The rapid phase of the apoprotein specific activity curve (or, for that matter, a number of other radioactive compounds introduced in the plasma compartment) is believed to be due to the mixing of the label into the IV and EV pools (Scott and Hurley, 1970; Langer *et al.*, 1972; Blum *et al.*, 1977). Since cholesterol is an integral part of lipoprotein, it would be expected to distribute into the IV and EV pools as rapidly as lipoproteins. As suggested earlier, if the rapid decline of plasma cholesterol specific activity is due to the mixing of the label between the IV and EV pools, and not to exchange, then conditions which affect the distribution of lipoproteins [especially that of low-density lipoproteins (LDL), since they are the major cholesterol-carrying lipoproteins] between the two pools would also affect the rapidly declining exponential of plasma cholesterol specific activity ($t_{1/2}$ days of exponential A). That such is the case is suggested by the data available in the literature, e.g., in homozygous hypercholesterolemic subjects there is a decrease in the EV pool of LDL apoB as compared to normolipemic subjects (Bilheimer *et al.*, 1979). The $t_{1/2}$ of the rapidly declining exponential of plasma cholesterol specific activity in the former type of subject is 6–9 days (Carter *et al.*, 1979), compared to 3–4 days in normolipemic subjects (Goodman *et al.*, 1973; Sodhi *et al.*, 1980). Similarly, in normal rabbits the $t_{1/2}$ of the rapid exponential of the plasma cholesterol specific activity curve, which is 1–2 days, is increased to 3–4 days upon cholesterol feeding (Hough and Basset, 1975). The extravascu-

lar pool of LDL apoB, which is 60 ± 9% ($N = 10$) in normal rabbits, is decreased to 30 ± 4% ($N = 8$) upon cholesterol feeding (unpublished observations). In fact, the data available in the literature suggest that the $t_{1/2}$ of the rapid exponential of plasma cholesterol specific activity is similar to the $t_{1/2}$ of the second exponential of the LDL apoB specific activity. Thus in humans the $t_{1/2}$ of the LDL apoB (second exponential) and of the plasma cholesterol (first exponential) specific activity are between 3 and 4 days in N and T, 4 and 6 days in C and CT, and 6 and 9 days in C_{xan} and CT_{xan} subjects. In rabbits on control diet it is 1–1.5 days, while in cholesterol-fed rabbits (1%) it is 3–4 days (Fig. 4). Long-term studies by Scott and Hurley (1970) have shown that it takes 3–4 weeks for LDL apoprotein to completely equilibrate with tissue pools. They found that while the equilibration between the plasma, liver, and spleen was rapid and complete in 3–4 days (rapidly equilibrating pool), that with muscle,

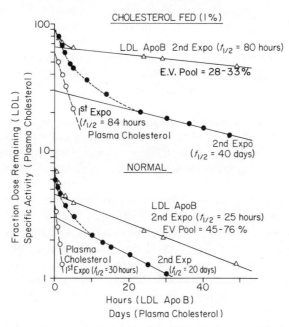

FIG. 4. Relationship between the biological half-lives of LDL apoB and plasma cholesterol. Specific activity time curves of LDL apoB (Δ) and plasma cholesterol (● and ○) after administration of a single intravenous dose of ^{125}I-labeled LDL ($d = 1.019–1.063$ g/ml) and plasma lipoproteins labeled *in vitro* with [^{14}C]cholesterol. Note that $t_{1/2}$ (hours) of the second exponential of LDL apoB is nearly equal to the $t_{1/2}$ (hours) of the first exponential of plasma cholesterol specific activity both in normal rabbits and in rabbits fed 1% cholesterol. [The data on LDL apoB specific activity are from our unpublished study, while those on plasma cholesterol are based on data from Hough and Basset (1975).]

aorta, and skin was slow and required 3-4 weeks (slowly equilibrating pool). It is interesting to note that cholesterol rapidly exchanges between the plasma, liver, and spleen (rapidly exchangeable pool), and it takes 3-4 weeks to equilibrate with most other tissues (slowly exchangeable pools) before a linear exponential phase is attained.

2. The rapidly exchangeable pool, or pool A, is believed to comprise mostly of cholesterol in the liver, blood, bile, proximal intestines, spleen, and lungs (Goodman and Noble, 1968; Wilson, 1970; Smith et al., 1976). Rapid equilibration of specific activity between blood, liver, and bile is expected, since the liver is the primary site for the removal of plasma cholesterol and its excretion into the bile. Equilibration with the proximal part of the intestines is expected because of the absorption of radioactive biliary cholesterol. Equilibration with the spleen can also be expected since the spleen is the catabolic site for the radioactive red blood cells. All of these, however, are metabolic events. Rapid equilibration with the lungs, however, is questionable. While the tissue cholesterol specific activity data obtained after IV labeling of plasma cholesterol (most of these are unphysiological methods) suggest that cholesterol in the lung is in rapid equilibrium, those available (Green et al., 1976) after feeding (physiological labeling) suggest that cholesterol in the lung may not be in rapid equilibrium with plasma cholesterol specific activity.

That the cholesterol in pool A may consist primarily of liver–bile–blood cholesterol, and the "extravascular" pool and may not include cholesterol in other tissues (intestines, spleen, lung), is suggested by comparison of the chemically measured cholesterol mass in the blood, bile, and liver and the calculated mass in the EV pool with the cholesterol mass in pool A as determined by isotope kinetics. In normolipemic subjects, the addition of calculated [assuming that cholesterol distributes itself similarly as lipoprotein, and that the distribution of lipoproteins (LDL and HDL) between the IV and EV pools is 60:40] EV pool cholesterol (5 g) to the chemically measured mass of cholesterol in the plasma (IV) pool (7 g), red blood cells (3 g), bile (1 g), and liver (6 g) (total: 22 g) leaves no room for "additional cholesterol," since it is almost equal to that estimated by the tracer kinetic method (18–23 g).

That part of the pool A cholesterol is the cholesterol in the EV pool is strongly suggested by the reanalysis of pool A data obtained in various types of hyperlipoproteinemic subjects, and especially those data obtained in hypercholesterolemic subjects with xanthomatosis. Low-density lipoprotein apoB turnover studies in these types of subjects suggest that in these subjects, the extravascular distribution of LDL apoB is reduced. Nearly 80% of the LDL apoB is found intravascularly (plasma pool)

(Bilheimer et al., 1979). In these types of subjects, the plasma cholesterol pool increases from the normal 25–35% of the calculated mass in pool A to 50–65% of pool A. The mass of cholesterol in pool A tissues other than plasma (M_{AX}) decreases compared to normals [185 mg/kg (N) versus 135 mg/kg (C_{xan})]. The value of M_{AX} also tends to have a negative correlation ($r = -0.40$, $N = 24$) with the pool of cholesterol in plasma (IV pool). Since there is no reason for the cholesterol content of liver, red blood cells, and other tissues in pool A to decrease [and other studies have shown that it, in fact, is increased (Buja et al., 1979)], these data suggest that an increase in the IV pool cholesterol levels in C and especially in C_{xan} subjects may be due to a shift of cholesterol from the EV to IV pool, or to the inability of cholesterol from the IV pool to shift to the EV pool. Compared to C and CT subjects, C_{xan} and CT_{xan} subjects had nearly 30–35 mg/kg more cholesterol in their plasma. Calculations of the total (IV + EV) pool, however, showed that it was the same (185–190 mg/kg) among these subjects (Table XIV).

Table XIV
EVIDENCE SHOWING THAT THE MASS OF CHOLESTEROL IN TISSUE POOL A (M_A) MAY NOT INCLUDE CHOLESTEROL IN TISSUES OTHER THAN LIVER–BILE–BLOOD (INTRA- AND EXTRAVASCULAR)

Parameter[a]	Type of subject						
	N	T	CT	C	C_{xan}	$C_{y,xan}$[b]	CT_{xan}
M_A (mg/kg; Isotope kinetics, two-pool)	280	290	310	310	290	410	410
M_{AP} (mg/kg; chemical)	95	95	122	122	155	255	150
M_{AX} (mg/kg; $M_A - M_{AP}$)	185	195	188	188	135	155	260
Percentage IV distribution of lipoproteins[c]	56	56	65	65	80	80	80
Calculated plasma IV + EV pool (mg/kg)	170	170	187	187	193	318	187
Extravascular pool (mg/kg)	75	75	66	66	39	63	38
M_{AX} − EV (mg/kg)	110	120	121	121	96	92	222
Red blood cells (mg/kg; chemical)[d]	42	42	42	42	42	42	42
(M_{AX} − EV) − red blood cells (mg/kg)	68	78	79	79	54	50	180
Liver (mg/kg; chemical)[d]	60–85 (+biliary cholesterol, 15–20)						

[a] M_{AP}, mass of cholesterol in the plasma pool; M_{AX}, mass of cholesterol in the tissues of pool A excluding the plasma pool ($M_{AX} = M_A - M_{AP}$).
[b] Subscript "y" indicates that the subjects were young, i.e., under 20 years.
[c] Based on the distribution of LDL in the intra- and extravascular pools.
[d] Based on the literature value (type of lipemia not known).

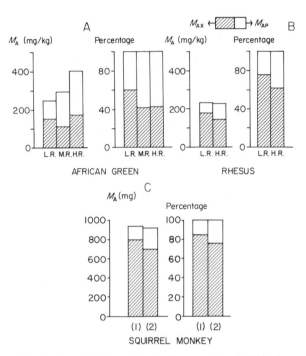

FIG. 5. The distribution of cholesterol in the plasma and pool AX in the rapidly exchanging pool A. When the mass of cholesterol in the total rapidly exchangeable pool (M_A) is the same (B and C), the increase in the plasma pool (M_{AP}) of cholesterol seems to be due to a corresponding decrease in the mass of cholesterol in the remaining tissues of pool A (M_{AX}). When the mass of cholesterol in the total pool is increased (A), the increase seems to be restricted to cholesterol in the plasma pool. (A) Based on data from Parks et al. (1977); (B) Based on data from Bhattacharyya and Eggen (1980) (L.R., M.R., and H.R. represent low-, moderate-, and high-responding monkeys); (C) based on data from Lofland et al. (1970) [(1) safflower oil + high cholesterol; (2) lard + high cholesterol].

Analysis of the data on pool A available in animals (Fig. 5) lends further support to the previous suggestion. There is no difference in the total cholesterol mass in pool A between hypo- and hyperresponding monkeys studied by Bhattacharyya and Eggen (1980). The increase of 30 mg/kg cholesterol in the plasma (IV) pool of HR monkeys was balanced by a corresponding decrease in the cholesterol mass in M_{AX} (part of which may be EV pool). Similarly, the mass of cholesterol in pool A (938 mg) in squirrel monkeys fed cholesterol with safflower oil was similar to that (926 mg) in monkeys fed cholesterol with lard. The increase in the plasma cholesterol pool (72 mg) of monkeys fed cholesterol with lard was balanced by a corresponding decrease in the cholesterol mass in pool AX (Lofland et al., 1970). Also, in African green monkeys (Parks et al., 1977)

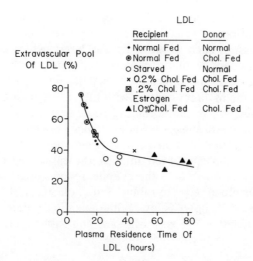

FIG. 6. Relationship of the extravascular pool of LDL (percentage) with the plasma residence time of LDL in rabbits. In normal-fed rabbits, the plasma residence time was negatively correlated with the amount of LDL in the extravascular pool. Although such a relationship was not seen in starved and cholesterol-fed rabbits, there was a marked decrease in their extravascular pools of LDL. Since other studies suggest that starvation (Stoudemier et al., 1982) and cholesterol feeding (Mahley et al., 1981) decrease receptor activity, the data suggest that the distribution of LDL between the IV and EV pools may be dependent on cellular receptor activity. (B. J. Kudchodkar, unpublished data.)

the increase in M_A upon cholesterol feeding was entirely due to an increase in the plasma cholesterol pool. The ability to shift cholesterol from the IV to the EV pool may, therefore, be an important factor in determining plasma levels of cholesterol.

The factors affecting the distribution of lipoproteins between the IV and EV pools are not known. Available data on LDL apoB turnover, as well as our preliminary studies (Fig. 6) in rabbits, suggest that conditions which decrease the EV pool also decrease the fractional catabolic rate (k_{FC}) of LDL apoB ($r = -0.63$, $N = 19$). Brown and Goldstein (1979) have suggested that the k_{FC} of LDL apoB is a reflection of the cellular LDL receptor activity. If this is true, then these analyses, which show a close relationship between the $t_{1/2}$ of LDL apoB and the $t_{1/2}$ of the rapid (first) exponential of the plasma cholesterol specific activity curve, and a close relationship between the distribution of LDL apoB into the IV and EV pools and the distribution of cholesterol between the plasma and pool AX (of which the EV pool may be a major part) suggest that the equilibration of labeled plasma cholesterol with tissue cholesterol (as well as its distribution between the IV and EV pools) may not be due to exchange, as is believed at present, but may be due to cellular receptor and nonreceptor activity. If equilibration was due to exchange, then the specific activity of cholesterol in tissues such as the aorta and heart, which comes in direct contact with radioactive blood cholesterol, should be in a more rapid equilibrium with plasma cholesterol specific activity than the tissues which do not come in direct contact with blood. The fact that the specific activity of cholesterol in the aorta and heart is not in rapid equilibrium with plasma cholesterol suggests that exchange may not play an important part in the equilibration of cholesterol between plasma and tissues. It was also found that in cholesterol-fed animals, the calculated mass of cholesterol in pool A was lower than the chemically measured mass of cholesterol of the liver alone, e.g., in rabbits fed cholesterol (0.1%), the calculated M_A (2165 ± 628 mg) was lower than chemically determined mass of the liver alone (2450 ± 791 mg) (Massaro and Zilversmit, 1977).

V. Is There an Exchange of Plasma Cholesterol with Tissue Cholesterol?

Strong evidence suggesting that *in vivo* plasma unesterified cholesterol may not exchange with tissue cholesterol other than red blood cells is given by the following observations. When plasma cholesterol is labeled either by feeding or by injecting a colloidal suspension of radioactive cholesterol, the labeled cholesterol from the plasma also disappears in a series of exponential rates (Grundy and Ahrens, 1969; Bhattacharyya *et al.*, 1976; Sodhi and Kudchodkar, 1973a). Again, the rapid exponential or the exponential A is believed to be due to the exchange of labeled plasma cholesterol with unlabeled cholesterol in tissues comprising the rapidly exchanging pool, that is, cholesterol in the liver, red blood cells, intestines, lungs, spleen, etc. Under both of these conditions, however, liver cholesterol specific activity is initially higher than plasma cholesterol specific activity, since both the absorbed cholesterol transported to the plasma by chylomicrons and the colloidal suspension of cholesterol injected in the plasma are rapidly removed by the liver. Furthermore, the specific activity of cholesterol in the intestines (upon feeding) and spleen (upon injection as colloidal suspension) is also initially higher than plasma cholesterol specific activity (Fig. 7; also Fig. 1 of Green *et al.*, 1976; Fig. 3 of Viikari *et al.*, 1976; Nilsson and Zilversmit, 1972; Sodhi and Kudchodkar, 1973a). Red blood cell cholesterol specific activity also reaches equilibrium with plasma cholesterol by the time it reaches a peak (24–48 hours in man, 6–8 hours in rats). The rapid decline of plasma cholesterol specific activity in these situations therefore cannot be ascribed to the exchange of plasma

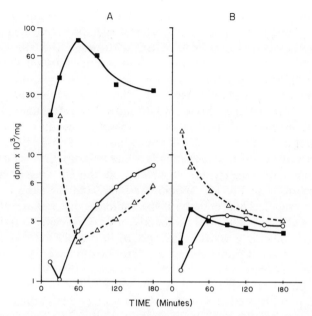

FIG. 7. Specific activity of plasma (○), liver (■), and biliary (△) free cholesterol after iv injection of [^{3}H]cholesterol (A, particulate cholesterol) and [^{14}C]mevalonate (B, endogenous cholesterol) in alcoholic saline. Under both conditions, the specific activity of biliary and liver free cholsterol was higher than the specific activity of plasma free cholesterol. The rapid decline of plasma free cholesterol specific activity in these situations, therefore, cannot be ascribed to the exchange of plasma cholesterol (with lower specific activity) radioactivity with the cholesterol in the liver (with higher specific activity), the tissue containing the major portion of cholesterol of tissue pool A. (B. J. Kudchodkar and H. S. Sodhi, unpublished.)

cholesterol with cholesterol (with higher specific activity) in the liver, red blood cells, intestines, spleen, etc., that is, the cholesterol in tissues belonging to pool A.

When plasma UC is labeled by injecting labeled precursors such as acetate or mevalonate, the plasma UC specific activity, after reaching a peak (2–4 hours), begins to decline rapidly. This rapid decline has also been interpreted to be due to the rapid exchange of labeled UC with unlabeled tissue UC. In this situation, liver cholesterol specific activity also is higher than plasma UC specific activity. Red blood cell cholesterol is also in equilibrium with plasma UC specific activity 1–2 hours after the peak (i.e., between 5 and 7 hours). Therefore, the rapid decline in plasma UC specific activity 1–2 hours after the peak cannot be ascribed to the exchange of labeled UC between lipoprotein–liver–red blood cell cholesterol.

One of the facts that has been neglected in the interpretation of the data on the early part of the plasma cholesterol specific activity curve is that besides the postulated exchange of UC with tissue UC, the UC is also converted to esterified cholesterol (CE) in plasma through the action of plasma lecithin:cholesterol acyltransferase (LCAT) (Glomset, 1968). After labeling the plasma cholesterol by the various methods previously mentioned, if the specific activity of UC and CE is determined, it is found that the UC specific activity is higher than the CE specific (Hellman et al., 1954; Nestel and Monger, 1967; Kudchodkar and Sodhi, 1976). The CE specific activity begins to rise gradually while that of UC falls (after reaching the peak) until equilibrium, at which time the CE specific activity is slightly greater than the UC specific activity, and then the two curves decline in parallel. Theoretically, if the radioactivity was not lost to the tissues by exchange, the total radioactivity [(UC specific activity × UC pool) + (CE specific activity × CE pool)] in the plasma should remain more or less constant from the time of secretion of labeled UC into plasma until its conversion to CE, that is, the decrease in the radioactivity in the UC pool should be balanced by an increase in the radioactivity in the plasma CE pool. Indeed, in humans when radioactivity in the plasma total cholesterol pool at the peak UC specific activity (2–4 hours) or after its equilibration with red blood cells (6–7 hours) is calculated, it seems to be nearly the same as the total radioactivity at the point of intersection of the UC and CE specific activities, which in normal humans is anywhere between 2 and 3 days. In other words, until the equilibration of plasma UC and CE specific activities, that is, for 2–3 days, no significant amount of radioactivity is lost from the plasma compartment. This observation, which was noted as early as 1954 by Hellman et al. (Table III of the reference), has remained essentially unnoticed.

Data on plasma CE formation available in other studies also support the just-mentioned suggestion. Thus, upon mevalonate injection in two normolipemic subjects (Goodman, 1964), the total radioactivity (dpm) in the plasma pool was 90.8×10^3 at 7 hours and 85.5×10^3 at 48 hours, while in the other it was 76.4×10^3 at 3 hours and 75.8×10^3 at 49 hours (plasma UC and CE concentrations in these normolipemic subjects were assumed, respectively, to be 60 and 140 mg per 100 ml plasma, and the plasma volume was assumed to be 4.5% of the body weight). In another study, 30 hours after feeding labeled cholesterol (peak of UC specific activity) the total radioactivity in the plasma cholesterol pool was 6.87×10^6 dpm, and at 72 hours it was 6.92×10^6 dpm (Hellman et al., 1955). Our preliminary studies in rabbits support these observations. In each of the four rabbits on control diet the total radioactivity in the plasma cholesterol pool at 4 hours after the injection of labeled mevalonate ($131 \pm 24 \times 10^3$ dpm) was nearly identical to the total radioactivity at 14 hours ($132 \pm 25 \times 10^3$ dpm), the time at which the declining UC specific activity curve intersected the increasing CE specific activity curve. The results in three cholesterol-fed and three starved rabbits were also similar. In the latter two cases, the peak CE specific activity occurred between 25 and 48 hours (Kudchodkar et al., 1981).

There are a number of other observations scattered throughout the literature which suggest that the decline in plasma UC specific activity may be primarily due to the formation of CE in the plasma and not to the exchange of labeled UC with unlabeled tissue UC. In 1952, Biggs et al. fed [^{3}H]cholesterol to four patients, one of whom was comatose. Except for labeled cholesterol (dissolved in oil and emulsified in milk) which was fed by stomach tube, the comatose patient received all nourishment intravenously. In all patients, including the comatose subject, the plasma total cholesterol specific activity reached a peak between 24 and 48 hours, and thereafter began to decline. However, in the comatose subject there was virtually no decline in plasma cholesterol specific activity for a 30-day period (length of the study; Fig. 8). Also, in one subject with severe hypercholesterolemia (total cholesterol, 550 mg/100 ml plasma), the rate of decline after the peak was slower than in the subjects with normal plasma cholesterol levels. Studies have shown that the fractional turnover rate of plasma CE is decreased in subjects with hypercholesterolemia (Nestel and Monger, 1967; Kudchodkar and Sodhi, 1976). Similarly, as compared to fed rabbits, in fasting rabbits the rate of decline in plasma UC specific activity is very slow (see Fig. 2 of Klauda and Zilversmit, 1974). In fasted

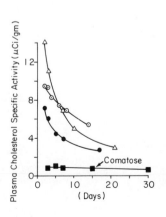

FIG. 8. The rate of plasma cholesterol specific activity decline upon feeding of [^{3}H]cholesterol. The plasma total cholesterol (TC) specific activity declined rapidly in the subject with normal plasma cholesterol levels (△; TC = 195 mg/100 ml plasma, free cholesterol = 57 mg/100 ml plasma) slowly in the subject with hypercholesterolemia (●; TC = 270 mg/100 ml plasma, free cholesterol = 76 mg/100 ml plasma) and very slowly in the subject with severe hypercholesterolemia (○; TC = 550 mg/100 ml plasma, free cholesterol = 196 mg/100 ml plasma). In the comatose subject (■), with a normal cholesterol level (TC = 201 mg/100 ml plasma, free cholesterol = 56 mg/100 ml plasma), there was virtually no decline in plasma cholesterol specific activity during the study period. [Redrawn from Figs. 1–4, Biggs et al. (1952).]

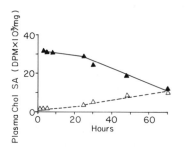

FIG. 9. Plasma free and esterified cholesterol specific activity (SA) changes in a rabbit injected with Triton WR-1339 and [³H]mevalonate. The rate of decline of plasma free cholesterol (▲) specific activity was low during the time (≈ 24 hours) when its rate of esterification was low (△, cholesteryl esters). As the rate of esterification increased, the rate of decline of plasma free cholesterol specific activity was also increased. (B. J. Kudchodkar, unpublished observations.)

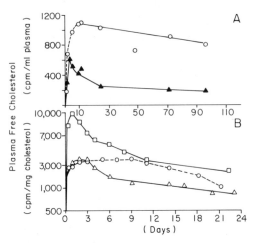

FIG. 10. The rate of decline of plasma free cholesterol specific activity after intravenous administration of [³H]mevalonate (A) and after feeding [³H]cholesterol (B). The rate of decline of plasma free cholesterol specific activity was very slow in subjects with plasma LCAT deficiency [IS (O———O), AR (O--O), MR (△)] compared to the rate of decline of plasma free cholesterol specific activity in normal subjects [HT (▲) and KN (□)]. [Data from Norum and Gjone (1967); (A) based on data in Table IV; (B) redrawn from Fig. 2.]

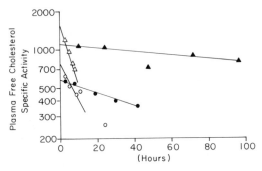

FIG. 11. Semilogarithmic plot of time versus plasma free cholesterol specific activity after intravenous administration of labeled mevalonate or acetate to various types of sub-

rabbits there is also a decline in the rate of formation of plasma CE (Klauda and Zilversmit, 1974).

Our preliminary studies (Kudchodkar *et al.*, 1981) in fasted rabbits, rabbits fed 1% cholesterol, and rabbits treated with Triton WR-1339 lend further support to the last-mentioned suggestion. The $t_{1/2}$ of the rapidly declining plasma UC specific activity was prolonged in all conditions where the fractional turnover rate (k_{FT}) of plasma CE was decreased. The $t_{1/2}$ of UC specific activity was negatively correlated ($r = -0.75$, $N = 11$) with the CE k_{FR}. In Triton-treated animals, in whom the formation of CE was markedly inhibited (CE $k_{FT} = 0.012$/day, compared to 0.111/day in normals), there was a marked decrease in the rate of fall of plasma UC specific activity ($t_{1/2} = 104$ hours, compared to 12 hours in normals). Twenty-four hours after treatment, the rate of plasma CE formation began to increase and it was associated with a corresponding increase in the rate of decline of plasma UC specific activity (Fig. 9). Data available in Triton-treated rats (Fourcans *et al.*, 1974) also support these observations. Thus, there is virtually no loss of radioactivity from plasma UC when its conversion to CE is inhibited. This is best illustrated by the reanalysis of data on plasma UC specific activity available in LCAT-deficient subjects studied by Norum and Gjone (1967). Upon injection of labeled mevalonate (Fig. 10A) or upon feeding of radioactive cholesterol (Fig. 10B), the results show that plasma UC specific activity does not decline rapidly in the absence of esterification.

All these observations strongly support the suggestion that the *in vivo* exchange of plasma UC with tissue (other than red blood cell) UC, if it occurs at all, is minor. The decline in plasma UC radioactivity is due to its conversion to radioactive CE (Fig. 11). The results also suggest that cholesterol (newly synthesized, absorbed, and resecreted in the plasma by the liver) enters the plasma in the unesterified form and leaves the plasma mostly as CE (some UC is needed for the lipoprotein structure). The formation of esterified cholesterol may be necessary for the physiological removal of cholesterol from the plasma. As noted later, the time at which the UC specific activity intersects the CE specific activity is nearly identical to the plasma residence time ($1/k_{FC}$) of LDL apoB.

---

jects. The biological half-life (hours) of plasma free cholesterol specific activity was prolonged when plasma free cholesterol esterification was absent [IS (▲; LCAT deficient, $t_{1/2}$ = 164 hours) from Norum and Gjone (1967)] or its rate was reduced [CHT-1 (●; $t_{1/2}$ = 56 hours, total cholesterol = 802 mg/100 ml plasma, free cholesterol = 301 mg/100 ml plasma) from Hellman *et al.* (1955)] compared to normal subjects [N (△; $t_{1/2}$ = 7 hours, total cholesterol = 124 mg/100 ml plasma, free cholesterol = 32 mg/100 ml plasma) from Barter (1974); HT (○; $t_{1/2}$ = 10 hours) from Norum and Gjone (1967).]

VI. Relationship between Plasma Cholesteryl Ester Turnover and LDL ApoB Turnover

Our analysis of data on plasma CE turnover and LDL apoB turnover suggested that the two parameters were closely linked. Thus it was found that the residence time of LDL apoB, which was 2–2.5 days in N and 3–4 days in C_{xan} subjects, was identical to the crossover point of UC and CE specific activities in similar types of subjects. This belief is strengthened by our preliminary studies in rabbits, in whom LDL apoB (d = 1.09–1.063 g/ml) turnover and plasma CE turnover was studied after varying the plasma esterification rate by starvation and by feeding cholesterol at different levels (Table XV). A significant direct correlation between the k_{FC} of LDL apoB and of plasma CE (r = 0.83, N = 11) was observed. Most importantly, in each rabbit *the time* at which the plasma CE specific activity reached its peak or crossed UC specific activity was very nearly the same as the plasma residence time ($1/k_{FC}$) of LDL apoB (r = 0.99, N = 11).

There are also a number of metabolic conditions where the k_{FC} of LDL apoB and the k_{FC} of plasma CE are affected in the same direction. For example, the k_{FC} of LDL apoB is increased upon carbohydrate feeding (Nestel *et al.*, 1979), cholestyramine treatment (Sheperd *et al.*, 1980), or in subjects with hypertriglyceridemia (Type IV) (Sigurdsson *et al.*, 1976).

Table XV

RELATIONSHIP OF THE FRACTIONAL CATABOLIC RATE OF LDL APOB AND THE FRACTIONAL TURNOVER RATE OF THE PLASMA CHOLESTERYL ESTERS IN RABBITS

Rabbits	Number	LDL apoB (d = 1.019–1.063 g/ml)		Plasma cholesteryl esters		
		k_{FC} (hours^{-1})	$1/k_{FC}$ (hours)	k_{FT} (hours^{-1})	$1/k_{FT}$ (hours)	UC and CE specific activity crossover point (hours)
Control	5	0.0611	16.8	0.111	9.0	14.8
Starved	4	0.0332	30.5	0.0559	17.9	30.0[a]
Cholesterol fed						
1%	2	0.0132	75.7	0.0177	56.5	78
0.2%	1	0.0226	44.2	0.0336	29.8	43.0
0.2%; + estrogens	1	0.0556	18.0	0.0862	11.6	17.0

[a] Peak of CE specific activity. In two out of four starved animals the CE specific activity did not cross the UC specific activity.

The k_{FC} of CE is also increased in these conditions (Kudchodkar and Sodhi, 1976; Clifton-Bligh et al., 1974; Barter, 1974). Similarly, in conditions where the k_{FC} of LDL apoB is decreased (Bilheimer et al., 1979; Walton et al., 1965; Eisenberg and Levy, 1975) (hypercholesterolemia, hypothyroidism) the k_{FC} of CE is also decreased (Nestel and Monger, 1967; Kudchodkar and Sodhi, 1976). Since in a steady state the rate of catabolism is equal to the rate of synthesis, the data suggest that the rate of synthesis of plasma CE (CE turnover) is a reflection of the formation of plasma LDL (LDL turnover). If the rate of LDL apoB turnover is a reflection of cellular LDL receptor activity (Brown et al., 1979), then the data would suggest that the k_{FC} of plasma CE is a reflection of cellular LDL receptor activity. Since UC leaves the plasma only after the formation of CE, which, in turn, represents the formation of LDL in the plasma, it is suggested that under normal conditions the equilibration of plasma cholesterol with the tissues takes place via the receptor and nonreceptor pathway and not by exchange. The different rates of equilibration of plasma cholesterol with tissue cholesterol may therefore represent the rates of cellular membrane turnover since LDL receptor activity (Brown et al., 1979), as well as cholesterol synthesis (Sodhi and Kudchodkar, 1974; Sodhi, 1975), under normal conditions is believed to be associated with cellular membrane turnover.

In vivo cholesterol metabolism studies in Hruza (1971a,b; Hruza and Zbuzkova, 1975) in young and old rats, and that of Zilversmit and Hughes (1973) in young and old rabbits, lend strong support to the suggestion that *in vivo* the equilibration of plasma cholesterol with tissue cholesterol takes place primarily via the receptor (and nonreceptor) pathway and not by exchange. These workers have shown that the uptake of radioactive cholesterol by tissues such as muscle, skin, adipose tissue, brain, aorta, etc., but not RBC, is considerably higher in young, actively growing animals as compared to old animals. Interestingly, tissues showing differences in exchange have LDL receptors and the one (RBC) showing no differences in exchange has no LDL receptors (Brown et al., 1979). Also the fractional rate of plasma-free cholesterol esterification is markedly reduced in old rats compared to young rats (Carlile and Lacko, 1981). In addition, Hruza (1971b) has found that the uptake of plasma cholesterol by the various tissues of young as well as old rats was decreased upon hypophysectomy and thyroidectomy, and increased upon treatment with insulin and thyroxine. These data suggest that the mechanism of the equilibration of plasma cholesterol with the cholesterol in tissues possessing receptors is under metabolic control.

Recent *in vivo* and *in vitro* studies with cells in culture have suggested that the uptake of plasma lipoproteins (and hence cholesterol) is dependent on cellular receptor activity (Brown et al., 1979). These studies have shown that cellular receptor activity varies with age and tissue, and is under hormonal and other metabolic influences (Brown et al., 1979). Both thyroxine and insulin treatments have been shown to increase the catabolism of plasma LDL in vivo. Conversely, hypothyroidism and insulin deficiency have been shown to decrease the in vivo catabolism of both plasma LDL apoB and cholesterol (Walton et al., 1965; Mazzone et al., 1982; Lehner et al., 1972; Mathe and Chevallier, 1979). Similarly, studies with cells in culture have shown a marked increase in LDL receptor activity upon treatment of the cells with insulin and thyroxine (Chait et al., 1979a,b).

Observations of Bing and Sarma (1975) also cast doubt on "exchange" being the major mechanism for the equilibration of plasma free cholesterol with tissue free cholesterol. These workers observed a reduction of nearly 90% in the uptake of radioactive plasma free cholesterol by arteries perfused in the presence of 7-ketocholesterol. This could be explained on the basis of other studies (Goldstein and Brown, 1976) which show a marked decrease in cellular LDL receptor activity of cultured cells grown in the presence of 7-ketocholesterol. It is interesting to point out that the uptake of covalently bound sucrose-labeled LDL, which is assumed to be a reflection of in vivo LDL receptor activity (Steinberg et al., 1979), is moderate in tissues which "exchange" their cholesterol slowly, low in those which exchange their cholesterol very slowly, and high in those which exchange their cholesterol rapidly with labeled plasma cholesterol. Recent studies of Dietschy and his colleagues (Jeske and Dietschy, 1980; Turley et al., 1981) also cast doubt on exchange being the major mechanism for the equilibration of plasma cholesterol specific activity with cholesterol in peripheral tissues.

At present it is believed that the curvilinear portion of the plasma cholesterol specific activity time curve (obtained upon intravenous injection of a single dose of labeled cholesterol) is due to the input of unlabeled cholesterol as well as to the exchange of radioactive plasma cholesterol with the nonradioactive cholesterol in the various tissues of the body. Once the specific activity of cholesterol in different tissue pools reaches an equilibrium with the specific activity of plasma cholesterol, the plasma cholesterol specific activity begins to decline linearly. This linear decline is believed to be mostly due to the entry of unlabeled cholesterol into the total body pool (Hellman et al., 1957; Grundy and Ahrens, 1966; Goodman and Noble, 1968). Increased entry of unlabeled cholesterol, therefore, will be reflected by an increased rate of decline (increased slope) of plasma cholesterol specific activity and, conversely, a decrease in the entry of unlabeled cholesterol will cause a decreased rate of decline (decreased slope).

In the body, unlabeled cholesterol can enter through biosynthesis and through the absorption of dietary cholesterol. Theoretically, the dietary absorbed cholesterol should not affect the slope of the plasma cholesterol specific activity curve if the amount absorbed from the diet is equal to or less than the amount of cholesterol synthesized, since it will not increase the entry of unlabeled cholesterol due to a compensatory decrease in an equal amount of unlabeled cholesterol synthesized (negative feedback inhibition). Only if the amount of dietary absorbed cholesterol exceeds the amount synthesized under basal conditions, or is not compensated for by a decrease in synthesis, will the dietary absorbed cholesterol affect the decline of plasma cholesterol specific activity. Under such conditions, the absorbed dietary cholesterol will be expected to increase the slope of plasma cholesterol specific activity because of the increased entry of unlabeled cholesterol into the body.

That factors other than the entry of unlabeled cholesterol in the body are responsible for the control of the rate of plasma cholesterol specific activity decline, both during the early rapid phase and the later slower phase, is suggested by our analyses of plasma cholesterol specific activity data in animals fed low and high cholesterol diets. These analyses showed that cholesterol feeding had a marked effect on the slopes of both exponentials (A and B), and especially on that of exponential B (Table XVI). The effect was not related to the amount of unlabeled dietary cholesterol entering the body, but seemed to be related to the response of the animals to cholesterol feeding. In species such as rats, dogs, and baboons, which are known to be resistant to the effect of dietary cholesterol, the slopes tended to increase, while in species responsive to dietary cholesterol feeding (rhesus monkey, rabbit, guinea pig), the slopes tended to decrease inspite of the fact that the entry of unlabeled cholesterol was increased 2–5 times over the basal entry of unlabeled cholesterol through synthesis by the liver (or 1–3 times through whole body synthesis; Table XVI). In cholesterol-fed guinea pigs studied by Green *et al.* (1976), our analysis showed that the decrease (compared to the control period) in the slope of the exponential B ($0.693/t_{\mathrm{mean}}$) was positively correlated with the accumulation of cholesterol in the liver ($r = 0.75, N = 6$).

These data are in accord with our earlier suggestion that the equilibration of plasma cholesterol specific activity with tissue cholesterol may not be due to the phenomenon of exchange, but may be due to cellular receptor activity. This suggestion is further supported by comparison of the

Table XVI

EFFECT OF DIETARY CHOLESTEROL ON THE SLOPES OF EXPONENTIAL A AND EXPONENTIAL B OF THE PLASMA CHOLESTEROL SPECIFIC ACTIVITY CURVE ANALYZED BY THE TWO-POOL MODEL

Species		Cholesterol input (mg/kg/day)			Slope of the exponential[b]	
		Endogenous		Dietary absorbed	A	B
		Total[a]	Synthesis			
Rat	LC[c]	90	56		0.407	0.0553
	HC			80	0.462	0.0693
Dog	LC	90	13		0.243	0.0243
	HC			58	0.223	0.0459
Baboon	LC	20	10		0.210	0.0184
	HC			22	0.257	0.0228
Squirrel monkey[d]	LC	94	52			0.0365
	HC			28	NA[e]	0.0444
Squirrel monkey	LC	94	52		0.531	0.0630
	HC			126	0.465	0.0537
Insulin deficient	HC			119	0.315	0.0459
Hypothyroid	HC			95	0.239	0.0099
Rhesus	LC	18	9		0.239	0.0343
	HC			29	0.169	0.0131
Rabbit	LC	45	25		0.541	0.0343
	HC			130	0.197	0.0160
Guinea pig	LC	26	15			0.0307
	HC			16	NA	0.0231

[a] Total = synthesis + endogenously reabsorbed.
[b] Slope = $0.693/t_{1/2}$ days of the exponential.
[c] LC, low cholesterol; HC, high cholesterol.
[d] These squirrel monkeys did not respond to cholesterol feeding.
[e] NA, not available.

slopes of the two exponentials (A and B) of the plasma cholesterol specific activity curve obtained in squirrel monkeys (Table XVI). In the study in which squirrel monkeys did not respond (plasma cholesterol response) to cholesterol feeding (Eggen, 1974), the slope of exponential B was increased in spite of the fact that the amount absorbed was lower than the basal synthesis and was compensated for by decreased synthesis (excretion). Data on exponential A were not available in this study. In other studies (Lofland et al., 1970; Raymond et al., 1976; Lehner et al., 1972), where cholesterol feeding increased the plasma and body cholesterol mass, the slopes of both exponentials (A and B) were markedly decreased, especially in monkeys made insulin deficient or hypothyroid, in spite of the fact that the amounts of unlabeled dietary absorbed cholesterol entering the body were similar in all groups. *In vivo* (Mazzone et al., 1982) and *in vitro* (Chait et al., 1979a) data suggest that cellular receptor activity is under hormonal control. In hypothyroidism cellular receptor activity (as reflected by decreased fractional turnover of LDL apoB) has been found to be decreased (Thompson et al., 1981). Analysis of data available in humans suggests that endogenously reabsorbed cholesterol (extrahepatic synthesis?) may have a marked effect on the slopes of exponentials A and B of the plasma cholesterol specific activity time curve. The $t_{1/2}$ (days) of both the exponentials tended to show a positive correlation with the amounts reabsorbed. The relationship, however, improved when the absorbed cholesterol was expressed on the basis of its contribution (percentage) to the total input (exponential A, $r = 0.48$; exponential B, $r = 0.76$; $N = 22$). The $t_{1/2}$ of exponential A (7.5 days), as well as that of exponential B (81.3 days), was markedly increased in young subjects with hypercholesterolemia and xanthomatosis (this includes two subjects with very low cellular receptor activity and others with a strong family history of coronary heart disease), in spite of the fact that the contribution of endogenously reabsorbed cholesterol to the total cholesterol input was either similar to or even lower than that in subjects with hypercholesterolemia alone ($t_{1/2}$ of exponential A = 5.1 days; $t_{1/2}$ of exponential B = 67.0 days). Cellular receptor activity (Brown et al., 1979), as well as the fractional turnover rate of plasma cholesteryl esters (Nestel and Monger, 1967; Moutafis and Myant, 1969; Kudchodkar and Sodhi, 1976), was low in the former subjects.

A critical look at the data in the literature on the plasma cholesterol specific activity curves and the slopes of the two exponentials suggests that the slopes of the two exponentials, especially that of exponential B, are increased in almost all conditions where there is an increase in the synthesis (excretion) of cholesterol (unlabeled), especially if it is accompanied by an increased catabolism to bile acids. On the other hand, if the input of unlabeled cholesterol is increased through dietary cholesterol absorption, the effect is variable in spite of a compensatory decrease in synthesis. This paradox could also be related to cellular receptor activity. Whereas dietary absorbed cholesterol leads to an inhibition of cellular receptor activity (Brown et al., 1979; Mahley et al., 1981), conditions which stimulate cholesterol synthesis and its catabolism seem to stimulate cellular receptor activity (Kovanen et al., 1981; Shepherd et al., 1980).

VII. Quantitative Relationship of Cholesterol Synthesis, Absorption, Esterification, and Catabolism with the Metabolism of Plasma Lipoproteins, Especially the Low-Density Lipoproteins

On the basis of LDL apoB turnover and the cholesterol : protein ratio, it has been suggested that in humans between 25 and 30 mg/kg of LDL cholesterol is catabolized daily. Since the daily excretion (synthesis) of cholesterol is 10–12 mg/kg, these workers have suggested that LDL cholesterol is reutilized at least 2–3 times (Bilheimer *et al.*, 1979). Workers (Nestel and Monger, 1967; Kudchodkar and Sodhi, 1976; Angelin *et al.*, 1981; Myant, 1982) studying plasma CE turnover have found that the daily turnover of CE is about 25–30 mg/kg, and since the daily synthesis of cholesterol is 10–12 mg/kg, some of these workers have also suggested that plasma cholesterol is reutilized (Angelin *et al.*, 1981; Klauda and Zilversmit, 1974; Myant, 1982). The impression one gets from these suggestions is that cholesterol transported to the plasma is newly synthesized cholesterol only, and that this cholesterol is reused for the secretion of new lipoproteins (the liver → plasma → tissue → plasma → liver → plasma cycle) before being excreted in the bile. The fact that cholesterol is reutilized is well known, but this reutilization is through the enterohepatic cycle, that is, the liver → plasma → tissue → plasma → liver → bile → intestines → plasma → liver → plasma cycle. The relationship between the input of unesterified cholesterol in the plasma, its esterification, and its catabolism can be explained by the model presented in Fig. 12. The model is based on the enterohepatic cycle and assumes that the cholesterol synthesized by the liver and the absorbed cholesterol (which initially may have been synthesized in extrahepatic tissues) removed by the liver is transported to the plasma first before being excreted, as such and after its conversion to bile acids, in the bile. Ample evidence for this suggestion was provided earlier.

The true body cholesterol turnover (which equals cholesterol transport in plasma) in a normolipemic adult was shown to be 20–21 mg/kg/day. At present cholesterol is believed to be transported in the plasma as UC, mostly through very low density lipoproteins (VLDL) and some through high-density lipoproteins (HDL) (Glomset and Norum, 1973; Eisenberg and Levy, 1975). During the catabolism of VLDL triglycerides, the UC is esterified (Schumaker and Adams, 1969; Kudchodkar and Sodhi, 1976) and then becomes part of LDL, the catabolic product of VLDL (Eisenberg and Levy, 1975). All of the UC transported to the plasma, however, is not esterified, since some of it is needed for the structural needs of lipoproteins. Assuming that 70% of the UC entering the plasma is esterified (since 70% of plasma cholesterol is esterified cholesterol), about

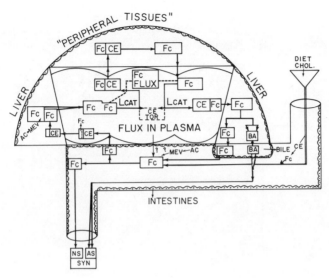

FIG. 12. A model depicting true body cholesterol turnover. Cholesterol from two sources (newly biosynthesized and the absorbed removed from plasma) is quantitatively secreted into the plasma as free cholesterol (Fc) in hepatic lipoproteins. Upon entering the plasma, a portion is esterified (CE) and transported to peripheral tissues where CE is hydrolyzed and the Fc is utilized. In a steady state, an equivalent amount of free cholesterol from peripheral tissues is transported back to the plasma where a portion is reesterified before being transported back to the liver for catabolism. The liver excretes cholesterol, as such and after its conversion to bile acids (BA), in the bile. During its passage through the GI lumen, part of the cholesterol and most of the biliary bile acid pool is reabsorbed, and the remaining is lost through the feces. The amount lost from the body is made up by new synthesis (SYN). Thus the total cholesterol utilized daily by the body (true body cholesterol turnover) is equal to the sum of cholesterol (and bile acids) excreted plus the cholesterol reabsorbed from the gastrointestinal tract. AC, acetate; MEV, mevalonate; TOR, turnover rate; AS, acidic steroids; NS, neutral steroids.

14–15 mg/kg of cholesterol will be esterified upon entering the plasma. According to the present concept (Brown and Goldstein, 1976; Brown *et al.*, 1979), the CE along with UC will be taken up by the peripheral tissues, via the receptor and nonreceptor pathways, and utilized for their structural (and other) needs. In a steady state, in order to keep the tissue cholesterol content constant, 20–21 mg/kg of cholesterol transported to tissues has to be transported to the liver, via plasma lipoproteins, for catabolism. For this, the UC from the peripheral tissues reentering the plasma has to be reesterified before removal (Glomset, 1968). The total UC esterified will therefore be 14–15 mg/kg, transported to the plasma from the liver, plus 14–15 mg/kg transported to the plasma from peripheral tissues, i.e., a total of nearly 28–30 mg/kg/day. Total unesterified cholesterol transport in the plasma based on this model will be 40–42

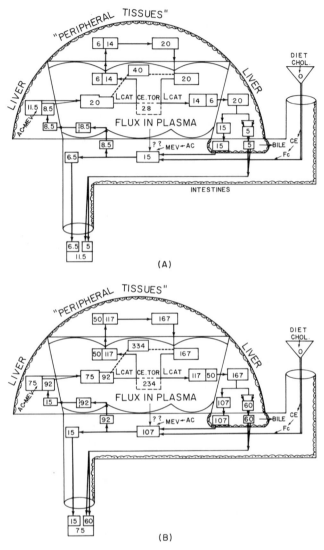

FIG. 13. (A) Quantitative relationship in humans between the input of free cholesterol and its esterification in the plasma. The total flux of free cholesterol (Fc) from the liver and from the peripheral tissues (40 mg/kg/day), and the amounts of free cholesterol esterified in the plasma (28 mg/kg/day), are calculated based on the model shown in Fig. 12. The turnover of plasma esterified cholesterol is calculated assuming that 70% of the free cholesterol transported to the plasma is esterified. The calculated values for the plasma CE turnover rate (TOR) were nearly similar to the ones measured by *in vivo* or *in vitro* techniques (see Table XVII). Similarly, the values calculated for free cholesterol flux based on the model were similar to those calculated by tracer kinetics as the flux of cholesterol through the rapidly

mg/kg/day. Data presented in Table XVII and Fig. 13 show that the values calculated on the basis of the model are nearly identical to the *in vivo*- and *in vitro*-determined values for plasma CE turnover in humans (Fig. 13A), as well as to the flux of UC through the plasma [calculated by tracer kinetics as the product of the mass of cholesterol in pool A (M_A) and its fractional turnover rate (α, which equals $0.693/t_{1/2}$ days of the first exponential); in this calculation, the input and output of UC is assumed to occur through the rapidly equilibrating pool A]. Data available in animals also give similar results. In rabbits (Table XVII, Fig. 13B), for example, the amount of free cholesterol secreted daily in the plasma (as measured after injection of Triton WR-1339 by Klauda and Zilversmit, 1974) is similar to the amount which is biosynthesized + endogenously absorbed, as measured by the sterol balance method or by tracer kinetics. Similarly, the values for plasma cholesteryl ester turnover calculated on the basis of the model are similar to those measured *in vivo* by Klauda and Zilversmit (1974). The model is not valid when liver secretes CE-rich lipoproteins.

Although the total CE turnover in humans is 28–30 mg/kg/day, the total cholesterol catabolized is still 20–21 mg/kg/day, the amount originally secreted in the plasma. The cholesterol secreted in the intestinal lumen (biliary cholesterol + synthesized in the intestines + transported to the intestines via plasma lipoproteins) is about 14–15 mg/kg/day, and that catabolized to bile acids is about 5–6 mg/kg/day, that is, a total of 20–21 mg/kg/day of cholesterol is excreted daily in the intestinal lumen, part of which is then irreversibly lost in the feces.

Since plasma cholesteryl esters are formed only in the plasma, the identity of the calculated values with those actually measured suggests that cholesterol, both the newly biosynthesized and the total absorbed (exogenous and/or endogenous), removed by the liver enters the plasma as unesterified cholesterol before being removed by peripheral tissues as esterified cholesterol for utilization. The unesterified cholesterol released into the plasma by the peripheral tissues is reesterified and is then re-

---

equilibrating pool A ($M_A \times \alpha$). (B) Quantitative relationship in rabbits between the input of free cholesterol and its esterification in the plasma. The daily true body cholesterol turnover in rabbits (body weight = 4–4.5 kg) as measured by modified sterol balance [167 mg: excretion (75 mg) + reabsorbed (92 mg)] or calculated by modified tracer kinetics ($R_{AB} = 168$ mg) was nearly the same as that measured chemically as free cholesterol secreted in plasma lipoproteins upon injection of Triton WR-1339 (168 mg/day). Similarly, the daily plasma esterified cholesterol turnover (234 mg) calculated based on the model was similar to that measured *in vivo* (230 mg) and *in vitro* (240 mg). Also, the values for the total free cholesterol input in the plasma [liver + peripheral tissues (334 mg/day)] calculated based on the model were similar to those calculated as the flux of cholesterol through the rapidly equilibrating pool A ($M_A \times \alpha = 345$ mg/day). AC, acetate; MEV, mevalonate.

Table XVII

DATA SHOWING THAT THE VALUES CALCULATED FOR PLASMA CHOLESTERYL ESTER TURNOVER BASED ON THE MODEL FOR CHOLESTEROL METABOLISM ARE SIMILAR TO THE *in Vivo*- AND *in Vitro*-MEASURED VALUES[a]

Species	Input of UC in plasma from		Total flux of UC in plasma (liver + peripheral tissues)	Flux of cholesterol in pool A[d] ($M_A \times \alpha$)	Plasma cholesteryl ester turnover		
	Liver[b]	Peripheral tissues[c]			Calculated[e]	Measured	
						In vivo[f]	*In vitro*[g]
Humans							
N	20.2	20.2	40.4	42.2	28.0	27.8	27.1
C	17.8	17.8	35.6	44.7	25.0	37.1	29.8
T	24.4	24.4	48.8	50.7	34.2	35.9	32.6
CT	25.3	25.3	50.6	50.6	35.4	46.7	35.3
$C_{y,xan}$[h]	32.2	32.2	64.4	NA[i]	45.0	43.0	NA
Rabbit[j]	41.8	41.8	83.6	86.2	58.5	57.5	57.2

[a] All values in milligrams per kilograms per day.
[b] Value for true body cholesterol turnover as measured by tracer kinetics. All the cholesterol is assumed to enter the plasma from the liver as unesterified cholesterol (UC).
[c] It is assumed that cholesterol transported to the plasma from the liver is utilized by other tissues. In a steady state, the same amount of cholesterol enters the plasma in unesterified form.
[d] Calculated by tracer kinetics of plasma cholesterol specific activity. The flux of cholesterol in pool A = mass of cholesterol in pool A (M_A) × daily fractional turnover rate (α, which is $0.693/t_{1/2}$ days of the exponential A.
[e] Calculated values are based on the model for cholesterol metabolism presented in Fig. 12.
[f] Kinetics of plasma unesterified and esterified cholesterol specific activity; mean of pooled data.
[g] *In vitro* determination of plasma LCAT activity; mean of pooled data.
[h] Subjects under 20 years of age.
[i] NA, not available.
[j] In rabbits the amount of UC secreted daily in the plasma was 40 mg/kg after injection of Triton WR-1339 (Klauda and Zilversmit, 1974).

moved by the liver which, in turn, secretes it into the gastrointestinal lumen as cholesterol and as bile acids (the catabolic product of) cholesterol). Some cholesterol may enter the gastrointestinal lumen directly through the uptake of plasma lipoproteins by the intestines. Furthermore, the identity of the values for biosynthesis + absorbed measured by sterol balance (fecal analysis) with those calculated by tracer kinetics (the methods which assume that cholesterol enters the plasma before secretion into the bile), and, most importantly, the identity of the calculated values for total body cholesterol mass with those measured chemically, support the validity of the model presented here for cholesterol metabolism.

At present it is believed that peripheral tissues, after removal of the cholesteryl esters from LDL, also catabolize the LDL apoB (Brown and Goldstein, 1976; Brown et al., 1979). Our model (Fig. 13A) is in full support of the former suggestion, but not of the latter. Transport of CE to peripheral tissues can account for only 14–15 mg/kg/day of LDL apoB CE turnover out of the total of 28–30 mg/kg/day of LDL apoB CE catabolized. The other half can be accounted for only if we assume that LDL apoB is reutilized for the reverse transport of the cholesteryl esters (14–15 mg/kg/day) formed in the plasma from the UC released by the peripheral tissues. That this assumption may be true is suggested by the fact that the metabolic clearance rate of cholesterol from the plasma [calculated as input of cholesterol in plasma (mg/day)/plasma pool of cholesterol (mg)] is similar to the clearance rate of LDL apoB from the plasma ($t_{1/2}$ days of second exponential; Table XVIII). Evidence available in the literature also suggests that the catabolism of cholesterol to bile acids

Table XVIII
SIMILARITY BETWEEN THE HALF-LIFE OF LDL ApoB (SECOND EXPONENTIAL) AND THE HALF-LIFE OF PLASMA CHOLESTEROL

Group	Biological half-life ($t_{1/2}$ days)	
	LDL apoB (second exponential)	Plasma cholesterol[a]
N	3.5	3.6
C	4.7	5.1
T	3.0	3.0
CT	—	4.4
$C_{y,xan}$ (homozygous)	6.0	6.7
CT_{xan}	—	5.8

[a] Calculated as $0.693/k_{FT}$ where k_{FT} = input in plasma (mg/kg) = (synthesized + total absorbed)/plasma pool of cholesterol (mg).

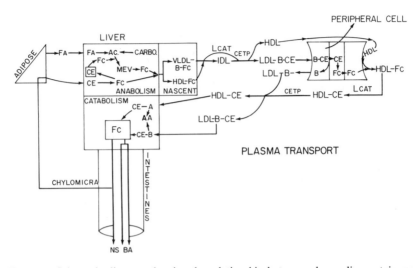

FIG. 14. Schematic diagram showing the relationship between plasma lipoprotein metabolism and cholesterol metabolism. Free cholesterol (Fc; synthesized + absorbed) is transported into the plasma mostly (85–90%) in triglyceride-rich very low density lipoproteins (VLDL) and some in high density lipoproteins (HDL). Upon entering the plasma, VLDL, through its interaction with endothelial lipases (lipoprotein lipase, hepatic lipase), plasma LCAT, and HDL [cofactors and cholesteryl ester transfer protein (CETP)], is catabolized to cholesteryl ester-rich LDL. Low density lipoproteins thus formed bind to the specific cell surface receptors and transport their CE (and other nutrients) to the cell. (It is assumed that LDL apoB, the carrier protein, is not catabolized by the peripheral cells.) In the cell, the CE is hydrolized to free cholesterol and utilized. An equivalent amount of free cholesterol is released by the cell surface. This free cholesterol is accepted by HDL, where it is esterified to CE by LCAT. Cholesteryl esters thus formed are then transferred to the LDL apoB ''skelton'' via CETP and are transported back to the liver (some cholesterol comes back to the liver via HDL) where protein is hydrolyzed to amino acids (AA) and the CE is hydrolized to free cholesterol and excreted, as such and after its conversion to bile acids (BA), into the bile. NS, neutral steroids; FA, fatty acids; AC, acetate; MEV, mevalonate.

[especially the rate of formation of bile acids calculated as daily excretion of bile acids (mg/day)/the plasma pool of cholesterol (mg), assuming that bile acids are formed from plasma cholesterol, indirect evidence for which was provided earlier in this discussion] is directly related to the fractional catabolic rate of both LDL apoB and plasma CE. Since bile acids are formed only in the liver, these observations suggest that LDL apoB may be the primary lipoprotein involved in the reverse transport of cholesterol to the liver. This is further supported by the fact that the catabolism of cholesterol through the turnover of HDL can account for only 250–300 mg/day [HDL cholesterol pool (1350 mg) × HDL protein k_{FC} (0.20/day) =

HDL cholesterol catabolism (270 mg/day)] out of the daily total of 1400–1500 mg cholesterol excreted in the gastrointestinal lumen.

These analyses therefore suggest the possible existence of an *in vivo* mechanism whereby peripheral cells, after binding LDL apoB to their receptors, selectively take up the lipoprotein cholesteryl esters (and other core lipids) with or without internalization of the LDL molecule. The LDL "skeleton" [ApoB–UC–phospholipid released from the surface of the cell or from the interior through reverse endocytosis (Bierman *et al.*, 1974)] then accepts the CE formed on HDL (through the action of LCAT on UC removed by HDL from peripheral tissues) via the cholesteryl ester transfer protein (Zilversmit *et al.*, 1975; Chajeck and Fielding, 1978) and transports it to the liver for its catabolism and excretion into bile (Fig. 14). Some cholesterol will be transported to liver by HDL. Based on their studies in perfused heart and endothelial cells in culture, Fielding and Fielding (1979) have postulated a mechanism similar to this.

VIII. Significance

If cholesterol uptake by tissues is by physiological (receptor and nonreceptor pathways) means and not by exchange, and if there is no significant loss of newly synthesized radioactive cholesterol from the plasma during the first 2–3 days, then more simpler and rapid methods could be developed for the quantitation of the flux of the newly synthesized cholesterol in the plasma, e.g., the method of Nestel and Kudchodkar (1975) could be used to determine the flux of newly synthesized cholesterol in the plasma *within 6–8 hours*. The method is based on the precursor–product relationship between squalene and cholesterol. In man, the specific activity of squalene reaches a plateau within 2 hours after the start of infusion of labeled mevalonate. The rate of formation of cholesterol (from squalene) can be calculated as follows: Influx into plasma of newly synthesized cholesterol (mg/hour) = (change in unesterified cholesterol radioactivity per hour × plasma volume)/specific activity of plasma squalene. These values, however, may primarily reflect cholesterol synthesized by the liver only (Table XIX).

In animals (or humans if sampling of tissues could be achieved) it may also be possible to determine the *in vivo* uptake of plasma cholesterol by different extrahepatic tissues. If the radioactivity found in different tissues (free of blood) is due to physiological uptake, as suggested by these analyses, then the available data would suggest that different tissues utilize (not exchange) plasma cholesterol at different rates, which may represent the

Table XIX
Cholesterol Influx Derived from Squalene during Constant Infusion of Radiomevalonate[a]

Subjects	Weight (kg)	Plasma lipids (mg/100 ml) Cholesterol	Plasma lipids (mg/100 ml) Triglycerides	Special procedure	Plasma squalene (μg/100 ml)	Cholesterol influx[b] (mg/24 hours)
1a	75	332	98	—	34	596
1b	75	262	117	Colestipol (15 g/day, 12 weeks)	59	1530
2a	79	261	160	—	43	429
2b	79	200	168	Colestipol (15 g/day, 12 weeks)	81	1172
3a	74	254	108	—	32	580
3b	74	198	166	Colestipol (15 g/day, 12 weeks)	58	1276
4a	95	136	150	Low cholesterol diet (<100 mg/day, 3 weeks)	81	1363
4b	95	139	141	High cholesterol diet (1000 gm/day, 3 weeks)	52	595
5a	62	142	75	Low cholesterol diet (<100 mg/day, 3 weeks)	66	823
5b	63	154	72	High cholesterol diet (1000 mg/day, 3 weeks)	45	177
6a	79	225	243	—	43	584
6b	79	189	156	Clofibrate (2 g/day, 8 weeks)	28	351
7a	68	241	171	—	43	790
7b	68	208	114	Clofibrate (2 g/day, 8 weeks)	26	534

[a] Based on data from Nestel and Kudchodkar (1975).
[b] These values may primarily reflect the cholesterol input from the liver.

time of tissue cell turnover. Tissues with rapid cell turnover (e.g., the intestines) or specialized tissues such as the adrenals will then be expected to take up cholesterol much more rapidly than tissues with slower cell turnover, for example, the brain and aorta. Tissue culture studies have shown that LDL receptor activity is high in faster as compared to slower proliferating cells (Brown and Goldstein, 1976; Brown *et al.*, 1979).

These analyses have shown that it may be possible to obtain the total body cholesterol synthesis, as well as the synthesis of cholesterol by hepatic and extrahepatic tissues, and the total body cholesterol mass, as well as the sizes of the tissue (or cell) cholesterol pools, accurately and as rapidly as in 4–6 weeks. Thus, means are now available for the complete description of cholesterol metabolism in normal and pathological states with greater precision than ever before. It should, therefore, be possible to assess with confidence the effects of physiological, dietary, and pharmacological manipulations on the previously mentioned parameters. Indeed, our analyses (to be published separately) of the available data in humans and in animals show that endogenously reabsorbed cholesterol (or the extrahepatic synthesis?) plays a major role in the regulation of plasma cholesterol and the deposition of cholesterol in the body. The data suggest that the effect of dietary cholesterol on the plasma (hypo- and hyperresponse) and other body tissue cholesterol pools may not be due to the amount of dietary absorbed cholesterol *per se*, but may be due to its effect on the secretion and reabsorption of *endogenous* cholesterol. Similarly, it appeared that the "endogenous" hypercholesterolemia and cholesterol deposits in tissues, seen under a variety of conditions, may also be a consequence not of overproduction but of an imbalance between excretion and reabsorption.

ACKNOWLEDGEMENTS

The author gratefully acknowledges the skillful assistance of Mrs. Beverly Baird, Ms. Joye Wesley, and Mrs. Sarita B. Kudchodkar in the preparation of this manuscript.

References

Anderson, J. M., Turley, S. D., and Dietschy, J. M. (1979). *Proc. Natl. Acad. Sci. U.S.A.* **76**, 165.
Angelin, B., Einarsson, K., Leijd, B., and Wallentin, L. (1981). *J. Lab. Clin. Med.* **97**, 502.
Barter, P. J. (1974). *J. Lipid Res.* **15**, 234.
Bell, F. B. (1976). *In* "Low Density Lipoproteins" (C. E. Day and R. S. Levy, eds.), p. 113. Plenum, New York.
Bhattacharyya, A. K., and Eggen, D. A. (1980). *J. Lipid Res.* **21**, 518.
Bhattacharyya, A. K., Conner, W. E., and Spector, A. A. (1976). *J. Lab. Clin. Med.* **88**, 202.

Bierman, E. L., Stein, O., and Stein, Y. (1974). *Circ. Res.* **35,** 136.
Biggs, M. W., Kritchevsky, D., Colman, D., Gofman, J. W., Jones, H. B., Lindgren, F. T., Hyde, G., and Lyon, T. P. (1952). *Circulation* **6,** 359.
Bilheimer, D. W., Stone, N. J., and Grundy, S. M. (1979). *J. Clin. Invest.* **64,** 524.
Bing, R. J., and Sarma, J. S. M. (1975). *Biochem. Biophys. Res. Commun.* **62,** 711.
Blum, C. B., Levy, R. I., Eisenberg, S., Hall, M., Gobel, R. H., and Berman, M. (1977). *J. Clin. Invest.* **60,** 795.
Borgstrom, B. (1969). *J. Lipid Res.* **10,** 331.
Brown, M. S., and Goldstein, J. L. (1976). *Science* **191,** 150.
Brown, M. S., Kovanen, P. T., and Goldstein, J. L. (1979). *Ann. N.Y. Acad. Sci.* **348,** 48.
Buja, M. L., Kovanen, P. T., and Bilheimer, D. W. (1979). *Am. J. Pathol.* **97,** 327.
Bullock, B. C., Lehner, N. D. M., Clarkson, T. B., Feldner, M. A., Wagner, W. D., and Lofland, H. B. (1975). *Exp. Mol. Pathol.* **22,** 151.
Carlile, S. I., and Lacko, A. G. (1981). *Comp. Biochem. Physiol.* **70B,** 753.
Carroll, K. K. (1971). *Atherosclerosis* **13,** 67.
Carter, G. A., Conner, W. E., Bhattacharyya, A. K., and Lin, D. S. (1979). *J. Lipid Res.* **20,** 66.
Chait, A., Bierman, E. L., and Albers, J. J. (1979a). *J. Clin. Invest.* **64,** 1309.
Chait, A., Bierman, E. L., and Albers, J. J. (1979b). *J. Clin. Endocrinol. Metab.* **48,** 887.
Chajeck, T., and Fielding, C. J. (1978). *Proc. Natl. Acad. Sci. U.S.A.* **75,** 3445.
Chevallier, F. (1967). *Adv. Lipid Res.* **5,** 209.
Chobanian, A. V., Burrows, B. A., and Hollander, W. (1962). *J. Clin. Invest.* **41,** 1738.
Clifton-Bligh, P., Miller, N. E., and Nestel, P. J. (1974). *Metabolism* **23,** 437.
Deckelbaum, R. J., Lees, R. S., Small, D. M., Hedberg, S. E., and Grundy, S. M. (1977). *New Engl. J. Med.* **296,** 465.
Dietschy, J. M., and Siperstein, M. D. (1967). *J. Lipid Res.* **8,** 97.
Eggen, D. A. (1974). *J. Lipid Res.* **15,** 139.
Eggen, D. A. (1976). *J. Lipid Res.* **17,** 663.
Eisenberg, S., and Levy, R. I. (1975). *Adv. Lipid Res.* **13,** 2.
Feldman, E. B., Russel, B. S., Schnare, F. H., Miles, B. C., Doyle, E. A., and Moretti-Rojas, I. (1979a). *J. Nutr.* **109,** 2226.
Feldman, E. B., Russel, B. S., Schnare, F. H., Moretti-Rojas, I., Miles, B. C., and Doyle, E. A. (1979b). *J. Nutr.* **109,** 2237.
Fielding, C. J., and Fielding, P. E. (1979). "Atherosclerosis V" (A. M. Gotto, Jr., L. C. Smith, and B. Allen, eds.), p. 379. Springer-Verlag, Berlin and New York.
Fourcans, B., Breillot, J., Meliv, B., Piot, M. C., Alcindor, L. G., and Polonovsky, J. (1974). *IRCS Libr. Compend.* **2,** 1187.
Glomset, J. A. (1968). *J. Lipid Res.* **9,** 155.
Glomset, J. A., and Norum, K. R. (1973). *Adv. Lipid Res.* **11,** 1.
Goldstein, J. L., and Brown, M. S. (1976). *Curr. Top. Cell. Regul.* **2,** 147.
Goodman, D. S. (1964). *J. Clin. Invest.* **43,** 2026.
Goodman, D. S. (1965). *Physiol. Rev.* **43,** 747.
Goodman, D. S., and Noble, R. P. (1968). *J. Clin. Invest.* **47,** 231.
Goodman, D. S., Noble, R. P., and Dell, R. B. (1973). *J. Lipid Res.* **14,** 178.
Goodman, D. S., Smith, F. R., Seplowitz, A. H., Ramakrishnan, R., and Dell, R. B. (1980). *J. Lipid Res.* **21,** 699.
Green, M. H., Crim, M., Traber, M., and Ostwald, R. (1976). *J. Nutr.* **106,** 515.
Grundy, S. M. (1978). *West. J. Med.* **128,** 13.
Grundy, S. M., and Ahrens, E. H., Jr. (1966). *J. Clin. Invest.* **45,** 1503.

Grundy, S. M., and Ahrens, E. H., Jr. (1969). *J. Lipid Res.* **10**, 91.
Grundy, S. M., and Metzger, A. L. (1972). *Gastroenterology* **62**, 1200.
Grundy, S. M., Ahrens, E. H., Jr., and Davignon, J. (1969). *J. Lipid Res.* **10**, 304.
Grundy, S. M., Ahrens, E. H., Jr., and Salen, G. (1971). *J. Lab. Clin. Med.* **78**, 94.
Gurd, F. R. N. (1960). *In* "Lipid Chemistry" (D. J. Hanahan, ed.), p. 260. Wiley (Interscience), New York.
Hellman, L., Rosenfeld, R. S., and Gallagher, T. F. (1954). *J. Clin. Invest.* **33**, 142.
Hellman, L., Rosenfeld, R. S., Eidinoff, M. K., Fukushima, D. K., Gallagher, T. F., Wang, C. I., and Aldersberg, D. (1955). *J. Clin. Invest.* **34**, 48.
Hellman, L., Rosenfeld, R. S., Insull, W. R., and Ahrens, E. H., Jr. (1957). *J. Clin. Invest.* **36**, 898 (Abstr.).
Hough, J. C., and Basset, D. R. (1975). *J. Nutr.* **105**, 649.
Hruza, A. (1971a). *Exp. Gerontol.* **6**, 103.
Hruza, A. (1971b). *Exp. Gerontol.* **6**, 199.
Hruza, A., and Zbuzkova, V. (1975). *Mech. Age. Dev.* **4**, 169.
Huff, M. W., and Carroll, K. K. (1980). *J. Lipid Res.* **21**, 546.
Illingworth, D. R., Conner, W. E., and Lin, D. S. (1980). *Gastroenterology* **78**, 68.
Jeske, D. J., and Dietschy, J. M. (1980). *J. Lipid Res.* **21**, 364.
Julian, P., Downar, E., and Angel, A. (1981). *Circ. Res.* **49**, 248.
Kekki, M., Miettinen, T. A., and Wahlstrom, B. (1977). *J. Lipid Res.* **18**, 99.
Kellogg, T. F. (1974). *J. Lipid Res.* **15**, 574.
Klauda, H. C., and Zilversmit, D. B. (1974). *J. Lipid Res.* **15**, 593.
Kovanen, P. T., Bilheimer, D. W., Goldstein, J. L., Jaramillo, J., and Brown, M. S. (1981). *Proc. Natl. Acad. Sci. U.S.A.* **78**, 1194.
Kudchodkar, B. J., and Sodhi, H. S. (1976). *Eur. J. Clin. Invest.* **6**, 285.
Kudchodkar, B. J., Kushwaha, R. S., and Albers, J. J. (1981). *Arteriosclerosis* **1**, 367 (Abstr.).
Langer, T., Strober, W., and Levy, R. I. (1972). *J. Clin. Invest.* **51**, 1528.
Lehner, N. D. M., Clarkson, T. B., Bell, F. P., St. Clair, R. W., and Lofland, H. B. (1972). *Exp. Mol. Pathol.* **16**, 109.
Lei, I. M., and Lei, K. Y. (1981). *J. Nutr.* **111**, 450.
Lofland, H. B., Clarkson, T. B., and Bullock, B. C. (1970). *Exp. Mol. Pathol.* **13**, 1.
McNamara, D. J., Proia, A., and Edwards, K. D. G. (1982). *Biochim. Biophys. Acta* **711**, 252.
Mahley, R. W., Hui, D. Y., Innerarity, T. L., and Weisgraber, K. H. (1981). *J. Clin. Invest.* **68**, 1197.
Massaro, E. R., and Zilversmit, D. B. (1977). *J. Nutr.* **107**, 596.
Mathe, D., and Chevallier, F. (1976). *Biochim. Biophys. Acta* **441**, 155.
Mathe, D., and Chevallier, F. (1979). *J. Nutr.* **109**, 2076.
Mathe, D., Lutton, C., Rautureau, J., and Chevallier, F. (1977). *J. Nutr.* **107**, 466.
Mazzone, T., Foster, D. W., and Chait, A. (1982). *Clin. Res.* **30**, 400 (Abstr.).
Miettinen, T. A., Proia, A., and McNamara, D. J. (1981). *J. Lipid Res.* **22**, 485.
Mok, H. Y. I., Bergmann, K. V., and Grundy, S. M. (1979). *J. Lipid Res.* **22**, 485.
Morris, M. D., Chaikoff, I. L., Felts, J. M., Abraham, S., and Fansal, N. O. (1957). *J. Biol. Chem.* **224**, 1039.
Mott, G. E., Jackson, E. M., and Morris, M. D. (1980). *J. Lipid Res.* **21**, 635.
Moutafis, C. D., and Myant, N. B. (1969). *Clin. Sci.* **37**, 61.
Myant, N. B. (1982). *Clin. Sci.* **62**, 261.
Nervi, F. O., and Dietschy, J. M. (1975). *J. Biol. Chem.* **250**, 8704.

Nervi, F. O., Weis, H. J., and Dietschy, J. M. (1975). *J. Biol. Chem.* **250**, 4145.
Nestel, P. J., and Kudchodkar, B. J. (1975). *Clin. Sci. Mol. Med.* **49**, 621.
Nestel, P. J., and Monger, E. A. (1967). *J. Clin. Invest.* **46**, 967.
Nestel, P. J., Havel, R. J., and Bezman, A. (1963). *J. Clin. Invest.* **42**, 1313.
Nestel, P. J., Whyte, H. M., and Goodman, D. S. (1969). *J. Clin. Invest.* **48**, 982.
Nestel, P. J., Reardon, M., and Fidge, N. H. (1979). *Metabolism* **28**, 531.
Nilsson, A., and Zilversmit, D. B. (1972). *J. Lipid Res.* **13**, 32.
Norum, K. R., and Gjone, E. (1967). *Scand. J. Clin. Lab. Invest.* **20**, 231.
Parks, J. S., Lehner, N. D. M., St. Clair, R. W., and Lofland, H. B. (1977). *J. Lab. Clin. Med.* **90**, 1021.
Pertsemlidis, D., Kirchman, E. H., and Ahrens, E. H., Jr. (1973a). *J. Clin. Invest.* **52**, 2353.
Pertsemlidis, D., Kirchman, E. H., and Ahrens, E. H., Jr. (1973b). *J. Clin. Invest.* **52**, 2368.
Quintao, E., Grundy, S. M., and Ahrens, E. H., Jr. (1971). *J. Lipid Res.* **12**, 221.
Raymond, T. L., Lofland, H. B., and Clarkson, T. B. (1976). *Exp. Mol. Pathol.* **25**, 344.
Reiser, R., Henderson, G. R., O'Brien, B. C., and Thomas, J. (1977). *J. Nutr.* **107**, 453.
Samuel, P., and Lieberman, S. (1973). *J. Lipid Res.* **14**, 189.
Samuel, P., Lieberman, S., and Ahrens, E. H., Jr. (1978). *J. Lipid Res.* **19**, 94.
Schumaker, V. N., and Adams, G. H. (1969). *Annu. Rev. Biochem.* **38**, 113.
Scott, P. J., and Hurley, P. J. (1970). *Atherosclerosis* **11**, 77.
Shepherd, J., Packard, C. J., Bicker, S., Lawrie, T. D. V., and Morgan, H. G. (1980). *New Engl. J. Med.* **302**, 1219.
Sigurdsson, G., Nicoll, A., and Lewis, B. (1976). *Eur. J. Clin. Invest.* **6**, 151.
Smith, F. R., Dell, R. B., Noble, R. P., and Goodman, D. S. (1976). *J. Clin. Invest.* **57**, 137.
Sodhi, H. S. (1975). *Perspect. Biol. Med.* **18**, 477.
Sohdi, H. S., and Kudchodkar, B. J. (1973a). *J. Lab. Clin. Med.* **82**, 111.
Sodhi, H. S., and Kudchodkar, B. J. (1973b). *Metabolism* **22**, 895.
Sodhi, H. S., and Kudchodkar, B. J. (1973c). *Lancet* **I**, 513.
Sodhi, H. S., and Kudchodkar, B. J. (1974). *Proc. Asia Oceania Congr. Endocrinol., 5th,* **2**, 484.
Sodhi, H. S., Kudchodkar, B. J., Varughese, P., and Duncan, D. (1974). *Proc. Soc. Exp. Biol. Med.* **145**, 107.
Sodhi, H. S., Kudchodkar, B. J., and Mason, D. T. (1979). "Clinical Methods in Study of Cholesterol Metabolism," Monographs on Atherosclerosis (D. Kritschevsky and O. J. Pollak, eds.), Vol. 9. Karger, Basel.
Sodhi, H. S., Kudchodkar, B. J., and Mason, D. T. (1980). *Adv. Lipid Res.* **17**, 107.
St. Clair, R. W., Wood, L. L., and Clarkson, T. B. (1981). *Metabolism* **30**, 176.
Steinberg, D., Pittman, R. C., Attie, A. D., Corew, T. E., Pangburn, S., and Weinstein, D. (1979). *In* "Atherosclerosis V" (A. M. Gotto, Jr., L. C. Smith, and B. Allen, eds.), p. 800. Springer-Verlag, Berlin and New York.
Stoudemire, J. B., Renaud, G., and Havel, R. J. (1982). *Circulation* **66**, II-100.
Thompson, G. R., Soutar, A. K., Spengel, F. A., Jadhav, A., Gavigan, S. J. P., and Myant, N. B. (1981). *Proc. Natl. Acad. Sci. U.S.A.* **78**, 2591.
Tomkins, G. M., Sheppard, H., and Chaikoff, I. (1953). *J. Biol. Chem.* **201**, 137.
Turley, S. D., Anderson, J. M., and Dietschy, J. M. (1981). *J. Lipid Res.* **22**, 551.
Turley, S. D., Spady, D. K., and Dietschy, J. M. (1983). *Gastroenterology* **84**, 253.
Viikari, J., Saarni, H., Ruuskanen, O., and Nikkari, T. (1976). *Acta Physiol. Scand.* **100**, 200.
Vuoristo, M., Tarpila, S., and Miettinen, T. A. (1980). *Gastroenterology* **78**, 1518.
Walton, K. W., Scott, P. J., Dyke, P. W., and Davies, J. W. L. (1965). *Clin. Sci.* **29**, 217.

Wilson, J. D. (1964). *J. Lipid Res.* **5**, 409.
Wilson, J. D. (1970). *J. Clin. Invest.* **49**, 655.
Wilson, J. D. (1972). *J. Clin. Invest.* **51**, 1450.
Zilversmit, D. B. (1960). *Am. J. Med.* **29**, 832.
Zilversmit, D. B. (1972). *Proc. Soc. Exp. Biol. Med.* **140**, 862.
Zilversmit, D. B., and Hughes, L. B. (1973). *Atherosclerosis* **18**, 141.
Zilversmit, D. B., Hughes, L. B., and Balmer, J. (1975). *Biochim. Biophys. Acta* **409**, 393.

Lecithin : Cholesterol Acyltransferase and the Regulation of Endogenous Cholesterol Transport

MILADA DOBIÁŠOVÁ

Institute of Nuclear Biology and Radiochemistry
Czechoslovak Academy of Sciences
Prague, Czechoslovakia

I.	Introduction	107
II.	Reaction Pattern of LCAT	109
	A. Activation	109
	B. Hydrolytic (Phospholipase) Activity of LCAT	112
	C. Transfer of the Acyl Group to the Acyl Acceptor	115
	D. Reverse LCAT Reaction	116
III.	Sources, Isolation, and Properties of LCAT	117
	A. Sources	117
	B. Isolation and Properties of LCAT	121
IV.	Methods of LCAT Estimation and Regulation of Its Activity	126
	A. Substrate and Conditions of the LCAT Reaction	126
	B. Methods	127
	C. Regulation of LCAT Activity	132
V.	Studies on LCAT in Experimental Animals	143
	A. The Rat as an Experimental Model	145
	B. LCAT in Some Other Experimental Animals	150
	C. Some Comments on Animal Models	151
VI.	Clinical Studies	151
	A. In Health	154
	B. In Disease	165
	C. Effect of Hypolipemic Drugs	180
VII.	Closing Remarks	183
	References	185

I. Introduction

Lecithin : cholesterol acyltransferase (LCAT, EC 2.3.1.43), an enzyme which catalyzes the transfer of fatty acids between lecithin and cholesterol on the surface of plasma lipoproteins (Glomset, 1962), performs several unique functions in the system of lipid and lipoprotein interconversions:

1. By maintaining the balance between free and esterified cholesterol (UC and EC, respectively), it stabilizes the normal shape and size of

lipoproteins, and a continuous exchange rate of lipids and apoproteins among them (Glomset and Norum, 1973; Glomset et al., 1980).

2. It creates the gradient necessary for the transfer of free cholesterol from tissues to the plasma (Tall and Small, 1980) and, at the same time, the transfer of cholesteryl esters to the tissue. In tissues, cholesteryl esters undergo hydrolysis and the resulting free cholesterol can be utilized (unless it is either catabolized or returned to the recirculation process) for the buildup or reconstruction of cellular membranes. Essential fatty acids released from cholesteryl esters can enter into other biologically important cycles, for example, as prostaglandin precursors.

3. Lecithin:cholesterol acyltransferase is a plasma component of phospholipid pathways in the conversion of lecithin to lysolecithin. However, the full extent of the versatile activity of lysolecithin, which can act as a biological detergent (Lucy, 1970), a medium accelerating the output of cholesterol from the plasma (Dobiášová et al., 1975, 1976; Portman and Alexander, 1976), and an arrhythmogenic (Sobel et al., 1978) and a cholinolytic (Zvezdina et al., 1978) heart muscle effector, has not yet been fully clarified.

4. The cholesterol esterification rate reflects the kinetics of interchanges, that is, the input and output of cholesterol. In view of the good agreement of reported data on the cholesterol esterification rate in vivo with the estimation in vitro based on methods assessing LCAT activity (Kudchodkar and Sodhi, 1976), the determination of molar and fractional esterification rates (MER and FER, respectively) offers a unique possibility for the detection and quantitation of the deviations accompanying various disorders of cholesterol metabolism which might have been obscured by relatively normal levels of plasma lipids.

Since Glomset (1962) first defined the function of LCAT, several important advances have contributed to the general understanding of the reaction mechanism and the physiological significance of this enzyme: the discovery of hereditary LCAT deficiency (Norum and Gjone, 1967) pointed to the importance of cholesteryl esters in the structure of lipoproteins, and disclosed new viewpoints applicable to studying this problem. The isolation of highly purified enzyme created the necessary prerequisite to the elucidation of the three-step reaction mechanism of LCAT and also of its participation in reversible reactions. A number of sophisticated studies using refined detection methods in a complete plasma system yielded new and substantial information about physiological regulations of LCAT activity in man, experimental animals, and other animal species.

This article is an attempt to summarize the current knowledge on the mechanisms of LCAT action, its endogenous regulation, and the prospects of prognostic merit in the timely detection of subtle derangements in the interchange of cholesterol between intra- and extravascular pools. However, such implications are presented with the reservation that the conclusions bearing on clinical aspects are mostly based on indirect and multifactorial evidence.

Extensive reviews on LCAT (Glomset, 1968, 1979; Glomset and Norum, 1973; Gjone et al., 1978; Glomset et al., 1980, 1983; Doi and Nishida, 1981; Norum et al., 1982; Simon, 1979; Rose, 1981) and associated problems (Brunzel et al., 1978; Nestel, 1970; Forte and Nichols, 1972; Tall and Small, 1980; Nicoll et al., 1980; Sodhi et al., 1980) have been published.

II. Reaction Pattern of LCAT

Yokoyama et al. (1980) conceived a reaction scheme for the action of LCAT in the esterification process of plasma cholesterol, and Subbaiah et al. (1980) described an until-that-time unknown function of the enzyme in the esterification of lysolecithin to lecithin. In essence, the LCAT reaction pattern proceeds in the following distinct steps: (1) activation of the phospholipid bilayer by protein or peptide; (2) hydrolysis of fatty acid ester by means of phospholipase A_2-like activity (formation of an intermediate acyl-enzyme); (3) transfer of the fatty acyl to the acyl acceptor; and (4) reverse LCAT reaction.

A. Activation

Fielding et al. (1972a) were the first to show that the LCAT reaction can be activated by some protein components of plasma lipoproteins, in particular the basic protein unit of high-density lipoprotein apoprotein AI (HDL apoA-I), whereas apoA-II inhibits the reaction. The activating effect was also demonstrated for apoC-I (Soutar et al., 1975; Sigler et al., 1976), while apoC-II (postheparin lipase-activating apoprotein), apoC-III, and apoD showed no activation; on the contrary, the last three components inhibited the reaction at the optimum activator concentration (Albers et al., 1979). On the basis of the amino acid sequence analysis of apoC-I (Jackson et al., 1974), peptide fragments were isolated and recombined with the dimyristoyl phosphatidylcholine (DMPC) unilamellar vesicle (Soutar et al., 1978). The properties of these complexes were studied by means of density gradient centrifugation, fluorescence, and circular dichroism spectra. Stable complexes with DMPC containing 25–40 amino acids in peptide residues exhibited a significant increase in α-helicity, and induced at the same time an almost complete LCAT activation comparable to that achieved by apoC-I. The notion (Day and Levy, 1969; Segrest et al., 1974; Edelstein et al., 1979) of the arrangement of the peptide chain into amphipathic α-helical segments with opposite sides of polar and nonpolar characters which determine the type of protein–lipid interaction necessary for the reaction of LCAT (Morriset et al., 1977) was confirmed by Yokoyama et al. (1980) in a sophisticated quantitative study. They constructed a synthetic docosapeptide carrying optimal α-helix characteristics of the amphiphilic (synonymous with amphipathic) fragment; however, its amino acid sequence was designed deliberately so that it differed from that of apoA-I, as well as from that of apoC-I, fragments. This peptide bound with simple Langmuir's isotherm to vesicle surfaces, either of egg lecithin or unilamellar lecithin–cholesterol (4:1) vesicles, and activated the esterification of cholesterol by highly purified LCAT. The apoA-I (or docosapeptide)–lecithin vesicles, shown by gel permeation chromatography and ultracentrifugation to be homogeneous, exhibited, respectively, dissociation constants (K_d) of 9.0×10^{-7} and 1.92×10^{-6}, and binding capacities of 574 mol of phosphatidylcholine per 1 of mol apoA-I, and 66 mol of phosphatidylcholine per 1 mol of peptide. Incorporation of cholesterol to the vesicle (lecithin:cholesterol = 4:1) reduced the molar ratio of protein (or peptide) to lecithin by approximately half. The maximum activation for cholesterol esterification achieved by the peptide was about 18% of that obtained by apoA-I.

Chung et al. (1979) reported fairly similar results on the binding capacity for maximum activation of purified LCAT by apoA-I–egg lecithin–cholesterol (4:1) vesicles: they found the ratio of 336 mol of phosphatidylcholine to 1 mol of apoA-I. However, the K_d estimated for apoA-I–egg lecithin–cholesterol (4:1) differed more substantially: 1.4×10^{-6} (Chung et al., 1979) compared to 3.0×10^{-7} (Yokoyama et al., 1980).

According to current knowledge, the nature of LCAT activation by apoprotein, or by any protein meeting the requirements of amphiphilic α-helicity, can be outlined as follows: Activation involves the unilamellar layer of lecithin and not the enzyme itself. It appears that the activator–lecithin interaction consists in a reorientation of the olefinic bonds of the hydrophobic residues from the surface to the hydrophobic regions of the monolayer, and thus in the rearrangement of the ester bonds exposed to the enzyme attack (Yokoyama et al., 1980). It may be that as a result of the protein–lecithin interaction, the tension at the site of the ester bond increases due to the straightening of the naturally bent olefinic chain, and that this change facilitates the enzyme action. No direct interaction between the enzyme and the activating protein takes place for the following two reasons:

1. The presence of the apoprotein activator is not necessary for the enzyme–lecithin vesicle linkage, as was confirmed by Aron *et al.* (1978); over 95% of the enzyme activity was bound to lecithin or to lecithin–cholesterol liposomes immobilized on Sepharose in the absence of an activator.
2. The enzyme is efficiently displaced from the complex with the phospholipid vesicle by means of apoA-I, and the actual protein–enzyme interaction is so weak that it cannot be detected by gel permeation chromatography (Furukawa and Nishida, 1979).

Neither is the activation induced by too tight a linkage of the activator to the phospholipid vesicle (Soutar *et al.*, 1975). On the contrary, with a stronger protein–phospholipid interaction LCAT is inhibited in a manner similar to what happens if the phospholipid vesicles are tightly packed. It has been found by Chung *et al.* (1979), using the ultracentrifugation flotation technique, that the apoA-II incubated with the apoA-I–lecithin–cholesterol complex displaces the apoA-I from the surface of the vesicles. This observation confirmed Assman and Frederickson's (1974) assumption of a higher affinity of apoA-II to phospholipid liposomes. Parallel to the displacement of apoA-I from the vesicles, the LCAT activity decreased; thus, the authors concluded that the displacement of apoA-I from the surface of phospholipid vesicles by proteins with higher binding affinity constituted the basis of the physiological regulation of LCAT activity.

Another factor which affects the optimum conformation of the LCAT activator is the architecture of the phospholipid vesicle, which is dependent upon the saturation of hydrocarbon chains in the lecithin molecule, as well as on the concentration and nature of the sterols incorporated into the vesicle. Consequently, an increase in cholesterol concentration in the lecithin vesicle, causing a reduction in the random thermal motion of phospholipids and a "condensing effect" (Hsia *et al.*, 1972), is accompanied by an inhibition of the LCAT reaction (Nichols and Gong, 1971). The dependence of LCAT activity on the fluidity of the hydrocarbon region of the vesicular substrate, determined by the degree of saturation, position, and length of the hydrocarbon chains, was first noted by Soutar *et al.* (1975).

It can therefore be inferred from the information available up to now that the protein activator induces changes at the surface of the lecithin–

cholesterol vesicle, unless its attachment to the vesicle is too strong, which expose the fatty ester bond of lecithin to the hydrolytic action of the enzyme.

B. HYDROLYTIC (PHOSPHOLIPASE) ACTIVITY OF LCAT

Piran and Nishida (1976) observed in a system of partially purified LCAT (100 times) that the presence of apoA-I activator, albumin, and a single bilayer of lecithin vesicles resulted in a release of fatty acid from position-2 of lecithin. In this study a mixture of labeled 2-[9,10-^{3}H]dipalmitoyl phosphatidylcholine (DPPC) and egg lecithin (1:5) was used. It was confirmed later (Piran and Nishida, 1979) that this mixture of saturated and unsaturated lecithins also facilitates the activation of saturated acyls from position-2 lecithin and makes them available for interaction with LCAT, in agreement with the requirement of the necessary fluidity of vesicles needed for an interaction with the protein activator. In the absence of cholesterol, partially purified LCAT released from DPPC about 5% of [^{3}H]palmitate, whereas in the presence of cholesterol in lecithin vesicles (1:4) only about 3% of free acid originated, and 14% was found incorporated into cholesteryl esters. The hydrolytic activity of LCAT appeared to have been inhibited by the same agents that inhibit the complete transesterification activity, i.e., heating, sulfhydryl group-blocking reactive compounds, and deoxycholate. This hydrolytic activity was defined by the authors as "phospholipase A_2-like activity," since unlike the genuine phospholipase activity, this type was independent of Ca^{2+} ions. The conclusion, i.e., the identity of LCAT and phospholipase activity, is supported by the results obtained by Aron *et al.* (1978) in a study using highly purified LCAT (10,000–17,000 times from the original plasma) and a mixture of lecithin or lecithin–cholesterol vesicles. Aron *et al.* showed a close similarity in kinetic properties of phospholipase and transferase regarding dependence on pH, absolute requirement for an activator (apoA-I in this case), and substrate concentration. Moreover, in a kinetic study on the inhibition of LCAT phospholipase and LCAT transferase activities by means of sulfhydryl inhibitors [5,5'-dithiobis-(2-nitrobenzoic acid), (DTNB)], as well as serine–histidine antiesterase inhibitory agents (diisopropyl fluorophosphate and diethyl nitrophenyl phosphate), they could demonstrate identical inhibition plots for both the phospholipase and transferase activities, which can be taken as an evidence that the two reactions are catalyzed by the same enzyme-protein and, furthermore, that they both bind to identical active-site residues. Yokoyama *et al.* (1980) studied the activation of LCAT by a synthetic docosapeptide, and observed likewise a release of fatty acid from lecithin even in the absence of cholesterol; if cholesterol was present, the rate of hydrolysis rose in parallel with the increasing intensity of esterification. These results may be interpreted as a confirmation of the assumption formulated by Aron et al. (1978) that "the reaction of hydrolysis and ester production share a common intermediary substance, possibly an acyl-enzyme whose production is rate-limiting."

Specificity of the Acyl Donor

The exclusive acyl donor for the LCAT reaction is lecithin (Glomset, 1968), a basic nitrogen atom-containing phospholipid. It cannot be substituted by any other phospholipid (Nichols and Gong, 1971), unless it is N-methylated (Fielding, 1974). Neither do synthetic-position isomers of phosphatidylcholine take part in the reaction with LCAT. Smith and Kuksis (1980) synthesized 2-[^{3}H]16:0,3-16:0-sn-glycerol 1-phosphorylcholine, 2-[^{3}H]18:2, 3-16:0-sn-glycerol 1-phosphorylcholine, and 1-[1-^{14}C]18:2,3-16:0-sn-glycerol 1-phosphorylcholine and compared the capacity of these compounds to serve as substrates for LCAT with that of likewise synthetic 3-sn isomers which are known to occur naturally as well. The results showed that in comparison to the 3-sn isomers, the 2-sn isomers yielded only 16% of LCAT activity, whereas no significant activity was detectable with 1-sn isomers. However, since this interesting experiment was performed in a complete plasma system, it remains to be clarified which particular step of the LCAT reaction is involved.

The validity of the original thesis that LCAT reacts merely with fatty acid acyl in the C-2 position of lecithin (Glomset, 1962; Sgoutas, 1972) was reconsidered (Glomset, 1968; Goodman, 1964) in order to accommodate the finding that the degree of saturation of cholesteryl esters resulting from the LCAT reaction was substantially higher than that which would correspond to the saturation of acyls in the 2 position of plasma lecithins. Incomplete position specificity of LCAT was experimentally confirmed by Aron et al. (1978) and by Assman et al. (1978a). In a system consisting of highly purified LCAT, protein activator apoA-I and lecithin or lecithin–cholesterol (10:1) vesicles with exactly defined composition (1-palmitoyl,2-oleyl phosphatidylcholine; 1-oleyl,2-palmitoyl phosphatidylcholine; 1-palmitoyl,2-linoleyl phosphatidylcholine), Aron et al. (1978) succeeded in demonstrating a predominant, yet never an exclusive, release of acyl groups from the C-2 position. In a system composed of partially purified LCAT, apoA-I activator, and sonnicated vesicles of specifically labeled phosphatidylcholines and cholesterol (molar ratio of 10:1), Assmann et al. (1978a) failed to find the LCAT-mediated transacylation specific for fatty acids in the 2 position of phosphatidylcholines. In the case of dilinoleyl phosphatidylcholine, 40% of cholesteryl ester fatty acids were derived from the 1-position and 60% were derived from the 2-position phosphatidylcholine. For substrates consisting of phosphatidylcholine with identical fatty acids in the C-1 and C-2 positions, the order of relative transesterification rates was estimated as follows: $18:2 > 20:4 \simeq 14:0 > 18:1 > 10:0 \simeq 12:0 > 8:0 > 16:0 > 18:3 \simeq 18:0$.

It could further be shown that the degree of unsaturation of the fatty acid in the 1-position phosphatidylcholine affects the release of fatty acid from the 2-position phosphatidylcholine. This effect has been attributed by several authors to the preferential interaction of unsaturated lecithins with the apoprotein activator at an increased fluidity of the phospholipid vesicle. Such a conclusion is in full agreement with the previously mentioned activation hypothesis. The question of the species specificity of LCAT to a certain type of fatty acid has not yet been definitely clarified. Portman and Sugano (1964) demonstrated in cross experiments using enzymatically active human plasma versus LCAT-inactive rat plasma substrate, and vice versa, that the fatty acid composition of newly formed cholesteryl esters depended primarily on the preexisting pattern of cholesteryl ester composition in the active plasma, rather than on the pattern of the substrate derived from a different species. An even more striking specificity for the type of preexisting pattern was found with horse serum (Yamamoto et al., 1979b). Contrary to this, similarly designed cross-experiments using LCAT in the plasma of reptiles and amphibians persuaded Gillet (1978) to conclude that the determining factor for the pattern of fatty acids in cholesteryl esters was the composition of plasma lecithin in the substrate. However, the LCAT specificity may be modified by interspecies differences; yet the human serum linoleic acid appears to be the most favored substrate for the LCAT reaction (Yao and Dyck, 1977).

It appears from what has just been said, therefore, that LCAT specificity with respect to the acyl donor is bound to lecithin; that the position specificity is incomplete, even though the interaction with fatty acids in the 2 position predominates; and that optimum transesterification is achieved with dilinoleyl phosphatidylcholine, whereas saturated fatty acids containing lecithins (16:0, 18:0) exhibit a low degree of reactivity. Utilization of saturated lecithin fatty acids in the LCAT reaction increases if unsaturated lecithins are present in mixed lipid vesicles (Piran and Nishida, 1979). The effectivity of the LCAT reaction is enhanced about four- to fivefold by the use of micellar complexes composed of apoA-I–egg yolk phosphatidylcholine–cholesterol, when compared to commonly used small unilamellar vesicles of egg yolk phosphatidylcholine and cholesterol in the presence of apoA-I (Matz and Jonas, 1982; Jonas and Matz, 1982).

C. TRANSFER OF THE ACYL GROUP TO THE ACYL ACCEPTOR

In general, sterols have a double function in the LCAT reaction: (1) as molecular factors which affect the intermolecular interactions of the lecithin vesicle, and thus change its affinity to protein activator; consequently lecithin becomes activated to a different degree; and (2) as preferred acceptors of fatty acyls released from the protein-activated lecithin bilayer.

The ordering effect of sterols upon the molecular motion of lecithin layers depends on unsaturation and the distribution of double bonds among acyl chains, as well as on the sterol type. For instance, cholesterol induces a condensation effect in 1-saturated, 2-unsaturated lecithin, whereas no such effect is observed for 1,2-diunsaturated lecithins, for example, 1,2-dilinoleyl (Demel et al., 1972a). As to the sterol configuration, the interaction with lecithin depends on the presence of a 3-hydroxy group, a planar steroid nucleus, and an intact hydrocarbon side chain (Demel et al., 1972b). Moreover, the presence of a saturated side chain on C-17 is required for the maximum ordering effect in egg lecithin bilayers (Hsia et al., 1972). Cholesterol as a physiological acyl acceptor in the LCAT reaction can be substituted by other sterols, provided the following two prerequisites of the interaction with the lecithin bilayer are met: that a 3-hydroxy group be present, and that the A/B ring be in the transposition (Piran and Nishida, 1979). Neither the position of the double bond in B ring nor its complete absence had any effect on LCAT activity; however, the presence of a side chain on C-17 reduced the capacity of the sterol to react with acyl. Androsterol, a substance which lacks the side chain entirely, appeared to be a more active acceptor than cholesterol (Table I), and sterols with more voluminous or more rigid chains on C-17 were markedly less active as compared to cholesterol. On the basis of these findings, a conclusion was drawn that the interaction between sterol and lecithin, which leads to a condensation effect and to permeability changes, is not an essential prerequisite to a transesterification reaction. A similar conclusion was derived from a study comparing the effect of the optimum acyl donor, that is, 1,2-dilinoleyl phosphatidylcholine, to that of 1,2-dipalmitoyl phosphatidylcholine, as has been explained. Also, 1-hexadecanol can serve as an acyl acceptor (Kitabatake et al., 1979), as can possibly a whole number of other hydroxy compounds whose molecular interactions with the lecithin vesicle do not virtually change its interfa-

Table I
STEROLS AS ACYL ACCEPTORS IN THE LCAT REACTION[a]

Effect of the configuration of the hydroxyl group on C-3 and the position of the double bond in ring B			
(β-3)	(%)	(α-3)	(%)
Cholesterol (cholest-5-ene-3β-ol)	100	Epicholesterol (cholest-5-ene-3α-ol)	1
Cholestanol (cholestan-3β-ol)	90	Epicholestanol (cholestan-3α-ol)	1
Lathosterol (cholest-7-ene-3β-ol)	83		
Androsterol (androstan-3β-ol)	128	Epiandrosterol (androstan-3α-ol)	1
Effect of A and B ring configuration			
(A/B trans)	(%)	(A/B cis)	(%)
Cholestanol (cholestan-3β-ol)	90	Coprostanol (coprostan-β-ol)	1
Effect of the side chain on C-17			
(β-3)	(%)		
Androsterol (androstan-3β-ol)	128		
Campesterol (24-isoergost-5-ene-3β-ol)	89		
Stigmasterol (stigmast-5,22-diene-3β-ol)	34		
β-Sitosterol (stigmast-5-ene-β-ol)	79		
β-Sitosterol (stigmast-5-ene-3β-ol)	82[b]		
Desmosterol (cholest-5,24-diene-3β-ol)	46		

[a] Production of sterol esters (cholesterol = 100%) in the presence of lecithin vesicles and purified LCAT (according to Piran and Nishida, 1979).
[b] According to Kitabatake *et al.* (1979).

cial properties related to fluidity, gel–liquid crystalline transition temperature, and possibly molecular surface area. This is actually also supported by the fact that in the absence of sterol in the lecithin vesicle the acyl acceptor is water.

D. REVERSE LCAT REACTION

The notion of a one-way transfer of fatty acid from lecithin to cholesterol has been revised. Subbaiah *et al.* (1980) showed that highly purified preparations of LCAT also had the capacity to catalyze the energy-independent acylation of lysolecithin to lecithin. Low-density lipoprotein (LDL) is the activator for this lysolecithin acyltransferase activity of LCAT, whereas the acyl transfer from lecithin necessarily requires the presence of apoA-I or a protein with amphiphilic α-helical configuration, as has been amply demonstrated. The course of this reverse LCAT reaction was in no way affected by the presence or absence of apoA-I. The identity of LCAT and lysolecithin acyltransferase was studied in an isolated system of LCAT–LDL, as well as in the plasma of patients suffering from familial LCAT deficiency. The effect of pH, temperature, and chemical inhibitors (p-hydroxymercuribenzoate, phenylmethylsulfonyl fluoride, sodium deoxycholate) was found to be similar on both the deacylation and acylation activities of LCAT. Blood plasma that was LCAT-deficient also lacked lysolecithin acyltransferase activity. It is not as yet certain what the origin is of the acyl residue transferred to lysolecithin. Subbaiah and Bagdade (1978, 1979) suggested originally that the energy-independent acylation of lecithin was associated with the combination of two lysolecithin molecules, and was accompanied by a release of glycerophosphatidylcholine, in a manner similar to the reaction described by Erbland and Marinetti (1965). However, this initial concept was later refuted by more detailed studies of the same group (Subbaiah et al., 1980). In a series of well-conceived experiments on the interaction of LCAT, LDL, and lysolecithin double-labeled with radioisotopes in the glycerol as well as in the fatty acid groups, no doubling of the specific radioactivity could be found in the lecithin acyl residues. The identity of the acyl donor, whether it is LDL lecithin, cholesteryl esters, or other glycerides, remains to be specified. It is believed that, like in the LCAT reaction, lecithin can be the acyl donor, and that the formation of an acyl-enzyme intermediate can be the common product of all reactions which are catalyzed by the enzyme and differ only in the acyl acceptor, be it water, sterols, or lysolecithin. It appears, however, that the notion of lecithin being the acyl donor is incompatible with the theory of lecithin bilayer activation by means of amphiphilic protein (Yokoyama et al., 1980), a process which is the prerequisite to the formation of the presumed acyl-enzyme intermediate. In the case of lysolecithin acylation, the presence of a protein activator of this type is not required. Clarification of the LCAT–LDL interaction in its reverse function may be of particular interest also from the standpoint of the physiological regulation of LCAT.

III. Sources, Isolation, and Properties of LCAT

A. Sources

1. Tissues

Evidence concerning the origin of LCAT is based primarily on perfusion studies: experiments with perfused liver (Osuga and Portman, 1971; Simon and Boyer, 1970; Marsh and Kashub, 1970) demonstrated that

LCAT was released from the parenchyma to the perfusate. These experiments were performed using isolated rat liver perfused with various perfusion fluids: rat blood (with citrate anticoagulants), heat-inactivated rat plasma with human erythrocytes, or synthetic medium containing electrolytes, albumin, and human red blood cells. After 2 hours of perfusion, LCAT activity was detectable in all types of perfusates tested, although the synthetic perfusate consistently contained the lowest concentration. It could further be shown that the liver tissue actively removed LCAT from the medium; this in fact prevented any reliable quantitative conclusion to be made from being drawn. Lecithin : cholesterol acyltransferase removal activity has already been described by Quarfordt and Goodman (1969) in similar perfusion studies. Direct evidence that liver parenchyma cells synthesize and secrete LCAT was presented by Nordby et al. (1976). The optimum rate of LCAT secretion from isolated hepatocytes was attained after 5 hours of incubation at pH 7.3–7.4; the release could be stimulated by addition of enzymatically inactive serum to the essential synthetic medium. Colchicine, an agent known to inhibit protein and lipoprotein secretion from the liver, markedly reduced the LCAT activity in the medium. Cycloheximide, an inhibitor of protein synthesis, exerted an effect similar to that of colchicine. The fact that hepatocytes released cholesterol and triacylglycerol in parallel with LCAT (Nordby et al., 1979) substantiated the conclusion that LCAT becomes associated with lipoproteins in the course of secretion. However, LCAT secretion was independent of the secretion of very low density lipoprotein (VLDL), as was established by Miller (1978) in an *in vivo* experiment on rats demonstrating a specific blocking effect of orotic acid on VLDL liver secretion. Since orotic acid had little or no effect on hepatic nascent HDL secretion by the perfused rat liver (Marsh, 1976), the previously mentioned finding could be taken as indirect evidence that LCAT was secreted by hepatocytes together with HDL. A direct interaction between LCAT and HDL was demonstrated in an experiment carried out by Hamilton et al. (1976) in which rat liver perfused with fluid containing the LCAT inhibitor DTNB secreted into the medium disk-shaped HDL composed primarily of polar lipids and proteins, and only a small amount of cholesteryl esters and triglycerides. The main apoprotein component was represented by "arginine-rich protein," that is, apoE, in contrast to plasma HDL, where apoA-I prevails. These discoidal HDL resembled the particles found in the plasma of subjects with familial LCAT deficiency (Glomset and Norum, 1973). In the absence of the LCAT inhibitor, the perfusate contained spherical particles corresponding in their composition of cholesteryl esters and apoproteins to the normal plasma HDL. On the other hand, the VLDL secreted by the liver were in no way affected by the presence or absence of LCAT-specific inhibitors.

The intracellular site of origin of LCAT has not as yet been precisely identified, even though it is assumed (Russell *et al.*, 1976) that LCAT is synthesized in the microsomal fraction; this contention is indirectly corroborated by the finding that the administration to rats of phenobarbital, a compound stimulating microsomal synthesis of proteins, results in an increase in plasma LCAT activity. Contrary to Russell *et al.*, Heller and Desager (1978) failed to confirm the induction of LCAT formation in the plasma of healthy subjects given phenobarbital orally for 7 days. In view of the fact that the regulation of LCAT activity is modified by a number of physiological and physical factors, the question of localization of LCAT synthesis continues to be an urgent research topic. A serious methodological obstacle is the difficulty in determining LCAT directly in the liver homogenate by means of the presently available technique, measurement of cholesteryl ester production, in spite of the fact that LCAT is secreted by hepatocytes. Nordby and Norum (1978) believe that either the cell homogenate contains a cholesteryl ester-hydrolyzing activity, or the enzyme is stored in an inactive form and becomes activated only after it has been secreted.

No precise data are available at present on the quantity of LCAT produced. Nordby *et al.* (1976) calculated that rat liver can produce the entire plasma pool of the enzyme within 1 day; other calculations cited are 5 hours (Osuga and Portman, 1971) or 7–13 hours (Soler-Argilaga *et al.*, 1977) as the likely time interval for a complete turnover of the LCAT pool.

The only other organ which might possibly be involved in LCAT secretion is the gut: this is indicated by the finding that the gut synthesizes HDL, and that the mesenteric lymph displays a certain degree of LCAT activity (Bennett-Clark and Norum, 1977). A noteworthy case is the presence of LCAT in edematous rat brain tissue; the enzyme probably plays a role here in the formation of cholesteryl esters characteristic for the process of demyelinization. An intact brain does not, however, contain any LCAT activity; it is therefore likely that in the case of brain edema LCAT infiltrates the tissue together with plasma lipoproteins through the blood–brain barrier (Amaducci *et al.*, 1978).

2. LCAT in the Plasma

In blood plasma, LCAT is associated predominantly with the heterogeneous group of high-density lipoproteins (Glomset, 1968; Akanuma and Glomset, 1968), particularly with the smaller subfractions of HDL (Fielding and Fielding, 1971) which are simultaneously substrates and regulators of LCAT activity.

A more accurate measurement of LCAT plasma distribution by means of radioimmunoassay (Albers et al., 1981a) showed that the $d = 1.063-1.21$ g/ml HDL fraction contained about 50% of the enzyme, whereas some 35% of LCAT could be recovered from the $d \geq 1.21$ g/ml fraction. Only a small amount of active LCAT (approximately 1% of the plasma LCAT mass and activity) is consistently associated with the low-density lipoprotein fraction, whereas very low density lipoprotein fraction does not exhibit any LCAT activity (Chen and Albers, 1982a). As already pointed out, LCAT is secreted by liver cells together with nascent discoidal HDL (Hamilton et al., 1976), and assists actively in the process of their maturation in the plasma by changing free cholesterol to esterified cholesterol, which changes the HDL volume and size from discoidal to spherical. High-density lipoprotein metabolism has been reviewed by Nicoll et al. (1980). It appears, however, that no quantitative relationship exists between HDL and LCAT, since the subjects with Tangier disease, in whom the level of plasma HDL is barely detectable, were shown to have operative LCAT and an only slightly reduced rate of cholesterol esterification (Clifton-Blight et al., 1974; Assmann et al., 1978b). The half-life of LCAT in the plasma, determined by Stokke (1974) in patients suffering from familial deficiency and transfused with normal plasma, appeared to be 4.6 days; this value is comparable to the half-life of HDL (Nicoll et al., 1980). Yet it is not at all clear whether this value also applies to normal human subjects. The quantitative estimation of LCAT concentration in human plasma is now believed to be around 6 μg/ml; a method of reliable determination is now available (Albers et al., 1981b).

3. LCAT in Extravascular Fluids

Lymph of certain animal species, including man, exhibits some LCAT activity. Contrary to the initial assumption (Glomset and Norum, 1973), the LCAT pool in the lymph does not make a significant contribution to the production of cholesteryl esters. Bennett-Clark and Norum (1977) studied the LCAT activity in rat mesenteric lymph as a possible source of chylomicron cholesteryl esters. Lymphatic activity accounted merely for 2–3% of the serum LCAT activity, increasing somewhat after separation of the $d < 1.006$ g/ml fraction, or at the time of feeding, yet never exceeding 10% of the serum activity. A comparably low LCAT activity was found in lymph collected from human feet (Reichl et al., 1973), or in canine cardiac and thoracic duct lymph (Stokke, 1974): in all cases it amounted to 7–18% of the activity detected in the serum. The origin of this lymphatic LCAT is not entirely clear; it is conceivable that it originates by capillary filtration of the plasma, even though it cannot be excluded that it is synthesized in the gut, paralleled by the well-documented synthesis of HDL. However, the findings reported by Norum et al. (1979) appear to have clarified the origin of lymphatic cholesteryl esters. The acyl-CoA : cholesterol acyltransferase activity found in the microsomal fraction of bioptic samples taken from intestinal mucosa of a normal man is high enough (3.6 ± 1.37 mmol of cholesteryl ester formed per milligram protein per hour) to account for all the cholesteryl esters in the intestinal lymph.

Another body fluid with detectable LCAT activity is pig ovarian follicular fluid (Yao et al., 1980). Lecithin : cholesterol acyl transferase probably passes from the plasma via capillary filtration to the follicular fluid together with HDL. Unlike in the lymph, the molar LCAT activity in the follicular fluid is substantially higher, approximately 20–25% of that of the plasma. Ovarian follicular fluid can serve as a perfect physiological model for studies on the interaction between LCAT and lipoproteins, since the fluid contains just HDL, while lipoproteins of larger molecular size (VLDL, LDL) are unable to pass through the follicular barrier.

B. Isolation and Properties of LCAT

1. Purification

The plasma concentration of LCAT is around 6 μg/ml; thus the mass ratio of the enzyme to all other plasma proteins is about $1:10^4$. This low concentration of the enzyme, combined with its instability apparent even in a complete plasma system, has long made the isolation of the enzyme and the definition of its molecular properties impossible. The first attempts at isolating LCAT by using hydroxyapatite adsorption, salt precipitation, ultracentrifugation, and affinity column chromatography techniques (Glomset and Wright, 1964; Akanuma and Glomset, 1968; Fielding et al., 1972b; Lacko et al., 1974) yielded only partially purified preparations. Since LCAT in the plasma is mostly associated with HDL (Akanuma and Glomset, 1968; Albers et al., 1981b), its separation from the contaminating apoA-I, apoA-II, apoD, and albumin is a critical step. Methods which make it possible to obtain highly purified samples of LCAT have been worked out in various laboratories. Apart from minor modifications in adsorption and affinity chromatography, particular isolation schemes differ mainly in using either centrifugation or precipitation. When sequential centrifugation is applied, the middle, clear-zone, $d =$ 1.21–1.25 g/ml fraction is separated (Albers et al., 1976, 1979; Chung et *al.*, 1979) and subjected to further purification by means of adsorption and affinity chromatography or polyacrylamide gel electrophoresis (Suzue *et al.*, 1980). Time-consuming and expensive ultracentrifugation methods can be obviated by precipitating the active fraction either with ammonium sulfate (Varma and Soloff, 1976) or with a combination of ammonium sulfate and butanol (Soutar *et al.*, 1974) or, alternatively, by way of a simplified method elaborated by Kitabatake *et al.* (1979). The latter method (reviewed by Doi and Nishida, 1981) involves adsorption of the enzyme from diluted human plasma on DEAE–Sephadex, treatment with 1-butanol in the presence of ammonium sulfate, DEAE–Sephadex chromatography, treatment with dextran sulfate in the presence of Ca^{2+}, and hydroxyapatite chromatography. Another precipitation method using polyethylene glycol (PEG; MW 6000) has been described by Chong *et al.* (1981). Combinations of precipitation and ultracentrifugation methods have also been proposed (Soutar *et al.*, 1974; Aron *et al.*, 1978). In the purification scheme, specific use is made of HDL-Sepharose affinity columns (Albers *et al.*, 1976), which have now been replaced by more efficient sorbents: phenyl-Sepharose (Albers *et al.*, 1981b), Affigel Blue, which removes albumin (Chung *et al.*, 1979), or anti-apoD antibody-Sepharose (Aron *et al.*, 1978; Albers *et al.*, 1976). In this way, the final purification of the enzyme fraction from the contaminating apoD is achieved. Chromatography on hydroxyapatite can be adequately substituted by preparative anion electrophoresis of LCAT on polyacrylamide gels (Suzue *et al.*, 1980). Selected representative modifications of the purification techniques are listed in Table II; included also are the relevant quantitative data for each particular purification step.

2. *Enzyme Stability*

According to various reports, the period of stability of the purified enzyme varied from a few hours (Chung *et al.*, 1979), to over a few days (Aron *et al.*, 1978), to up to several weeks (Chen and Albers, 1981; Kitabatake *et al.*, 1979). The thermolability of the enzyme at 57°C is well known (Glomset, 1968). Furukawa and Nishida (1979) studied the stability of purified LCAT in more detail and came to the conclusion that it depended primarily on low ionic strength of the medium. At an ionic strength of 0.01 at pH 7.4 in 0.4 mM phosphate buffer, the enzyme activity remained unchanged for 6 hours when incubated at 37°C. The stability of the enzyme decreased progressively with increasing ionic strength up to 0.1 in 39 mM phosphate buffer; as early as 30 minutes after the start of incubation about 90% of the enzyme activity was lost. The study further showed that inactivation of the enzyme mostly took place at the "air–water inter-

Table II
SUCCESSIVE STEPS IN LCAT PURIFICATION SCHEMES ELABORATED
BY DIFFERENT AUTHORS

Step	Total protein (mg)	Specific activity (units[a]/mg)	Yield (%)	Purification (n-fold)	Reference
Plasma	13,939	1.26	100		Aron et al.
First CsCl step	420	27	66	21	(1978)
Second CsCl step	52	215	63	171	
Sephadex G-100	11	610	39	477	
Hydroxyapatite	0.22	10,200	13	8,075	
Anti-apoD column	0.065	20,370	8	16,166	
Serum	58,400	0.05	100	1	Chung et al.
$d = 1.21$ g/ml middle fraction	458	5.96	93	118	(1979)
Affigel Blue	81.1	19.2	53	381	
DEAE–cellulose	9.9	132.8	45	2,635	
DEAE–Sephadex A-50	2.9	192.4	19	3,817	
Hydroxyapatite	0.307	850.2	8.9	16,889	
Plasma	91,740	0.12	100	1	Chen and Albers
$d = 1.21$–1.25 g/ml middle fraction	2,068	3.56	66.9	30	(1981)
Phenyl–Sepharose	68.3	100	62.0	833	
DEAE–Sepharose	28.3	176	45.2	1,467	
Hydroxyapatite	0.75	1,980	13.4	16,500	
Plasma	105,000	6	100	1	Kitabatake
DEAE–Sephadex	10,000	43	68	7	et al. (1979)
Butanol–$(NH_4)_2SO_4$ Precipitate,	1,481	149	35	25	
DEAE–Sephadex Dextran sulfate	464	366	27	61	
supernatant	44	2,840	20	473	
Hydroxyapatite	0.3	208,000	10	34,700	

[a] One unit of enzyme catalyzed the esterification of 1 nmol of cholesterol per hour at 37°C.

face." In an experiment in which reaction mixtures of different phosphate buffer ionic strengths at 0°C were bubbled with nitrogen, the extent of inactivation was substantially higher than that observed in the same sample kept at 37°C and agitated. It appears that inactivation of the enzyme at the air–water interface at higher ionic strength may be due to extensive unfolding of the enzyme at the interface. The effect of ionic strength probably accounts for the difference in half-life of LCAT stored in different buffer solutions: 24 days and 14 days, respectively, for LCAT stored in Tris buffer (10 mM Tris, 1 mM EDTA, and 140 mM NaCl, pH 7.4) and in phosphate buffer (35 mM sodium phosphate and 150 mM NaCl, pH 7.4) (Chen and Albers, 1981). It has been noted that less-purified preparations exhibited greater stability than highly purified ones: dialyzed dextran sulfate supernatant fractions remain stable for 6 months, whereas fractions separated on hydroxyapatite remain stable for only 4 weeks (Kitabatake *et al.*, 1979).

This effect appears to be caused by the removal of albumin or, alternatively, apoA-I, which together with the phospholipid vesicles markedly stabilized the purified enzyme (Furukawa and Nishida, 1979; Chen and Albers, 1981). Mercaptoethanol or dithiothreitol, compounds applied in the purification process to protect the sulfhydryl groups responsible for the catalytic activity of the enzyme, do not, at low concentrations, affect the stabilty of the enzyme; at higher concentrations, however, they significantly reduce the esterification activity (Furukawa and Nishida, 1979). The mechanism of this effect is not entirely clear, but it cannot be attributed to the cleavage of intramolecular disulfide bridges, because the presence of these reducing agents does not decrease the molecular weight of the active protein (Chung *et al.*, 1979).

The low stability of highly purified enzyme preparations is further evinced by differences in the number of units of activity related to the weight of enzyme protein, as reported by different laboratories; the differences are as high as two orders of magnitude (Table II). The discrepancies in the specific activity of the enzyme may be at least partly attributable to variations in methodology.

3. Molecular Characteristics of LCAT

The relative molecular weight of a highly purified enzyme preparation, made homogeneous by polyacrylamide gel electrophoresis in 8 M urea and sodium dodecyl sulfate (SDS), has been estimated by various laboratories to be in the range of 65,000–66,000 (Kitabatake *et al.*, 1979; Albers *et al.*, 1979). A somewhat lower value is obtained by calculations made on the basis of data derived from sedimentation equilibrium centrifugation. The discrepancy can probably be explained by the fact that the latter calculation of molecular weight also included carbohydrates, the content of which is relatively high, that is, 24% (w/w) (Chung *et al.*, 1979). Determinations of amino acid composition by several research groups were similar (Table III), with relatively high contents of glutamic acid, aspartic acid, glycine, and leucine being reported by all. The following carbohydrates

Table III
AMINO ACID COMPOSITION OF HUMAN LCAT

Amino acid carbohydrate	Mol/10^5 g protein[a]	Mol/10^3 g protein[b]	Mol/10^5 g protein[c]
Lys	30.9 ± 1.9	33 ± 2	30.5 (27.2–33.8)
His	26.4 ± 1.4	28 ± 4	25.6 (23.8–27.9)
Arg	37.9 ± 3.4	43 ± 5	41.2 (36.5–44.9)
Asp	73.9 ± 2.7	95 ± 5	82.0 (75.9–88.3)
Thr	43.8 ± 1.6	59 ± 1	55.0 (51.6–60.7)
Ser	50.3 ± 3.0	62 ± 4	55.7 (48.5–62.5)
Glu	91.9 ± 4.2	102 ± 11	92.8 (85.1–100.5)
Pro	75.9 ± 2.9	81 ± 5	77.5 (66.6–87.4)
Gly	87.5 ± 4.8	98 ± 14	85.2 (76.7–94.2)
Ala	51.1 ± 1.8	62 ± 4	59.9 (57.0–61.1)
Val	46.2 ± 2.5	65 ± 3	60.6 (56.1–63.8)
Met	16.3 ± 0.7	20 ± 2	16.3 (15.3–17.2)
Ile	28.1 ± 2.3	41 ± 2	39.1 (37.1–41.9)
Leu	91.6 ± 1.8	114 ± 11	102.1 (97.2–108.0)
Tyr	38.0 ± 1.2	41 ± 4	29 (27.1–33.6)
Phe	36.6 ± 0.3	44 ± 3	32 (28.1–35.9)
Half-cys	—	8 ± 3	8.4
Trp	—	—	14.5
Glucosamine	—	31 ± 4	—
Glucosamine			17 (12–22)[d]
Sialic acid			13 (7–19)[d]
Galactose			20 (17–23)[d]
Mannose			31 (26–35)[d]

[a] From Aron et al. (1978).
[b] From Albers et al. (1979).
[c] From Chung et al. (1979). Range in parentheses.
[d] Mol/5.9×10^4 g.

were identified: mannose, galactose, glucosamine, and sialic acid. Very little is known up to now about the structural arrangement of the enzyme, except that no interchain disulfide bonds are involved. The enzyme can probably form as many as five isoforms with the isoelectric point ranging from 5.1 to 5.5, as demonstrated by analytical isoelectric focusing (Albers et al., 1979); other measurements located the p*I* in a lower pH region, between 4.5–4.8 (Doi and Nishida, 1981). However, another enzyme preparation isolated by a different method and measured by the same procedure yielded just a single zone between pH 3.7 and 4.0 (Gustow et al., 1978).

4. Immunological Properties of LCAT

The first attempt to induce antibody production against LCAT made it clear that goats were by far the best producers of anti-LCAT antibodies (Varma et al., 1977, 1978), while rabbits were found to be much more inferior in that respect. Tentative analysis demonstrated that the anti-LCAT serum totally inhibited LCAT activity in a complete plasma system (Varma et al., 1977, 1978). Albers et al. (1981a) confirmed these initial findings, and in a carefully conceived and detailed study they defined the basic immunological properties of LCAT. The following noteworthy conclusions were drawn: Isolated LCAT possesses the same immunoreactive properties as the lipoprotein-bound enzyme; immunoreactive sites of LCAT are not masked by lipids, because their extraction from the plasma does not result in an increase in immunoreactivity, but rather leads to a reduction, possibly due to a conformational change of the LCAT protein; and LCAT immunoreactivity remains unchanged even after storage at 4°C for as long as 2 months. It was further demonstrated that plasma samples of different nonhuman species had the capacity to displace ^{125}I-labeled human LCAT from goat anti-human LCAT; surprisingly, sheep LCAT was found to be almost identical to human LCAT. The LCAT displacement capacity of other tested species ranked in the following order: nonhuman primates > cat or dog > pig > rabbit or guinea pig > mouse > rat. It remains to be decided whether immunoreactivity parallels enzyme activity; an indirect indication that this may be so can be seen in the finding that sera of certain patients with hereditary LCAT deficiency exhibited a definite degree of immunoreactivity, even though they were completely inactive enzymatically.

IV. Methods of LCAT Estimation and Regulation of Its Activity

A. SUBSTRATE AND CONDITIONS OF THE LCAT REACTION

It can be logically inferred from the reaction mechanism that the lecithin–cholesterol vesicles are the substrates in the LCAT reaction, provided the physicochemical conditions meet the requirements for an optimum interaction with the activating protein. However, under physiological conditions *in vivo*, or in a complete plasma system *in vitro*, when the reaction takes place on the surface of lipoproteins constructed from qualitatively and quantitatively variable components, the definition of the term "substrate" becomes broader by convention, encompassing classes and subclasses of lipoproteins, or even the entire plasma. In this way, the term is applicable to a number of basic as well as auxiliary factors which all influence the ultimate value of the esterification rate. It seems therefore appropriate to outline schematically the role of the main components in the reaction mechanism and to attempt, as far as it is possible, to consider separately their particular contributions to the course of the esterification process.

Enzyme
1. Concentration of the enzyme protein (assuming that it goes parallel with the enzyme activity, which in fact need not be so)
2. Effect of physical factors (temperature, water–air interaction, pH, ionic strength)
3. Effect of chemical factors (inhibitors blocking the specific binding sites of the enzyme protein; nonspecific effectors)

Activating protein
1. Concentration of activating protein
2. Ratio of the activating protein (apoA-I) to other proteins in the active surface, either inhibiting (apoA-II) or diluting

Lecithin–cholesterol
1. Concentration
2. Ratio of cholesterol to lecithin in the active surface
3. Length and saturation of lecithin acyl chains

Active surface
1. Area of the active surface
2. Physicochemical conditions of the contact with reaction components (ionic strength, salt concentration, pH, detergents)
3. Dilution effect of lipoproteins having inactive surface

Substrate–product equilibration
1. Continuous supply of fresh substrate
2. Removal of reaction products

B. Methods

A good deal of research effort has been devoted to the search for a suitable method of LCAT estimation, a method that would be adequately sensitive, would be reproducible, would allow comparisons to be made among different laboratories, and would be simple enough to be applicable as a supplementary diagnostic method in clinical practice. Some of the problems and methodological difficulties bearing on LCAT determination have been reviewed elsewhere (Norum, 1974; Lacko, 1976; Verdery and Gatt, 1981). In general, all of the methods applied up to now can be divided into three groups. The first one (the so-called "common substrate" or "homologous substrate" method) is based on the measurement of cholesteryl esters produced in a system where two of the active components of the LCAT reaction, i.e., the activating protein and the lecithin–cholesterol substrate, are constant and the variable component is represented by the enzyme. The second group of methods (the so-called "autologous substrate" or "initial rate of esterification" methods), believed to reflect the actual individual course of the esterification rate, tests the net production of cholesteryl esters in a system where all three reaction components are variable. Finally, a third method was developed (Albers et al., 1981a) which makes it possible to determine directly the enzyme protein concentration. The complex character of the LCAT reaction, in which three components interact, implies that each particular method can answer adequately only a certain set of questions concerning the rate of cholesteryl ester production in the plasma. It is evident that each of the methods has its advantages and its inherent technical difficulties; they will be discussed in turn in the following section.

1. Method of "Uniform (Common, Homologous) Substrate"

This method provides data on the participation of the enzyme in the esterification of free cholesterol. The uniform component in this reaction system is either whole plasma, serum with heat-inactivated enzyme (Glomset and Wright, 1964), partially delipoproteinated serum (Alcindor et al., 1977), or a mixture of activating protein, lecithin–cholesterol vesicles, and albumin (Morin and Piran, 1981). The advantage of using whole serum or plasma lies in the fact that the absolute values of the cholesterol esterification rate in the plasma of healthy people determined by this method (Glomset and Norum, 1973; Dobiášová et al., 1978a) correspond almost perfectly to the values found in healthy human subjects by the *in vivo* method of compartmental analysis of cholesteryl ester turnover (Nestel, 1970; Kudchodkar and Sodhi, 1976), as well as to the values found by the *in vitro* method of "autologous substrate" (Wallentin and Vikrot, 1975).

Because the method of uniform substrate is being used in many laboratories, it is appropriate to address it in more detail, particularly since numerous critical objections have been raised questioning its merit. In essence, the procedure involves heat inactivation of plasma or serum, labeling with radioactive cholesterol, and incubation for 1 to several hours with a small amount (10–20%) of enzymatically active plasma or serum. After determining the ratio of radioactivity in the fractions of free and esterified cholesterol by means of thin-layer chromatography (TLC), and measuring the total content of free cholesterol in the incubation mixture, the rate of cholesterol ester production is then calculated. The main objections concern (Norum, 1974; Lacko, 1976) the risk that heat inactivation might to some extent cause denaturation of lipoproteins, so that it is not at all certain that the mixture of enzymatically active and inactive sera really represents a homogeneous pool; consequently, the esterification might preferentially take place in the enzymatically active sample. Another objection concerns the labeling process of the serum with radioactive cholesterol. In fact, this is the most critical point for the autologous substrate method as well. According to our experience (Dobiášová *et al.*, 1978b) the side effects can be kept to a minimum by strictly adhering to certain rules which assure constant optimum conditions. To begin with, serum is used instead of plasma, because no denatured proteins precipitate from the serum after heating to 57°C for 30 minutes. The serum sample is pooled from the blood of at least four normolipemic male donors. Inactive serum is incubated on a disk of filter paper in which radioactive cholesterol is finely dispersed. The labeled cholesterol is applied to the paper disk dissolved in benzene, because any trace of toluene, which is the solvent in the commercially supplied preparation, inhibits the LCAT reaction. The enzyme activity is also inhibited if silica gel is substituted for filter paper as sorbent, apparently due to derangement of the lipoprotein surface. A similar effect is observed after exposure of the labeled substrate to ultrasound for just a few seconds; for the same reason, even mechanical stirring must be avoided.

The radioactive substrate was shown to be stable for at least 3 months if stored in an N_2 atmosphere in sealed vials at $-20°C$ and allowed to equilibrate at 37°C for about 30 minutes before use. With this arrangement, the radioactive cholesterol is distributed evenly in the lipoproteins, depending on the lipoprotein quantity, as can be demonstrated by agarose gel electrophoresis. Our experience, gathered over many years, allows us to predict that most of the differences in the results reported by different laboratories will turn out to be due to variations in these technical details related to the preparation of the serum substrate, rather than to the variability among normal human subjects. Even though this method may hardly be used for diagnostic purposes, it is presently the only method which makes it possible to determine in cross-experiments the contribution to the LCAT reaction of the enzyme, on one hand, and of th enatural lecithin substrate and activating protein on the other. New methods have been developed in which the natural serum substrate is replaced by an artificial one. Alcindor *et al.* (1977, 1978) introduced plasma samples delipoproteinated by means of coprecipitation with Intralipid, dextran sulfate, and calcium chloride. Delipoproteinated serum (lacking LDL and VLDL) is heat inactivated and labeled with a small amount of [^{3}H]cholesterol in an ethanolic solution. The substrate prepared in this way is incubated for

30 minutes with a threefold amount of enzymatically active delipoproteinated serum. The percentage yield of produced cholesteryl esters is several times higher with this arrangement, even though the calculated absolute value of the esterification rate in normal subjects is less than one-half of that estimated by other methods. Nevertheless, by eliminating from the system lipoproteins lacking the enzyme activity, any changes in enzyme activity, for example, differences between normal subjects and patients with liver disorders, become more apparent than with methods using the complete system.

The replacement of serum substrate with a mixture of lecithin–cholesterol vesicles and activating protein is the underlying principle of other methods using delipoproteinated plasma as a source of the enzyme as well. Morin and Piran (1981) labeled the artificial substrate with [^{3}H]cholesterol, and measured the radioactivity of the produced cholesteryl esters in the extract after precipitation of the proteins and free cholesterol with a solution of 1% digitonin in 98% ethanol. A mixture of HDL apolipoprotein was used as the source of the activating protein because its activating property, though substantially less than that of apoA-I, remained stable for a longer storage period.

A well-defined artificial substrate of proteoliposomes of apoA-I–phosphatidyl choline–cholesterol (molar ratio 0.8:250:12.5) prepared by cholate dialysis technique was described by Chen and Albers (1982b). The mean LCAT activity estimated by using this substrate was 95.1 nmol/hour/milliliter of plasma from normal subjects, that is, a value in excellent agreement with that obtained in plasma of healthy subjects by other methods (Table VI). Pownall *et al.* (1982) prepared a similar highly reactive substrate for a LCAT reaction (apoA-I–phospholipid) with the physical properties of HDL.

These new approaches obviously hold some promise for the prospect of clinical use, though it still remains to be seen how reproducible and comparable the results will be when performed in different laboratories.

2. Method of "Autologous (Self) Substrate

This method, which uses the initial esterification rate, makes it possible to determine *in vitro* the rate of cholesteryl ester production in a sample of complete plasma or serum. Since in this case all the active components of the LCAT reaction are contained in one single sample, the esterification rate is expected to be, under the optimum conditions of measurement, the same as *in vivo*. Different methodological approaches exist: The historically oldest one involves incubation of the plasma for 6–48 hours, followed by determination of the resulting decrease in free cholesterol.

However, because the linearity of the reaction is of short duration, and usually does not extend beyond 40 minutes, apparently because of the delay in substrate equilibration, product removal (Norum, 1974), and possibly enzyme denaturation, it is difficult during such a short time interval to assess precisely the difference in the amount of free cholesterol, which ranges normally from 1–3%. Stokke and Norum (1971) worked out a radioassay involving a transient inhibition of LCAT activity by means of DTNB, followed by equilibration of the plasma with radioactive cholesterol introduced either in ethanol or an acetone–albumin mixture; the effect of DTNB was then abolished by mercaptoethanol, and the radioactivity of the labeled cholesteryl esters was measured. This method is used with various modifications quite generally, although several objections have been raised even in this case. The first one concerns the question of the homogeneity of the radioactive label in the plasma lipoproteins, or possible damage to the active surface of the lipoproteins which may occur when introducing the label in a solvent. Another unsettled issue is the problem of possible side effects caused by the surplus of mercaptoethanol needed for reinhibition of LCAT. The former objection was refuted by Walentin and Vikrot (1975), who modified the method (by introducing the label into the serum in a mixture with albumin) and confirmed experimentally the equilibration of the label in lipoproteins in relation to their quantity. In an extensive series of measurements they obtained convincing evidence for the sensitivity and reproducibility of the method. Barter (1974b) and Kudchodkar and Sodhi (1976) tested the LCAT activity in plasma samples obtained from subjects whose plasma lipoproteins were labeled *in vivo* with radioactive cholesterol; a good agreement was found between the *in vivo* and *in vitro* methods of cholesterol esterification rate estimation.

Attempts to avoid the use of additionally introduced radioactive label, as well as additional chemical treatment of LCAT, led to the elaboration of methods based on direct measurement of the decrease in free cholesterol. Gas–liquid chromatography methods (Marcel and Vezina, 1973b; Blomhoff *et al.*, 1974) are sufficiently sensitive when performed with top quality equipment, but they are unfortunately very time consuming, as they necessarily require several parallel measurements. Enzyme tests were applied which made it possible to determine the free cholesterol in the cholesterol oxidase–peroxidase system, generally following the original Patsch *et al.* (1976) method. Nagasaki and Akanuma (1977) measured photometrically the difference in free cholesterol concentration by the red quinone produced in the reaction of cholesterol oxidase and peroxidase in the presence of phenol-4-aminoantipyrine. Iwasaki and Hamada (1981) developed a sensitive fluorometric method utilizing tyramine as substrate for the cholesterol oxidase–peroxidase reaction. In the last two methods quoted, lecithin liposomes were added to the reaction mixture in order to stabilize the plasma during storage at low temperatures and, at the same time, to increase the enzyme activation and to prolong the linearity of the reaction. Another method of enzymatic measurement of free cholesterol was reported by Bartholomé et al. (1981). Dieplinger and Kostner (1980) succeeded in greatly simplifying the procedure by making use of the monotest "System Cholesterol Enzymatic" produced by Merck; the set was applied to entirely untreated serum samples before and after incubation for 40 minutes. The data obtained were in good agreement with the normal values of LCAT determined by radioassay according to Stokke and Norum (1971). The reproducibility was high if sufficient care was taken to achieve the maximum accuracy of the dosage. Somewhat lower accuracy was noted for serum samples exhibiting lower LCAT activity.

3. Method of LCAT Radioimmunoassay

A competitive-displacement, double-antibody radioimmunoassay for human plasma LCAT was developed by Albers et al. (1981a) and utilized to arrive for the first time at valid figures for the amount of LCAT in the plasma of normolipidemic, hyperlipidemic, and LCAT-deficient subjects. By using a highly purified preparation of the enzyme as antigen, the authors induced antibody production in goat, isolated the IgG immunoglobulin from the anti-LCAT serum, labeled the LCAT with ^{125}I, and assembled an LCAT radioimmunoassay test set comprising the following active components: diluted sample of plasma, ^{125}I-labeled LCAT, goat antihuman LCAT antiserum, normal goat serum, and anti-goat IgG. The test is highly reproducible, the lower limit of sensitivity being 20 ng of enzyme protein per milliliter of plasma. The mean recovery after addition of a standard amount of LCAT was found to be around 100%. Normal values of enzyme protein concentration were higher in females (6.44 μg/ml) than in males (5.98 μg/ml).

C. REGULATION OF LCAT ACTIVITY

The way in which LCAT is regulated under physiological conditions has not yet been completely clarified, even though very detailed information is available on some limiting cases of dysfunction, such as familial LCAT deficiency or disorders associated with liver diseases, etc. For most of the effectors which modify the rate of esterification, it has not as yet been established with certainty whether the factors affect the enzyme itself or influence the availability of the substrate for the LCAT reaction. Studies *in vitro* referred to in this section make it possible to differentiate to some extent between these effects and to infer more about the physiological validity of particular regulatory mechanisms.

1. Enzyme

The bulk of available data concerning a direct influence on the enzyme protein are based on chemical inhibition of LCAT, and are therefore more important for understanding the reaction mechanism than the actual regulation of LCAT in the organism itself. Apart from irreversible heat inhibition, a direct effect on LCAT fatty acid release, and on transferase activity, is exerted by sulfhydryl inhibitors (Glomset, 1962, 1968; Piran and Nishida, 1976; Verdery, 1981). Among these agents, the apparently mildest and most specific inhibitor is DTNB (Norum, 1965). From the mercury derivatives acting through the mercuration of essential sulfhydryl groups in the enzyme molecule, the most widely used ones are *p*-hydroxymercuribenzoate and *p*-chloromercuriphenylsulfonic acid (Glomset *et al.*, 1970; Owen *et al.*, 1979). Inhibition of LCAT induced by *p*-hydroxymercuribenzoate and by DTNB can be completely reversed by mercaptoethanol (Stokke and Norum, 1971).

However, inhibitors that are highly reactive with sulfhydryl groups need not necessarily be specific solely for LCAT, but may interact also with other proteins (Owen *et al.*, 1979). Inhibition by *N*-ethylmaleimide is irreversible; so are the inhibitory effects of iodoacetate (Chen and Albers, 1981) and paraoxone (Rose, 1978). Lecithin : cholesterol acyltransferase is further blocked by inhibitors of serine–histidine antiesterase activity, such as diisopropyl fluorophosphate and diethyl nitrophenyl phosphate (Lacko *et al.*, 1973), as well as by an inhibitor reacting with the hydroxy group of serine in the enzyme protein, that is, phenyl sulfonyl chloride (Chen and Albers, 1981). Nakagawa *et al.* (1977) described an LCAT inhibition effected by some heavy metal cations such as Ag^{1+}, Cd^{2+}, and Zn^{2+}. Inhibition by zinc could be reversed by means of EDTA. The inhibitory effect of Cu^{2+} and Hg^{2+} as well as the effects of some complexing agents (cysteine, mercaptoethanol, penicillamine, thioglucose, thiourea, etc.) on LCAT activity *in vitro* have been described (Nakagawa *et al.*, 1982b). Calcium ions which are known to inhibit LCAT activity in the plasma (Glomset, 1968) have no effect on the activity of purified enzyme (Piran and Nishida, 1976; Aron *et al.*, 1978). Mercaptoethanol and dithiothreitol, which in low concentration protect the sulfhydryl groups against oxidation, inhibit LCAT when present in higher concentrations (Furukawa and Nishida, 1979). The different sensitivity to mercaptoethanol and sulfhydryl blocking agents observed for mouse LCAT indicates that the enzyme structure may be species specific (Owen et al., 1979). Also, as novel inhibitors of plasma LCAT activity carnitine esters react only in experimental animals but not in humans (Bell, 1983).

The activity of LCAT *in vitro* is also inhibited by certain organophosphorus insecticides (Nakagawa and Uchiyama, 1974; Nakagawa et al., 1982c); local anesthetics, such as dibucaine, benzocaine, tetracaine, and lidocaine (Bell and Hubert, 1980); certain purines, such as guanine, xanthine, and hypoxanthine (Solera, 1978); various penicillins (Bojensen, 1978); constituents of plastics, such as phthalate (Lagente et al., 1979); and others. However, it is not certain whether all these active compounds act primarily on the enzyme protein, nor is it certain which of the reaction steps is the actual target for the inhibitory effect. It appears that the inhibition of LCAT activity by deoxycholate and detergents (Glomset, 1968), dicarboxylic lecithins (Douset et al., 1978), natural lysolecithins (Nakagawa and Nishida, 1973), as well as enantiomeric lysolecithins (Smith and Kuksis, 1980) belongs to the category of those interactions where the actual target is not directly the enzyme molecule, but rather the phospholipid bilayer. It is further conceivable that the same applies also to the effect of heavy metal cations on the phospholipid–protein interactions (Dobiášová and Linhart, 1969).

2. Lecithin–Cholesterol Substrate

Plasma lipoproteins contain around 1.5 μmol lecithin and cholesterol per milliliter of plasma. However, taking into account the substantially reduced proportion of both components in the surface-shell HDL, where esterification mostly takes place because of the localization of LCAT, it becomes obvious that a rapid and steady supply of both substrates must be assured. It has been known for a long time that a very active exchange exists in the plasma between plasma lipoprotein phospholipids both *in vitro* (Kunkel and Bearn, 1954) and *in vivo* (Eder and Steinberg, 1955), and that phospholipids exchange in proportion to their content in lipoproteins (Rubenstein and Rubinstein, 1972). Only polyene phosphatidylcholine appears to exhibit a higher affinity to the HDL fraction, one which is essentially time independent (Zierenberg et al., 1979). Even though it is generally believed at present that phospholipid turnover proceeds spontaneously, Jackson et al. (1980) and Wilson et al. (1980) were able to show that the transfer of phosphatidylcholine from LDL to artificial phospholipid complexes (sphingomyelin–apoA-II) was facilitated by a phospholipid transfer protein isolated from human plasma (Ihm et al., 1980). These investigators also observed that the entire content of phosphatidylcholine in lipoprotein was available for the exchange.

The exchange of free cholesterol among plasma lipoproteins is also very rapid (Bell, 1973); here again it is proportional to the mass of unesterified cholesterol in the lipoproteins, as was demonstrated by Bjornson et al. (1975) in experiments both *in vitro* and *in vivo*. The transfer of cholesterol is most probably closely related to phospholipid transfer, because the amphiphilic phospholipids control specifically the interaction of cholesterol with membranes and proteins (Dobiášová and Linhart, 1969; Dobiášová and Faltová, 1974), as well as its clearance from the plasma (Dobiášová et al., 1976). This hypothesis is corroborated by the findings that bilayer lipid membranes are formed during lipolysis of chylomicrons (Blanchette-Mackie and Scow, 1976), and that these membranes can be transferred to HDL from chylomicrons and VLDL in a way analogous to artificial egg yolk lecithin-mixed liposomes (Chobanian et al., 1979). That the transport of phospholipids and cholesterol is not strictly dependent on movement of the apoproteins has been demonstrated by the results of a study (Eisenberg and Rachmilewitz, 1973) using double-labeled VLDL with the label attached to the protein and lipid moieties: the decay of labeled lipids was found to be substantially faster. Finally, evidence has been presented recently showing that practically the entire amount of cholesterol substrate for the LCAT reaction in surface HDL is derived from VLDL and LDL (Fielding and Fielding, 1981). Thus, since the transport of the substrate to the LCAT reaction site is so rapid, the speed of the supply by itself cannot be the limiting factor. On the part of the substrate, the limiting factors are the lecithin–cholesterol ratio [the optimum being approximately 4:1 (Fielding and Fielding, 1972a)] and the degree of saturation of the phosphatidylcholine chains, which determines the optimum surface for interaction with the activating protein and for the acyl transfer.

3. Reaction Products

Since the physicochemical characteristics of the lecithin–cholesterol vesicle are so narrowly defined in the LCAT reaction (see Section II), it is evident that doubling the weight and the size of the esterified cholesterol molecule incorporated in the lecithin bilayer must necessarily impair the lipid–protein contact essential for the reaction. Lysolecithin, as the second reaction product, can probably either disturb the hydrophobic association of the lecithin bilayer with the amphiphilic protein acting as a biological detergent, or can compete with lecithin for the binding of apoprotein. It could in fact be experimentally demonstrated that both products inhibit the LCAT reaction *in vitro* (Fielding et al., 1972b). The removal mechanism for these products can now be explained.

a. Cholesteryl Ester Removal. It had been thought that cholesteryl esters, being highly nonpolar components of the lipoprotein core, were almost reistant to transport or exchange within the plasma pool, contrary to the behavior of lecithin and free cholesterol (Steinberg, 1974). It is now clear, however, that a live and intense exchange takes place between cholesteryl esters and plasma lipoproteins; core cholesteryl esters are in equilibrium with the cholesteryl esters of the lipoprotein surfaces, and are permanently available for transfer (Janiak *et al.*, 1979). If LCAT is eliminated by an inhibitor, and the plasma or lipoprotein fractions are incubated at 37°C, an exchange of cholesteryl esters among individual lipoprotein classes takes place without a change in mass distribution. By means of labeled lipoproteins, it was possible to demonstrate the transfer of cholesteryl esters from LDL to VLDL (Zilversmit *et al.*, 1975), from LDL to HDL (Barter and Jones, 1979; Pattnaik *et al.*, 1978), from VLDL to HDL and LDL (Barter and Lally, 1978a), and from HDL or HDL_3 to LDL (Sniderman *et al.*, 1978). The rate of bidirectional transfer is high, reportedly attaining levels as great as 108–301 nmol/liter of plasma/hour (Barter and Jones, 1979), and the process is saturable by HDL, but not by LDL (Barter and Jones, 1980). The transfer of cholesteryl esters between human and bovine lipoproteins suggests that the process is not species specific (Pattnaik *et al.*, 1978).

The exchange of cholesteryl esters in the plasma pool does require, however, the presence of a protein factor first discovered by Zilversmit *et al.* (1975), the so-called "CHE exchange protein" contained in the $d >$ 1.25 g/ml HDL fraction. The factor was later detected also in the lipoprotein-free fraction (Barter and Jones, 1979), as well as in the 1.063 < $d <$ 1.21 g/ml HDL fraction (Chajek *et al.*, 1980). A discrepancy persists at present among various laboratories in how they define the molecular and functional characteristics of this protein. Zilversmit *et al.* (1975) and Pattnaik *et al.* (1978) characterize the CHE exchange protein as a glycoprotein with an isoelectric point of 5 and an apparent molecular weight of 80,000. Immunologically it differs from any known apoprotein, and, apart from the transport function, possesses no enzymatic activity. Chajek and Fielding (1978), on the other hand, designate as "CHE transfer protein" (apoD) the protein with MW 35,000 which they isolated from the HDL fraction (1.063 < $d <$ 1.21 g/ml) by means of affinity chromatography on immobilized anti-apoD antibody and fractionation on a concanavalin A-Sepharose column. The latter authors believe that their transfer protein is different from the exchange protein described by Pattnaik. The lipid transfer complex (LTC) isolated recently from lipoprotein-free human plasma exchanges both cholesteryl esters and phosphatidyl choline between the LDL and HDL of human plasma at equal rates (Ihm *et al.*,

1982). The protein of the lipid transfer complex was characterized as a glycoprotein with a molecular weight of 61,000.

All of those cholesteryl ester transfer or exchange proteins were LCAT independent.

Thus with inactivated LCAT, the cholesteryl ester mass equilibrium is preserved, and only the exchange process takes place, whereas if LCAT is reactivated, the removal mechanism is initiated according to the following scheme based on the findings reported by Chajek *et al.* (1980) and Fielding and Fielding (1981):

1. Due to the action of LCAT, the surface content of cholesteryl esters in HDL increases with a concomitant reduction in LDL and VLDL of cholesterol, which has been transferred to HDL in order to compensate for its loss.

2. The LCAT reaction is inhibited by physical change of the phospholipid bilayer in the HDL surface as soon as the saturation with cholesteryl esters reaches approximately 3%. In the presence of the transfer (exchange) protein, the produced cholesteryl esters are directed to LDL and VLDL until a maximum saturation is attained in these lipoproteins as well; the maximum increment is 10% in LDL, and as much as 100% in VLDL. Once saturation with cholesteryl esters is reached in all lipoproteins, the LCAT activity ceases.

3. Lecithin : cholesterol acyltransferase becomes reactivated under the following circumstances: (a) if more particles of cholesteryl ester acceptor are added to the medium. These may not only be fresh LDL and VLDL, but also any artificial phospholipid vesicles, such as sphingomyelin–cholesterol liposomes, which function as acceptors, though not as LCAT substrates; (b) if fresh substrate is added in the form of lecithin–cholesterol vesicles.

It still remains an open question whether the "net transfer" of cholesteryl esters, which takes place with the productive participation of LCAT, represents just a certain form of the basic exchange process maintaining the concentration balance of cholesteryl esters among lipoproteins. The hypothesis of net transfer is based mainly on the work of Nichols and Smith (1965), who demonstrated the transfer of cholesteryl ester mass from HDL and LDL to VLDL. This unidirectional transfer could, however, be detected only in plasma containing at least tenfold-additional VLDL. In normo- as well as hyperlipemic plasma with normal lipoproteins levels, only a proportional increase in cholesteryl esters, and a decrease in free cholesterol, could be found in every lipoprotein fraction. The high affinity of VLDL to cholesteryl esters can probably be explained by the fact that cholesteryl esters, being highly nonpolar molecules, may be more easily solubilized in triglyceride-rich particles such as VLDL. On the other hand, under normal conditions, VLDL are less important as acceptors of cholesteryl esters than LDL because of their low plasma level. Some new arguments in support of the hypothesis on net transfer of cholesteryl esters from HDL or HDL subfractions to VLDL have been presented by Marcel *et al.* (1980, 1981).

  b. *Lysolecithin Removal.* Leaving aside the contribution of tissue phospholipases to the small but extremely rapidly turning plasma pool of lysolecithin, the LCAT reaction alone produces an appreciable amount of lysophosphatidylcholine, i.e., approximately 100 μmol/liter/hour, which in a normal subject amounts to about 3 g/day. Of course, such a quantity of a potentially destructive substance requires an efficient removal mechanism. In fact, lysolecithin is an illustrative example of a biologically active substance whose dosage determines whether its effect will be beneficial or destructive. Essentially beneficial functions, such as induction of membrane fusion (Lucy, 1970), activation of lipoprotein lipase (Chung *et al.*, 1973), activation of cholesterol esterification in rats *in vivo* (Dobiášová *et al.*, 1976), promotion of cellular cholesterol release (Stein *et al.*, 1979), acceleration of cholesterol transfer from the plasma of rats and squirrel monkeys (Dobiášová *et al.*, 1975; Portman and Alexander, 1976), increased immune response in mice (Langer *et al.*, 1972), and others, can all be reversed with an "excess dose" of lysolecithin at which its cytolytic membrane-damaging action takes place.

Nevertheless, as the amount of information available concerning lysolecithin metabolism is scanty, particularly as relates to human beings. The main difficulty lies in the fact that even an extreme dose does not result in a substantial change in lysolecithin plasma concentration. For instance, in a model experiment on rats *in vivo* (Dobiášová *et al.*, 1975), evidence was presented that the administration of a small dose of phospholipase A results in an almost complete conversion of plasma lecithin to lysolecithin within a few minutes. Although the decrease in lecithin plasma concentration could be clearly identified, the corresponding increase of lysolecithin could not be detected because of its immediate clearance. As has been shown earlier, lysolecithin, as compared to lecithin, has a higher affinity to cellular membranes (Stein and Stein, 1976; Akino *et al.*, 1972; Dobiášová *et al.*, 1971), and this explains its rapid clearance from the plasma. This rapid release of lysolecithin also indices a removal of cholesterol from the plasma. The surprising conclusion of cholesterol release being associated with lysolecithin clearance is supported by several findings: (1) unlike lecithin–cholesterol liposomes, labeled lysolecithin–cholesterol liposomes are bound in the plasma only loosely and can therefore be extracted by the insoluble protein acceptor in as much as a 10-fold amount (Dobiášová and Faltová, 1974); (2) lysolecithin–cholesterol liposomes injected intravenously into rats displayed a decay curve identical for both compounds, whereas lecithin–cholesterol liposomes were released from the plasma at different and much lower rates (Dobiášová *et al.*, 1976); and (3) lysolecithin induced an accelerated release of cholesterol from rat plasma (Dobiášová *et al.*, 1975) and of LDL from squirrel monkey plasma, primary to the liver (Portman and Alexander, 1976).

Much more detailed information is needed before the relationship between the LCAT reaction and the removal pathway of lysolecithin coupled with cholesterol can be fully explained. What is known at present is that lysolecithin itself forms stable micellar aggregates which have molecular weights around 92,000 and which contain 181 monomeric molecules (Helenius and Simons, 1975), and has perhaps a higher affinity to the lipid than to the protein component of lipoproteins (Steim *et al.*, 1968; Stein *et al.*, 1978). On the other hand, it associates with albumin in a molar ratio of 2:1, and occasionally much lower (Klopfenstein, 1969; Nagakawa and Nishida, 1973; Groot and van Tol, 1978); the same applies to its association with apoproteins. Rosseneu *et al.* (1976) found a molar ratio of 78 mol lysophosphatidylcholine/mol apoprotein HDL. The molar ratio of phospholipid to apoprotein, as well as the binding enthalpy, increases with the elongation of the fatty acyl chain. In fact, the molar ratio of phosphatidylcholine to apoA-I is seven times higher (see Section II). Taking this information into account, it seems logical to expect that the conversion of lecithin to lysolecithin in the LCAT reaction is unavoidably accompanied by the dissociation of the bulk of the lysophosphatidylcholine molecules from the original apoprotein–lecithin complex. Even though the binding enthalpy of the apoA-I–lecithin complex is higher than that of apoA-I–lysolecithin, the formation of the former complex was shown nevertheless to be mediated by lysolecithin (Nichols *et al.*, 1974). Judging by its mobility in agarose gel (Phillips and Wille, 1973), lysolecithin in the plasma concentrates primarily in the pre-α-lipoprotein fraction, which is identified with the HDL $d > 1.21$ g/ml fraction containing lysolecithin complexed with apoA-I and albumin (Wille *et al.*, 1978). In addition to this fraction, LDL likewise display a high binding affinity and capacity for lysolecithin (Portman and Illingworth, 1974). Stein *et al.* (1979) concluded on the basis of some experimental evidence that precisely the low-molecular-weight (less than 100,000) complexes, such as lysolecithin–albumin–apoA-I, which are capable of passing the capillary endothelial barrier, could possibly promote cholesterol removal from peripheral cells.

It has been established (Fielding *et al.*, 1972b; Nakagawa and Nishida, 1973) that in a closed system the LCAT reaction is inhibited by lysolecithin. Since the same effect on the LCAT reaction was seen with 1-sn and 3-sn enantiomers of lysolecithin, as well as with a nonionic detergent (Triton X-100), it was suggested by Smith and Kuksis (1980) that the inhibition is nonspecific and is based on the detergent effect of the listed substances on the interaction between the lecithin bilayer and the activating protein. The inhibition effect of all these substances can be reversed (though this is true to a lesser extent for Triton X-100) if the medium is supplemented with albumin. Depending on its concentration, albumin associates with lyso compounds and with Triton, whose micellar volume is similar to those of lyso compounds. It may seem that these data are at variance with our own finding that the esterification of plasma cholesterol is stimulated by lysolecithin *in vivo*; in other words, that the rate of esterification of labeled cholesterol injected into rats in combination with lysolecithin is at least three times higher than when it is injected in combination with lecithin (Dobiášová *et al.*, 1976). The reason for this discrepancy in the results obtained *in vivo* and *in vitro* may be either the highly efficient removal mechanism of lysolecithin *in vivo*, or its favorable effect on the formation of the apoprotein–substrate complex, as was outlined earlier.

A less efficient mechanism of lysolecithin removal from the plasma consists in the acylation of lysolecithin in the reverse direction of LCAT action, a process which requires low-density lipoproteins as activators of the reaction (Subbaiah and Bagdade, 1978, 1979; Subbaiah *et al.*, 1980, 1982). Compared to the transacylation activity, the reverse direction has merely some 10% efficiency. Due to the high removal rate of lysolecithin *per se*, the latter mechanism may play an insignificant role under normal physiological conditions, although other unusual circumstances might possibly increase the importance of this process.

In spite of the fact that the evidence concerning lysolecithin metabolism in the plasma still lacks many details, we have attempted to draw up a hypothetical scheme based on available information indicating that the release of lysolecithin originates from the action of LCAT: Due to the inability of apoA-I to bind the same number of molecules of lysophosphatidylcholine and phosphatidylcholine, most of the lysophosphatidylcholine dissociates from the active surface of HDL in various forms. These are as follows: (1) associated with albumin it is cleared out from the plasma; (2) in the form of a low-molecular-weight complex of apoA-I–lysophosphatidylcholine–albumin which migrates to peripheral cells, it deprives them of cholesterol, and facilitates the formation of an active complex, apoA-I–phosphatidylcholine, after reassociation with HDL; (3) it may dissociate as an apoprotein-independent micellar complex, lysophosphatidylcholine–cholesterol, which binds and migrates with LDL toward the peripheral cells or, if the concentration of lysolecithin is high, to the liver; and (4) a further possibility exists that due to its low molecular weight, high mobility, and high affinity to cellular membranes, this micellar complex of lysophosphatidylcholine–cholesterol may be extracted by tissue cells directly. Concomitantly, cholesteryl esters are released from HDL and transferred with the aid of a transfer protein to VLDL and LDL.

4. Apoproteins and High-Density Lipoproteins in the LCAT Reaction

The optimum activator of the LCAT reaction is apoA-I, the principal protein constituent of HDL; apoC-I activates it to a lesser extent (see Section II). Apoprotein AII, being the second main component of HDL, inhibits the LCAT reaction (Fielding *et al.*, 1972a) due to its greater affinity to phospholipids. The relationship of apoA-I and apoA-II to LCAT activity is inversely correlated (Chung *et al.*, 1979). However, other apoproteins besides apoA-II affect the ability of apoA-I to activate lecithin. Apoprotein CII, apoC-III$_1$, apoC-III$_2$, and apoD all lack the activation property (Albers, 1978) and inhibit esterification when added to the medium containing apoA-I–phospholipid vesicle–LCAT. Esterification likewise decreases if the apoA-I concentration rises such that the optimum ratio between the activator and the vesicle is exceeded.

Whereas apoA-II appears to compete actively with apoA-I for binding the phospholipid vesicles, the remaining apoproteins, as well as an excess of apoA-I, are not, in the proper sense of the word, inhibitors; rather they reduce the optimum activator/vesicle ratio. As was shown by Yokoyama *et al.* (1980), the rate of the LCAT reaction is strongly proportional to the surface density of apoA-I. Whenever a decrease of apoA-I occurred in the surface, whether due to a specific extraction by antibody, a competitive exclusion by apoA-II, or a dilution with other proteins such as rabbit serum γ-globulin, the rate of esterification was always reduced. Similarly, the use of a mixture of HDL apoproteins containing about one-third apoA-II led to a decrease in the efficiency of the reaction down to about 55% of that characteristic for pure apoA-I (Akanuma *et al.*, 1978).

From a theoretical viewpoint, lipoproteins having a high ratio of apoA-I must be expected to be better substrates for LCAT. In reality, however, the subclass HDL$_3$ of high-density lipoproteins was shown to be a preferred substrate for the LCAT reaction (Fielding and Fielding, 1971; Jahani and Lacko, 1982), in spite of the fact that, as has been repeatedly shown, it has a relatively lower content of apoA-I than has the subclass HDL$_2$ with a greater volume (Kostner and Alaupovic, 1972; Curry *et al.*, 1976). Moreover, HDL$_2$ even supresses the LCAT reaction, as was demonstrated by Marcel and Vezina (1973a) and by Kostner (1978), and confirmed by others (Pinon *et al.*, 1980; Jahani and Lacko, 1982). The unfavorable ratio of apoA-I to apoA-II in HDL_2 may be compensated for by a more advantageous structural and compositional ordering of the active surface: for the same concentration of total cholesterol, HDL_3 possesses a lower ratio of free cholesterol to phospholipids. Even though in HDL_3 the ratio of esterified to free cholesterol falls to almost one-half, it is possible that this is precisely the ratio which is optimal for the physicochemical organization of the phospholipid bilayer. Further, HDL_3 particles have well-defined surface substructures, and molecular weights which are only one-half (175,000) of those estimated for HDL_2 particles (Scanu *et al.*, 1974; Anderson *et al.*, 1977; Schaefer *et al.*, 1978).

Small particles logically provide a greater active surface; by analogy, a favorable relationship between the volume and surface area may be the underlying reason for the claim that certain other particles act as "optimum" substrates for LCAT, e.g., discoidal HDL particles isolated from the plasma of HDL-deficient patients (Uterman *et al.*, 1974; Norum *et al.*, 1975; Glomset *et al.*, 1980), discoidal nascent particles found in rat liver perfusates (Hamilton *et al.*, 1976; Ragland *et al.*, 1978; Felker *et al.*, 1977), the lipoprotein component lipoprotein lipase HDL-C (Kostner, 1978), or the VHDL $d > 1.21$ g/ml fraction, which is a fraction with a high content of lysolecithin, apoA-I, and albumin (David *et al.*, 1976). High-density lipoprotein fractionated by means of hydrophobic and ion-exchange chromatography yielded four fractions (Jahani and Lacko, 1981); the fraction with the lowest molecular weight contained most of the endogenous LCAT activity, but was also the best substrate for LCAT of all four isolated fractions. This particular fraction also exhibits a lower ratio of apoA-I/apoA-II than all the other fractions, and is thought to be analogous to HDL_3. Despite HDL being the main carrier of LCAT, the concentration of HDL in the plasma is not a direct indicator of LCAT activity, because no correlation could be found between the levels of HDL and LCAT (Rose and Juliano, 1976; Wallentin and Vikrot, 1976; Leiss *et al.*, 1978). On the other hand, a slight positive correlation between LCAT activity and HDL or HDL cholesterol in healthy, physically active young men has been described by Marniemi *et al.* (1982).

Another question which remains to be clarified is whether HDL with a higher concentration of cholesterol is a "worse" substrate for LCAT, as has been claimed by Soloff and Varma (1978). It is conceivable that a higher cholesterol concentration impairs the physicochemical conditions of the enzyme reaction, even though with the particular experimental setup using the self-substrate method the effect of the substrate could not be distinguished from the actual enzyme activity. However, Albers *et al.*

(1981b) demonstrated a negative correlation between LCAT concentration and the content of HDL cholesterol.

Another point concerning interconversion between HDL classes due to LCAT activity emerged from a study of Daerr and Greten (1982) in which they showed that following the LCAT reaction *in vitro*, essentially all of the HDL material bore the flotation characteristic typical of HDL_2, even though the HDL morphology, that is, the size of the molecule, remained unaltered. Similar evidence of conversion of HDL_3 to HDL_2 was recently provided by Schmitz *et al.* (1982): during the prolonged incubation of fresh sera, HDL_3 disappeared completely and the newly formed HDL_2 exhibited flotation, electron-microscopical, and electrophoretic migration properties identical with those of the HDL_2 contained in fresh serum. The reaction was not affected by VLDL or lipoprotein lipase but was suppressed by LCAT inhibitors. The discrepancy between the two quoted studies consists of the morphology of the newly generated HDL_2.

Thus, it appears that the inhibitory effect of HDL_2 on LCAT activity— as previously described—might be attributed to the feedback action of the product.

V. Studies on LCAT in Experimental Animals

Most of the data on LCAT activity reported up to the present time concerns animals used primarily in conventional experimental models for studies on atherosclerosis. Although no systematic search for LCAT in different animal species has been performed as yet, it is evident that the presence of esterification activity in the plasma is no privilege of mammals and birds, but can also be found in reptiles and amphibians (Table IV). Yet the variability of quantitative data reported by different authors for the same species is considerable, as can be seen in the table. The discrepancies in the reported esterification rates are mainly attributable to the choice of either the self or common substrate methods. Whereas in healthy human subjects the "normal" values obtained by one or the other method are basically similar (Table VI), great discrepancies are not uncommon in the data on other animal species. The reason for this is not entirely clear, but it is possible that various factors may play a role, such as differing resistance of serum lipoproteins to heat inactivation, slower equilibration of the readioactive label in the plasma, or an opposite action of the reactivation agent (mercaptoethanol) on LCAT in different species; for instance, a positive effect was reported for rats (Takatori and Privett, 1974), but a negative effect was reported for mice (Owen *et al.*, 1979).

Apart from these methodological variations and differences in the basic

Table IV
MOLECULAR AND FRACTIONAL ESTERIFICATION RATES IN VARIOUS ANIMAL SPECIES DETERMINED BY DIFFERENT METHODS

Animal	Self-substrate MER (μmol/liter/hour)	Self-substrate FER (%/hour)	Common substrate MER (μmol/liter/hour)	Common substrate FER (%/hour)
Rat	53.3^a	19.0^a	42.5^a	15.2^b
	123.6^c	29.4^c	70^d	15^d
	56^e	11.1^e	93^f	13.7^f
	156^f	22.9^f	84.1^b	12.1^b
			66.3^g	12.3^g
Dog	71.8^a	7.11^a	37.0^a	3.7^a
	93^c	7.7^c		
	81^h	7.6^h		
Calf	9.2^a	3.7^a	8.6^a	3.4^a
Bovine	30^i	3.8^i		
Horse	30–90^j			
Pig	52.1^a	7.2^a	41.9^a	5.8^a
	43.6^c	6.4^c		
Rabbit	27.5^a	5.8^a	30.6^a	6.5^a
			45.0^c	10.2^c
			43.7^k	8.4^k
Guinea pig	21.2^a	10.1^a	21.8^i	10.3^a
	36.3^c	13.6^c		
	58^l	20.3^l		
	52.3^m	11.4^m		
Monkey	104.7^a	14.5^a	59.4^a	5.8^a
Cebus monkey	152^n	15.0^n		
Squirrel monkey	154^n	14.4^n		
Chicken (male)	194°	21.0^o		
Female (laying hen)	136^p			
	34.7^p	2.5–5^p		
	55°			
Iguana iguana	240–265^q			

^{a} Stokke (1974), ^{b}Cisternas (1979), ^{c}Lacko et al. (1974), ^{d}Miller (1978), ^{e}David et al. (1978), ^{f}Manniner et al. (1978), ^{g}Dobiášová et al. (1982a), ^{h}Blomhoff et al. (1978), ^{i}Noble et al. (1972), ^{j}Yamamoto et al. (1979b), ^{k}Pinon and Laudat (1978), ^{l}Heller (1979), ^{m}Cisternas et al. (1981), ^{n}Lichtenstein et al. (1980).
^{o} *In vivo*.
^{p} *In vitro*.
^{q} Gillet (1978).

inter- as well as intraspecies biological characteristics, the cholesterol esterification rate is known to depend strongly in certain animal species on sex (e.g., chicken), stage of postnatal development, variations in standard laboratory diet, and perhaps a host of other factors related to the conditions of breeding and maintenance. Quantitative information available up to now on LCAT activity in different animal species does not permit the definition certain groups characterized by common quantitative features, although it is known that, for instance, herbivora have relatively the lowest activity among mammals. On the other hand, the highest LCAT activity of all was found in common leguan *Iguana iguana* (Gillett, 1978).

A. THE RAT AS AN EXPERIMENTAL MODEL

The rat is at present the most thoroughly studied animal with regard to LCAT activity, even though it cannot be considered an optimum experimental model, comparable to humans, with respect to lipid metabolism and susceptibility to atherosclerosis. There are at least four distinct basic differences between rat and man as concerns the esterification of cholesterol.

1. It has been found (Stokke, 1974) that rat liver contains a significant activity of acyl-CoA cholesterol acyltransferase, which may be responsible for part of the production of plasma cholesteryl esters. Sugano and Portman (1964), comparing esterification rates *in vivo* and *in vitro,* believed that the pathway mediated by acyl-CoA cholesterol acyltransferase might account for as much as 50% of the cholesteryl esters produced. Contrary to this, Chevalier *et al.* (1971) interpreted their own experimental results as indicating that the site of rat plasma cholesteryl esters was the plasma itself. At any rate, the activity of acyl-CoA cholesteryl acyltransferase is about five times higher in rats than it is in dogs and pigs, and at least 10 times as high as in monkeys. It appears that human liver tissue lacks this activity entirely (Stokke, 1972).

2. A special protein responsible for the transfer of cholesteryl esters is absent in rat plasma (Ihm *et al.*, 1982); the mechanism of cholesteryl ester removal from the site of LCAT action in rats is currently obscure (Barter and Lally, 1978b).

3. Rat VLDL are metabolized predominantly in the liver, and not in the plasma by converting VLDL to LDL via IDL (intermediate-density lipoprotein), as occurs in humans (Eisenberg and Rachmilewitz, 1973).

4. The process of esterification of lysolecithin to lecithin through a reversed LCAT reaction, as demonstrated convincingly in human plasma, does not take place in rats (Subbaiah and Bagdade, 1978; Webster, 1965).

The molar esterification rate of cholesterol in rat plasma *in vitro* ranges from 40 to 80 μmol/liter/hour, and the fractional esterification rate, representing the turnover rate in the plasma pool of free cholesterol, ranges from approximately 10 to 20% per hour; this is the highest value obtained among the mammalian species tested so far (Table IV). We have attempted in our laboratory to analyze the reasons for the variability in esterification rates seen in control groups, and have been able to demonstrate that the activity of rat plasma LCAT depends on the degree of postnatal development (Fig. 1). The experiment involved the testing of LCAT activity in various age-groups, and comparing it with that of "common" heat-inactivated serum substrate obtained from adult rat males; a reciprocal test was also performed with the serum substrate of identical age-groups (M. Dobiášová, unpublished results). It could be shown that 10-day-old rats already had a concentration (or activity) of the enzyme sufficient to maintain a rate of esterification similar to that of 120-day-old adult rats. The maximum enzyme activity was found around day 30, that is, at the time of weaning. Effective regulatory factors of LCAT activity were, however, contained in the substrate: whereas serum substrate from a 10-day-old suckling reduced the LCAT activity by 80%, and that from 30-day-old weaned younglings reduced it by only about 30%, no evidence of an inhibition of this type could be detected in the serum of later developmental stages.

We believe that the inhibition effect of suckling sera is not directly related just to the high concentration of plasma cholesterol in the serum of 10-day- and 30-day-old rats, since heated suckling serum added to the adult enzyme–adult substrate system blocked the LCAT reaction, whereas serum from 30-day-old rats did not. No sex differences in enzyme activity or in substrate availability could be discerned for the examined developmental stages between male and female animals fed standard diet.

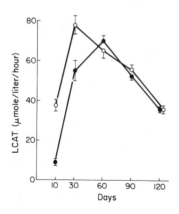

FIG. 1. Lecithin : cholesterol acyltransferase activity in rats during postnatal development measured by the common substrate method in the age-identical (autologous, ●) or adult (homologous, ○) serum. (M. Dobiášová, unpublished data.)

However, sex differences in LCAT activity do exist: Soler-Argilaga et al. (1977) demonstrated that the release of LCAT by perfused livers from fasted female rats exceeded that of the male rats. Also, the decay of serum LCAT activity with LCAT output blocked by means of cholchicine was more rapid in fasted female rats than in males. From the slope of the exponential decay curve they calculated the half-lives ($t_{1/2}$) for female and male rats as 6.9 and 13.2 hours, respectively. This phenomenon is of particular interest, because unlike male rats, female rats are apparently able to better regulate the esterification rate of cholesterol under different extreme conditions. For instance, loading with cholesterol-rich diet leads in males to a decrease in the fractional esterification rate, whereas females succeed in maintaining the standard rate under these conditions (Aftergood and Alfin-Slater, 1967).

Similarly, in lactating rats, where food consumption rises by as much as threefold (Babický et al., 1973) and metabolic activity increases sharply, the rate of plasma cholesterol esterification almost doubles; simultaneously, the enzyme activity and the substrate availability increase by 30 and 60%, respectively (our own unpublished results). These experiments also demonstrated that the triglyceride concentration in the substrate affected neither its availability for the LCAT reaction nor the enzyme activity; lactating rat females exhibited a plasma triglyceride concentration reduced to one-half, and the 10-day-old sucklings had a concentration which was doubled, while the respective esterification rates were as indicated above, that is, high in lactating and low in suckling rats. Even though the relationship between triglycerides, or triglyceride-rich VLDL, and the esterification rate is often taken as a metabolic correlate (Glomset and Norum, 1973; Norum et al., 1982), considerable discrepancies still persist, since an inhibition by orotic acid of VLDL secretion, accompanied by a plasma triglyceride reduction, had no effect on LCAT (Miller, 1978; Nordby and Norum, 1978). On the other hand, a positive correlation exists between the secretion by rat hepatocytes of LCAT and triglycerides (Nordby et al., 1979), and the turnover rates of serum cholesteryl esters and VLDL are likewise positively correlated (Soler-Agrilaga et al., 1977).

Table V summarizes some actions of *in vivo* effectors on the rate of cholesterol esterification in rat plasma *in vitro*. Most of the published data do not differentiate between the effect of enzyme activity and that of substrate availability. Nevertheless, it seems that certain conclusions can be safely drawn from these results: any increase in metabolic activity, even one that does not exceed the physiological range, results in increased enzyme activity. This applies to weaned younglings, exposure to cold, lactation, physical exercise, or any hormone-induced elevation of metabolism such as occurs after administration of thyroxine, norepinephrine, or

Table V
THE EFFECT OF SOME FACTORS *in Vivo* ON LCAT ACTIVITY *in Vitro* IN RATS

Factor	Activity of enzyme	Availability of substrate	Net esterification	Fractional rate	Reference
Age					
10 days (sucklings)	Unaltered	Reduced	Reduced	Reduced	Dobiášová et al. (unpublished results)
30 days (weaned)	Increased	Reduced	Increased	Unaltered	Dobiášová et al. (unpublished results)
60 days	Increased	Increased	Increased	Increased	Dobiášová et al. (unpublished results)
Lactation (10 days)	Increased	Increased	Increased	Increased	Dobiášová et al. (unpublished results)
Cold-exposed female	Increased	Reduced	Unaltered	Unaltered	Dobiášová et al. (unpublished results)
Exercise			Increased		Simko and Kelley (1979)
Diet					
Cholesterol (male)	Unaltered	Reduced	Reduced	Reduced	David et al. (1978)
Cholesterol (male)				Unaltered	Aftergood and Alfin-Slater (1967)
Cholesterol (female)					Aftergood and Alfin-Slater (1967)
Unsaturated oil	Unaltered	Reduced	Reduced		Larking and Sutherland (1977)
Vitamin E deficiency			Reduced		Takatori and Privett (1974)
Copper deficiency			Reduced		Lau and Klevay (1981); Harvey and Allen (1981)

Condition	Effect	Reference
Bile duct ligated		
2–6 hours after	Reduced	Ho et al. (1976)
8 hours after	Unaltered	Ho et al. (1976)
12–72 hours after	Unaltered	Williams et al. (1981)
Triglyceride turnover		
Accelerated	Increased	Soler-Argilaga et al. (1977)
Accelerated		Nordby et al. (1979)
VLDL blockade by orotic acid	Increased	Miller (1978); Nordby and Norum (1978)
Hypophysectomy	Unaltered	
7 days	Reduced	Takatori and Privett (1978)
2–4 weeks	Reduced	Manninen et al. (1978)
Ileal bypass	Increased	David et al. (1978)
Triiodothyronine	Increased	Dobiášová and Faltová (1976)
Norepinephrine	Unaltered	Dobiášová and Faltová (1976)
Isoprenaline	Reduced	Dobiášová and Faltová (1976)
Dexamethasone	Reduced	Dobiášová and Faltová (1976)
Clofibrate	Increased	Cisternas (1979)
Clofibrate + cholesterol diet	Increased	Cisternas (1979)
Clofibrate	Increased	Dobiášová and Faltová (1976)
DL-Methylthyroxine		Dobiášová et al. (1982b)
Phenobarbital		Russell et al. (1976)
Chlorpromazine	Reduced	Bell and Hubert (1981)
Ionizing radiation 48 hours after	Reduced	Dousset and Douste-Blazy (1975)

Additional effects noted:
- Triiodothyronine: Increased
- Norepinephrine: Increased
- Isoprenaline: Reduced
- Dexamethasone: Increased
- Clofibrate: Increased
- Clofibrate + cholesterol diet: Unaltered
- Clofibrate: Reduced
- DL-Methylthyroxine: Increased
- Phenobarbital: Increased

dexamethasone. On the other hand, a modification of the pituitary function, or administration of the tranquilizer chlorpromazine, induces a depression of activity. It is conceivable that the regulation of LCAT activity, or concentration, is mediated by humoral factors. Contrary to that, dietary manipulations or processes that would affect the spectrum of plasma lipoproteins, such as ileal bypass, are reflected rather in the availability of the substrate for enzyme action. That LCAT may be regulated by a subtle mechanism at the hormonal level is perhaps indicated also by the fact that any functional impairment of the liver, the source of LCAT, such as severe intoxication with praseodym salts (Godin and Frohlich, 1981), or biliary obstruction simulating cholestasis, reduces LCAT activity by a lesser extent than would be expected.

B. LCAT in Some Other Experimental Animals

The effects of various dietary factors on cholesterol esterification rates were examined in several studies using atherosclerosis-susceptible experimental animals. Lichtenstein *et al.* (1980) measured LCAT activity in two monkey species—one resistant to atherosclerosis (cebus monkey) and the other susceptible (squirrel monkey). After supplementing the animal diets either with coconut oil, which has a higher content of saturated fatty acids, or with corn oil and coconut oil combined with cholesterol, both monkey species exhibited a tendency toward increased cholesterol concentration and decreased esterification rate. However, there was a difference between the two species: whereas the enzyme activity in the squirrel monkey remained unchanged, it increased sharply in the other species. Consequently, the fractional esterification rate in the cebus monkey became reduced by a substantially smaller degree.

The significance of the fractional rate was first noted by Lacko *et al.* (1972); they concluded that FER appears to be a more useful tool for comparative studies because of the positive correlation of the cholesterol plasma level and LCAT activity under normal conditions. In this connection, it seems to be of particular interest that the atherosclerosis-susceptible squirrel monkey is unable to control the FER by means of increased LCAT activity. Neither is the rabbit, a species characterized by an easy inducibility of atherosclerosis, able to respond to elevated concentrations of cholesterol by increasing the activity of LCAT, and consequently the fractional rate decreases 4- to 10-fold (Pinon and Laudat, 1978). Another atherosclerosis-susceptible species, the guinea pig, reacts in a way described earlier (Glomset and Norum, 1973). A somewhat less intensive dietary regimen which reduces the rate of cholesterol esterification is a low protein diet. Its effect was tested in mice (Yashiro and Kimura, 1980); an interesting finding in this study was that exercising mice (revolving tread wheel), even when fed a low protein diet, maintained a significantly higher esterification rate compared to nonexercising animals.

C. Some Comments on Animal Models

The results of animal experiments brought to light many new facts and stimulated further studies and deliberations concerning the function of LCAT. Particularly noteworthy are the following points:

1. A possible hormonal regulation of LCAT concentration or LCAT activation.
2. The ability or inability of a particular organism to respond by a modification of enzyme activity to changes in substrate availability induced either by diet or by metabolic agents.
3. The developmental aspects of the rate cholesterol recycling between the plasma and the extravascular pools mediated by LCAT. Assuming that the reduction of cholesterol esterification in rat sucklings is not merely a response to the high-fat diet supplied by rat milk, then the shortcut recirculation process can be the consequence of a preferential utilization of cholesterol for the buildup of membranes of newly originating cells in physiologically immature rat younglings, particularly since the *de novo* synthesis of cholesterol is greatly reduced (Carrol, 1964).
4. The recruitment of extravascular lipid sources and the activation of metabolism associated with LCAT activation.

VI. Clinical Studies

The goal of clinical studies carried out in recent years has been to identify the abnormalities in LCAT activity and in the plasma cholesterol esterification rate in patients suffering from various diseases, as compared to normal healthy subjects, and in this way to discover useful criteria which would indicate the presence of regulatory disorders in cholesterol metabolism; or more specifically, to find indicators of an imbalance between the plasma pool of cholesterol and its transfer to extravascular pools. The detection of disturbances in the kinetics of lipid metabolism is much more informative than a simple measurement of plasma levels. Even though cardiovascular diseases are no doubt consequences of disorders in lipid metabolism, most epidemiological studies clearly demonstrate that an increased plasma concentration of lipids can be detected in only less than 50% of patients. The restricted validity for diagnosis of data on mere plasma concentrations can also be inferred from the conclusions of the review by Sodhi *et al.* (1980). This inference underlines the fact that there is no strong correlation between the plasma concentration of cholesterol and its total turnover, and that even the concentration of triglycerides, which otherwise positively affect the turnover, is not linearly related to the turnover rate.

On the other hand, a highly significant linear relationship could be confirmed between triglyceride and cholesteryl ester fractional turnover rates, i.e., between dynamic components of the metabolic pathway. Within the framework of the total turnover of body cholesterol, the cholesterol–cholesteryl ester pathway mediated by LCAT represents a part of the very rapid cycle (Fig. 2): the total daily turnover of cholesterol reaches 896 ± 393 mg in normal human subjects (Sodhi *et al.*, 1980), whereas the average production of cholesteryl esters solely in the plasma pool is 3 g, which is approximately three times as much as the amount of total cholesterol turnover. This quantitative relationship was confirmed by experiments both *in vivo* and *in vitro* (Glomset and Norum, 1973; Nestel and Monger, 1967; Kudchodkar and Sodhi, 1976; Dobiášová and Vondra, 1978). The purpose of this "futile cycle" of cholesterol–cholesteryl esters (Norum *et al.*, 1982) may be participation in several different functions, that is, storage of cholesterol in a mobile and harmless form for possible acute needs of the organism, extraction of cholesterol from peripheral tissues (or, conversely, compensation for the loss of cholesterol in membranes) (Glomset, 1968; Glomset and Norum, 1973), establishment of suitable conditions for the metabolism of VLDL (Schumacher and Adams, 1970), or simply redistribution of cholesterol among sites of actual surplus or shortage. Whatever the real purpose may be, leaving aside the

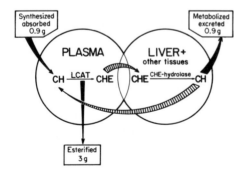

FIG. 2. Schematic comparison of total cholesterol turnover and LCAT-controlled recirculation of cholesterol. CH, cholesterol; CHE, cholesteryl ester. [Adapted from Dobiášová and Vondra (1982).]

physiological role of lysolecithin, a byproduct of the LCAT reaction, it appears that a constant recirculation rate of the plasma pool of free cholesterol is one of the prerequisites of a balanced cholesterol metabolism. This is why this parameter, fairly constant for healthy subjects, could serve as a basis for the evaluation of metabolic disorders.

No unified evaluation system has as yet been worked out for the assessment of individual cases. We have previously suggested (Dobiášová and Vondra, 1978) that the following parameters be necessarily taken into account: (1) the molar esterification rate depending on the concentration of active enzyme; (2) the fractional esterification rate depending on enzyme activity, substrate availability, and the concentration of plasma cholesterol; (3) the ratio of the molar esterification rate to the relative body weight index (MER/BI); and (4) the ratio of free-to-esterified cholesterol (UC/EC). The molar esterification rate is determined to at least 65% by the concentration of the enzyme (Albers *et al.*, 1981b) (more precisely by its enzymatically active form), as evidence of a functionally defective enzyme has been presented by Albers *et al.* (1981c). This strong correlation between the LCAT concentration and the rate of esterification seems obvious when comparing the low LCAT turnover number (not exceeding 200) with those found for most other enzymes (turnover numbers in the range 10^3–10^4). Both the fractional esterification rate, representing the rate of recirculation of the plasma pool of cholesterol, and the relationship between esterification rate and obesity, presumably rectify the meaning of the absolute value for LCAT activity in an individual. Also, the UC/EC ratio suggested by Saier *et al.* (1979) proved to be another useful indicator of disturbances in cholesterol metabolism.

The following sections will summarize the results concerning the effects of different physiological effectors on the rate of esterification of plasma cholesterol both in healthy normolipemic human subjects, comprising the control groups, and in patients suffering from diseases where a disorder of lipid metabolism is presumably involved. The effect of certain therapeutically effective drugs is also reviewed. Finally, we have attempted to summarize the available information on various effectors and to classify them according to the previously mentioned criteria in the form of a table. Since the numerical values are hardly comparable because of methodological differences both in LCAT estimation and in the measurement of total as well as of free cholesterol, accounting for approximately a 10% deviation in the calculation of the free-to-esterified cholesterol ratio, we have found it preferable to express the data as percentages of relative values, taking the results obtained in each appropriate control group as the reference value.

A. In Health

1. Normal Values

The rate of esterification of free cholesterol in the plasma of normal healthy human subjects is around 90 μmol/hour/liter of plasma (Table VI); the fractional esterification rate, that is, the fractional turnover of the intravascular pool of free cholesterol, is 6–7% per hour; and the ratio of molar esterification rate to relative body weight is 0.8–0.9. The given values were obtained by using different methods, that is, by measuring LCAT activity in normolipemic common substrate and by measuring esterification rate in self substrate or *in vivo* after administration of labeled cholesterol precursor ([^{3}H]mevalonate), and are in good absolute agreement. The similarity of the rate of cholesterol esterification assessed in the plasma *in vivo* and *in vitro* was reported earlier (Monger and Nestel, 1967; Glomset, 1968), and Barter (1974b) and Kudchodkar and Sodhi (1976) demonstrated the conformity of results obtained by *in vivo* and *in vitro* tests in identical subjects. Similarly, the two *in vitro* methods (common or self substrate), applied to large groups of healthy normolipemic subjects, yielded concordant results (Table VI). Data on LCAT protein concentration were first determined by radioimmunoassay (Albers *et al.*, 1981a); they are also included in the table.

2. Sex Differences

The question of whether sex differences exist in esterification activity has not yet been unequivocally decided. Many investigators described obvious sex differences, mostly regarding a decrease in the molar esterification rate in women, but others deny it with equal vehemence. Moreover, measurements of enzyme protein in a large group of examined subjects (Albers *et al.*, 1981a, 1982) indicated surprisingly that the LCAT concentration was higher in women than in men. Studies using the radioassay method in self substrate (with DTNB and mercaptoethanol) with a temporarily inhibited LCAT (Wallentin, 1977; Gillett and Silva, 1978) usually did not reveal any sex differences, although Sutherland *et al.* (1979), using the same method, found a significant decrease in the fractional esterification rate in women; however, no significant change could be detected in the molar rate. Other methods, such as measurement of the decrease in free cholesterol (Marcel and Vezina, 1973b; Patsch *et al.*, 1976) or assessment of the activity of the enzyme by the method of common heated substrate (Dobiášová and Vondra, 1978), usually make it possible to detect a reduced cholesterol esterification rate in healthy, nor-

Table VI
LCAT Activity and Concentration Determined in Healthy Subjects by Different Methods

Method	Number	Sex	BI[a]	TC (μmol/liter)	UC (μmol/liter)	TG (μmol/liter)	MER (μmol/liter/hour)	FER (%/hour)
Self-substrate				LCAT activity				
Radioassay[b]	85	Pooled	NG[c]	4990 (2900–7800)	1480 (800–2300)	970 (350–2100)	94.3 (60–146)	6.37[d]
	18	Pooled	108 (73–137)	6360 (4100–7700)	2030 (1300–2500)	1520 (800–2300)	136.7 (104–174)	6.73[d]
Common substrate[e]	83	M	96 ± 8	5587 ± 1137	1370 ± 297	1064 ± 640	97.9 ± 17.3	7.34 ± 1.34
	12	M	120 ± 4	6145 ± 1176	1498 ± 329	1724 ± 324	112.1 ± 25.9	7.59 ± 1.58
	12	M	159 ± 34	6475 ± 970	1597 ± 354	1666 ± 632	114.5 ± 13.0	7.50 ± 2.05
	57	W	92 ± 10	5581 ± 843	1416 ± 219	1000 ± 627	82.8 ± 12.6	6.00 ± 1.41
	13	W	171 ± 27	5837 ± 810	1489 ± 194	1430 ± 485	104.3 ± 21.9	7.10 ± 1.58
Self-substrate								
In vivo[f]	15	M	NG	3627–9171	1166–2617	757–5010	93[d]	126.2 ± 41.9[g]
In vitro[h]	15	M	NG	3627–9171	1166–2617	757–5010	82[d]	110.5 ± 25.5[g]
				LCAT concentration				
Radioimmunoassay[i]	44	M	NG	5518 ± 932	NG	1410 ± 633	5.98 ± 1.04[j]	
	22	W	NG	4870 ± 648	NG	1096 ± 520	6.44 ± 0.81[j]	

[a] Abbreviations: BI, body weight index; TC, total cholesterol; UC, unesterified cholesterol; TG, triglycerides; M, men; W, women; SD, standard deviation. Values expressed as mean ± standard deviation, or range (in parentheses).
[b] Data reported by Wallentin (1977).
[c] NG, numerical values not given.
[d] Value not given by author, but calculated on the basis of values listed in the publication.
[e] Data reported by Dobiášová et al. (1978a), and unpublished results).
[f] Production of cholesteryl esters in vivo. Data reported by Kudchodkar and Sodhi (1976).
[g] Net esterification (mg/hour).
[h] An in vitro test using plasma labeled in vivo. Data reported by Kudchodkar and Sodhi (1976).
[i] Data reported by Albers et al. (1981a).
[j] LCAT concentration (μg/ml).

molipemic, premenopausal females. There may be several reasons for this discrepancy: in some cases it is due to the lack of conformity in compared groups with regard to relative body weight, cholesterolemia, and triglyceridemia; further, if the examined groups are too small in size, uncontrolled factors may inadequately modify the results, such as intrapersonal variations or dependence of LCAT activity on the menstrual cycle (Wallentin and Vikrot, 1975), showing a reduced activity in the first decade.

The notion that DTNB with mercaptoethanol might be able to eliminate a hypothetical factor which would play a role in the regulation of LCAT activity in female plasma is purely speculative. The existence of such a hypothetical blocking regulatory factor would then explain the seemingly paradoxical observation that LCAT activity in normal female subjects is reduced under normal conditions, even though the plasma concentration of LCAT is higher in females than in males.

3. Age

Most studies fail to detect any correlation between esterification rate and age. If any correlation was found at all (Albers et al., 1981b) it was only partial, and the partial correlation index was not significant if the relative body weight coefficient was kept constant.

4. Relative Body Weight

A positive correlation between LCAT activity and relative body weight was reported by Akanuma et al. (1973); we were able to confirm the relationship in a group of normolipemic subjects (Dobiášová and Vondra, 1978). There is also a positive correlation between LCAT concentration and relative body weight (Albers et al., 1981b). In our quoted study, it was shown that the value of the correlation coefficient was higher in men than in women, and that the significance of this relationship is limited by the degree of obesity (Table VII). In males exhibiting an obesity index exceeding 125, the correlation was insignificant, indicating that the range of linear increase in LCAT activity with obesity is restricted and ends at a certain maximum.

Relative body weight also positively affects the production of VLDL triglycerides and the plasma concentration of triglycerides. However, this can be demonstrated only in males (Bernstein and Bernstein, 1978); in this respect, there is a certain parallel in the weak correlation of LCAT and obesity in females, as compared to the much stronger correlation found in men. Contrary to this, obesity in both sexes leads to a reduction in plasma levels of HDL cholesterol and apoA-I (Avogaro et al., 1978). This fact

Table VII

Estimates of Correlations between MER (and LCAT Concentration), on the One Hand, and Relative Body Weight, Total Cholesterol, Unesterified Cholesterol, Triglycerides, and Phospholipids, on the Other, in Healthy (Mostly) Normolipemic Subjects

Subjects	Sex[a]	Number	BI	TC	UC	TG	log TG	PL	Reference
					MER				
BI < 115	M	83	0.364	0.426	0.515	0.232[b]			
BI = 115–125	M	12	0.535	0.626	0.672	0.280[b]			
BI > 125	M	12	−0.398[b]	0.371[b]	0.019[b]	0.370[b]			
BI < 115	W	57	0.294	0.041[b]	0.062[b]	0.279[b]			
Pooled	{M	36}					0.63	0.56	Wallentin and Vikrot (1975)
	W	28}							
Pooled	{M	20}	0.563	0.454	0.447	0.397			Akanuma et al. (1973)
	W	13}							
				LCAT concentration					
Pooled, 11 hyperlipemic	{M	20}	0.619	0.608	0.562		0.247[b]		Albers et al. (1981b)
	W	5}							
44 normolipemic	M	90		0.384		0.131[b]			Albers et al. (1981a)
22 normolipemic	W	34		0.519		0.512			Albers et al. (1981a)

[a] Abbreviations as in Table VI; PL, phospholipids.
[b] Correlations not significant.

further confirms the idea that the concentration of apoA-I is not limiting, even though it is practically the only activator of the LCAT reaction *in vivo*.

5. Concentration of Plasma Cholesterol

A positive relationship between LCAT activity and the concentration of plasma cholesterol was established by many laboratories (Akanuma *et al.*, 1973; Fabien *et al.*, 1973; Soloff *et al.*, 1973; Wallentin and Vikrot, 1975; Sutherland *et al.*, 1979; Albers *et al.*, 1981a,b). In most studies, however, the data obtained for men and women were pooled. We have evaluated separately 99 men and 58 women, and found a highly significant positive correlation between LCAT and cholesterol for men only, while under the same conditions of normal cholesterolemia, the data on women displayed almost no correlation at all (Table VII). It will be shown later, however, that in certain situations associated with hypercholesterolemia (pregnancy, diabetes), there is a highly significant positive correlation between LCAT and cholesterol in women.

6. Plasma Triglycerides

Despite the fact that the mechanism of cholesterol esterification is closely related to the intravascular transformation of VLDL and chylomicrons (Glomset and Norum, 1973; Norum *et al.*, 1982), and that hypertriglyceridemia in humans is often associated with increased LCAT activity (Akanuma *et al.*, 1973; Blomhoff *et al.*, 1974; Kudchodkar and Sodhi, 1976; Wallentin, 1977; Kuczynska and Sznajderman, 1980), a significant relationship between the concentration (or activity) of LCAT and the concentration of triglyceride could not be regularly confirmed (Table VII; Rose and Juliano, 1976). Neither is the decrease in plasma lipoproteins induced by heparin administration directly related to triglycerides; reduction of LCAT is, rather, attributable to the increase in the concentration of free fatty acids which inhibited LCAT *in vitro* (Rutenberg *et al.*, 1973).

7. Gravidity

Gillett and Silva (1978) followed the changes in plasma lipid concentration and in the cholesterol esterification rate in a group of young women during the period of pregnancy. Plasma lipid levels, LCAT activity, and esterification rate were found to increase successively, but subsequently fell postpartum (Table VIII). Plasma levels of cholesterol, phospholipids,

Table VIII
Rate of Esterification in Young Pregnant and Nonpregnant Women Determined by the Self-Substrate Method[a]

Subject	Number	TC (μmol/liter)	UC (μmol/liter)	TG (μmol/liter)	MER (μmol/liter/hour)	FER[b] (%/hour)
Nonpregnant	28	4440 ± 800	1330 ± 240	1650 ± 490	73.6 ± 11.3	5.53
First trimester	27	4590 ± 850	1350 ± 220	2111 ± 610	76.2 ± 14.0	5.64
Second trimester	25	6040 ± 1150	1790 ± 300	2980 ± 730	93.6 ± 19.5	5.23
Third trimester	17	6550 ± 1170	1950 ± 370	3930 ± 740	116.1 ± 26.5	5.95
Postpartum	21	4820 ± 900	1460 ± 280	2280 ± 650	93.5 ± 22.8	6.40

[a] Data reported by Gillett and Silva (1978). Abbreviations as in Table VI; values expressed as mean ± standard deviation.
[b] Not given by the author, but calculated on the basis of values listed in the publication.

triglycerides, and LDL, which tend to increase during pregnancy, were positively correlated with LCAT activity. The levels of HDL did not change during pregnancy; no correlation between LCAT and HDL was found, however, in young, nonpregnant women either. It appears that the course of LCAT activation during pregnancy is an excellent example of the regulatory mechanism which maintains a practically constant rate of exchange of plasma cholesterol despite markedly increased concentrations. This fact can be illustrated by expressing the obtained values as relative quantities if the data pertaining to the first trimester are taken as a reference value of 100% (Table XII). The higher activity of LCAT found 1 month postpartum may be considered as analogous to the increased LCAT activity found in lactating rats (see earlier), a phenomenon which may be connected with the recruitment of stored fat during the lactation period. This is also in keeping with the interpretation of the results of a study on lipid metabolism in pregnant women published by Warth *et al.* (1975).

8. Infancy

Table IX presents data on LCAT activity found in umbilical cord neonatal serum and in the serum of children aged 3 days and 3 months. The results were obtained by using either the common heated substrate of adult normolipemic subjects or pooled common substrate of donnors of identical age (Dobiášová *et al.*, unpublished results). Lecithin : cholesterol acyltransferase activity in healthy children determined immediately after birth (umbilical cord serum) exhibited the lowest values. A significant increase was detectable on the third day of life, and at the age of 3 months the babies had LCAT activity roughly comparable with that of adults. This development of LCAT activity was equally significant when estimated by either of the two methods using different substrates, although the identical-age substrate appeared to be preferable to the adult-subject substrate (the difference being about 20%, and as much as 40% in the 3-day-old babies). The fractional esterification rate values generally approximated those found in normal adult subjects. There are up to now very few data on the metabolism of lipoproteins in children in the early neonatal period, but some results (McConathy and Lane, 1980; Stožický *et al.*, 1982) indicate that LCAT activity and differences in the content of particular plasma lipoproteins (or apoproteins) are not correlated.

9. Physical Activity

It has been reported that physically well-trained men, as compared to sedentary men, have lower levels of plasma cholesterol, lower levels of

Table IX
Human LCAT Activity in Early Postnatal Development[a]

Age	Number	TC (μmol/liter)	UC (μmol/liter)	TG (μmol/liter)	MER-A[b] (μmol/liter/hour)	MER-I (μmol/liter/hour)	FER-A (%/hour)	FER-I (%/hour)
Newborn	11	1819 ± 325	478 ± 127	500 ± 100	32.3 ± 6.0	40.6 ± 11.2	7.18 ± 2.15	8.59 ± 2.49
3 days	14	3897 ± 622	1258 ± 152	1600 ± 600	50.2 ± 8.4	83.5 ± 21.1	4.11 ± 0.99	6.83 ± 1.91
3 months	5	3342 ± 865	1127 ± 227	—	86.3 ± 4.3	102.3 ± 8.0	7.86 ± 1.50	9.09 ± 0.96
Adult		5363	1541	1400	106.5		6.91	

[a] From Dobiášová et al. (unpublished results).
[b] "A" signifies common heated substrate of adult normolipemic subjects; "I" signifies common heated substrate of donors of identical age. Otherwise, abbreviations as in Table VI; values expressed as mean ± standard deviation.

serum total triglycerides and VLDL triglycerides, and significantly higher levels of HDL cholesterol (Nikkilä et al., 1978). These investigators examined a group of long-distance runners, and observed low concentrations of plasma triglycerides accompanied by increased activities of skeletal muscle and adipose tissue lipases. In a comparable group of athletes (cross-country skiers), we have measured the activity of LCAT and the concentration of plasma lipids (Dobiášová et al., 1978a). In concordance with the results of the Finnish group, we also have observed lower level of triglycerides as well as of cholesterol. The molar rate of esterification did not differ from that of control groups, but due to the lower content of plasma cholesterol the fractional rate was significantly elevated both in females and in males. Lopez et al. (1974) found a significant decrease in serum triglycerides, a moderate reduction in serum cholesterol, and an increase in the activity of LCAT among volunteers who participated in an exercise program lasting 7 weeks. Increased LCAT activity, as presented by Marniemi et al. (1982), was found in 26 active sportsmen and in 35 military conscripts 2 months after military training. In addition, a stimulatory effect on the endogenous exchange of cholesterol has been described in experimental animals given exercise (Simko and Kelley, 1979; Yashiro and Kimura, 1980).

10. Diet

a. Alimentary Lipemia. The response of LCAT to high-fat diet has been examined in various laboratories (Rutenberg et al., 1973; Marcel and Vezina, 1973a; Rose and Juliano, 1977; Kostner, 1978). In all the quoted studies, the total cholesterol esterification rate was assessed either by radioassay in self-substrate or by measuring the decrease in free cholesterol after incubation; consequently, data on the activity or concentration of the enzyme are lacking. Rose and Juliano (1977) examined eight subjects and found a significant positive correlation between percentage increase in plasma triglycerides and esterification rate; the average increments in triglyceride concentration and in molar esterification rate were 181 and 37%, respectively. However, the increases in esterification rate and triglyceride concentration were not parallel in time. Esterification attained a rather broad optimum after 5–7.5 hours, followed by moderate trend of decrease until ten hours after food intake. On the other hand, the triglyceride concentration was substantially increased as early as 2.5 hours after feeding, reached a maximum in the fifth hour, and fell sharply after 7.5 hours; a normal postabsorptive level was attained after 10 hours.

In the course of these changes, the concentration curves for triglycerides, VLDL, and chylomicrons run parallel. No obvious explanation for the time shift in the changes of the two former parameters is available, but a possible hypothesis has been forwarded: The reason is not an augmented secretion of the enzyme by liver cells, because the regulatory response based on this mechanism would be too slow due to the relatively long half-life of LCAT in the plasma. The clue is rather to be found in the supply of a better substrate for LCAT, which is supposedly the increase in the concentration of HDL which persits even after the reduction of the triglyceride concentration. However, since the increase in HDL is mostly due to the HDL_2 fraction (Havel et al., 1973), which, as was explained earlier, inhibits rather than stimulates the LCAT reaction (see Section IV,C), it seems unlikely that this hypothesis will be verified. Kostner (1978) observed a parallel in time in the increases in esterification rate and triglyceride concentration. At the same time, he presented data on the concentration of cholesterol in the postprandial period; when the data on MER and UC are converted to FER, it can be deduced that the magnitude of FER remains essentially unchanged because the MER increases concomitantly with cholesterol concentration. This constancy of fractional turnover rate might reflect an obvious regulatory tendency in healthy subjects aiming at a constant exchange rate of the plasma pool of free cholesterol. Some circumstantial evidence, i.e., that the activity of LCAT by itself (estimated by the method of common substrate) is not affected by alimentary load, can be inferred from our study on a group of volunteers subjected to sleep deprivation stress (Dobiášová et al., 1979a). Data obtained by determining the selected parameters in samples taken in the morning and evening at 3-hour intervals after food intake showed that no changes in LCAT activity could be detected which would correspond to the observed changes in triglyceride concentration (Fig. 3).

b. *Composition of Dietary Fat.* Gjone et al. (1972b) demonstrated a reduction in the cholesterol esterification rate *in vitro* if saturated fat was replaced by unsaturated fat in the diet of the examined subject. The finding was later confirmed by Miller et al. (1975) in an experiment in which the effect of a swich from saturated fat [polyunsaturated/saturated (P/S) ratio of 0.2] to polyunsaturated fat (P/S ratio of 2.4), fed for a period of 7–9 days, on LCAT activity and esterification rate *in vitro* was studied in the same individuals. The unsaturated fat diet resulted in a decrease in the concentration of cholesterol and triglycerides in parallel with a reduction in LCAT activity and in net rate of cholesterol esterification. The ingestion of saturated fat during the postprandial period led to a transient increase in both the molar and fractional esterification rates (Wallentin and Vikrot, 1976). After converting Miller's reported data to comparable relations, it became obvious (Table XII) that changing the diet does not affect the esterification rate and the concentration of plasma lipids independently of

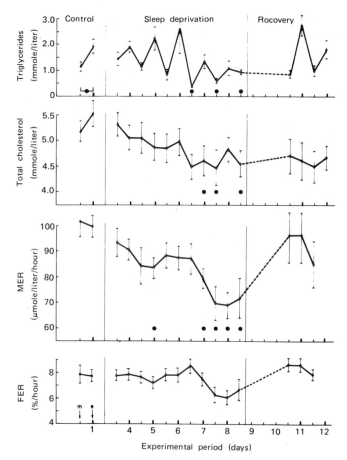

FIG. 3. Plasma lipids and esterification rate of cholesterol in stress induced by sleep deprivation. Black dots indicate statistically significant differences compared to initial control values, corresponding to morning (m) or evening (e) blood sampling. [Data adapted from Dobiášová et al. (1979a).]

each other, so that it seems impossible to induce by dietary means a reduction of solely the lipid level and at the same time maintain constant the desirable normal esterification rate.

11. Stress

Although more than 20 years ago Klein and Dahl (1961) reported that the cholesterol esterification rate was remarkably sensitive to environmental stress, not much attention has been paid to this problem since. In a model stress situation induced by sleep deprivation lasting for 120 hours (Kuhn, 1968), we attempted to assess in 7 healthy, young, male volunteers the effect of stress upon the cholesterol esterification rate and to relate it to changes in plasma lipid levels (Dobiášová et al., 1979a). The arrangement of the experiment was such that during the entire test period the subjects received mixed diets in equienergetic doses given at 3-hour intervals; blood sampling was performed always at 9 AM and 9 PM, 3 hours after food intake. With this arrangement of periodical alimentation the extent of the daily biorhythmic changes in lipid levels could also be assessed. Whereas the triglyceride levels were profoundly affected by the biorhythm until the last day of the test, no effect of this factor was seen on other parameters, such as LCAT activity and cholesterol concentration (Fig. 3). After approximately 3 days of sleep deprivation, a significant reduction was noted in triglyceride and cholesterol levels, as well as in LCAT activity. The decrease in LCAT and cholesterol was simultaneous in time and parallel in extent, so that the average fractional esterification rate was not significantly reduced as compared to the initial values. Individual response to stress was, however, variable: one subject exhibited practically no response of plasma lipids and LCAT concentration, whereas two other subjects responded by the progressive lowering of not only the molar but also the fractional esterification rate. In the recovery phase, all LCAT values returned to normal. Further, the results of this experiment indicate that the enzyme activity (tested by the common substrate method) is not causally related to the actual concentration of plasma triglycerides.

B. In Disease

The primary disease which impairs the mechanism of cholesterol esterification in the plasma is the hereditary deficiency of LCAT. In addition to several Scandinavian families, in whom this defect was detected first (Norum and Gjone, 1967; Hamnström et al., 1969), other LCAT-deficient families were discovered on Sardinia (Uterman et al., 1972), England—a family of Indian ancestry (Bron et al., 1975), Canada (Frohlich et al., 1978), France (Chevet et al., 1978), and Japan (Iwamoto et al., 1978). Typical clinical symptoms of this disease, i.e., corneal opacities, proteinuria and glomerular tufts of the kidney, and anemia with target cells, have been described (Gjone, 1973). Abnormalities of plasma lipoproteins, defects in the transfer of apoprotein and lipid moieties among plasma lipoproteins, as well as the defective removal of lipoprotein from plasma have been reviewed (Glomset and Norum, 1973; Glomset et al., 1980). A new aspect in the study of this disease has come to light, namely, the nonidentity of enzyme activity and enzyme concentration found in certain individuals suffering from LCAT deficiency (Albers *et al.*, 1981b); this finding suggests the possibility of an as yet unknown mechanism of regulation of enzyme activity.

Other diseases which obviously affect the mechanism of cholesterol esterification belong to hepatocellular pathologies. Certain liver diseases impair enzyme production, as was independently shown by Gjone and Norum (1970) and Simon and Scheig (1970). It has been known that this pathology is characterized by an increased ratio of unesterified to esterified cholesterol; the discovery of LCAT provided a logical explanation of this observation. Simon (1979) demonstrated that the reduced esterification rate of plasma cholesterol in parenchymal liver diseases is attributable neither to the presence of bile acids in the plasma, which are known to inhibit LCAT *in vitro* (Glomset, 1968), to a decreased availability of the substrate, to the inhibitory effect of abnormal lipoproteins (LpX), or to the activity of cholesteryl ester hydrolase, but rather it is attributable to defect in the production of LCAT by damaged liver cells. On the other hand, the cause of an increased UC/EC ratio in cholestatic (obstructive) liver disorders is more complex, since LCAT activity mostly remains normal, so that other factors are likely to be involved (Simon, 1979).

Possibly more attention should be paid to the changes in LCAT activity and rate of cholesterol recirculation in the course of those diseases which until recently have not been considered in connection with a disorder in cholesterol metabolism because of relatively normal levels of plasma lipids, for example, renal insufficiency or hypertension, pathologies with high degrees of intrinsic risk for the development of atherosclerosis and with grave cardiovascular consequences. In this context, it is not inconceivable that LCAT activity is likely to play a role in the pathogenesis of ischemic heart disease.

1. Hypertension

The relationship of hypertension, lipids, and lipoproteins to atherosclerosis has been reviewed by Lewis and Naito (1978). It is well known that hypertension is accompanied by a considerable risk for the development of atherogenous complications such as cerebrovascular, coronary, and arotic lesions, and in particular a special manifestation of ischemic disease, myocardial infarction. It remains to be clarified, however, whether hypertension is the initiating or the accelerating factor in the development of atheroclserosis. Since there is no correlation between hypertension and plasma lipid levels, various possibilities relating atherosclerosis to hypertension are being considered: increased compression of plasma components into the vessel wall (Duncan et al., 1965), which may be suggested by a more frequent incidence and a greater extent of atherosclerotic lesions in localities with higher pressure and turbulence (Dustan, 1974); increased vascular permeability for plasma lipoproteins induced by the effect of certain hormones (Giese, 1973); or the possibility of a modified affinity of the vessel wall to lipoproteins (Bretherton et al., 1976). Thus, in spite of the apparent relationship between hypertension and lipid metabolism disturbances, the mere concentration of cholesterol has only a limited diagnostic or prognostic value as far as a particular patient is concerned (Proudfit et al., 1966; Vaurik et al., 1974).

Guided by the idea that, apart from the instances just mentioned, impairment of the influx and efflux rates of plasma cholesterol might take part also in the development of atherosclerosis in hypertensive patients, we have embarked upon a study investigating the activity of LCAT in 91 hypertensive subjects, of which 51 are men and 40 are women (Dobiášová et al., 1979b). Features common to the two subgroups (males and females) were increased levels of cholesterol and triglycerides (Table X). The molar esterification rate did not differ significantly from the control value in males, but it was slightly higher in females. The most remarkable difference between the control subjects and the hypertensive patients concerned in the reduction in the fractional esterification rate: it was highly significant in the men, but significant only at the level of 95% confidence in the women. Unlike that in the females, the ratio of unesterified to esterified cholesterol was significantly increased in the males. A positive correlation between LCAT, on one hand, and relative body weight, triglyceride, and cholesterol concentrations, on the other, was found in the men, whereas in the women the correlation was restricted to just the plasma cholesterol level (Table XI). A highly significant correlation was found in both groups between the fractional esterification rate and the UC/EC ratio (percentage of UC in the table).

Sex differences in the regulation of LCAT activity, however, were more apparent when the patients were subdivided according to the criteria set up by WHO for the classification of the severity of the disease (Table XII). Women classified in the first and second stages of the disease displayed increased LCAT activity concomitant with increased plasma cholesterol, so that the fractional esterification rate remained unchanged; only in the third stage was FER reduced below 80% of the control level. In men, however, LCAT activity rose only slightly compared to the plasma cholesterol concentration, and the values for FER were reduced both in the second and in the third stages by as much as 40%. The fractional esterification rate was found low also in those cases in which the cholesterol level remained normal while the LCAT activity decreased. Further, it was

Table X
PLASMA LIPIDS AND ESTERIFICATION RATE IN HYPERTENSIVE SUBJECTS[a]

Subject	Number	BI	TC (μmol/liter)	UC (μmol/liter)	UC (%)	TG (μmol/liter)	MER (μmol/liter/hour)	FER (%/hour)
Men, healthy	99	103 ± 21	5722 ± 1147	1401 ± 306	24.5 ± 3.0	1257 ± 681	100.2 ± 18.6	7.34 ± 1.48
Men, hypertensive	51	106 ± 13	6869 ± 1893[b]	1903 ± 637[b]	27.8 ± 4.9[b]	2172 ± 2365	94.2 ± 22.0	5.28 ± 1.45[c]
Women, healthy	57	92	5581 ± 834	1416 ± 219	25.3 ± 2.3	1000 ± 626	82.8 ± 12.6	6.00 ± 1.41
Women, hypertensive	40	113 ± 19[b]	6613 ± 1519[d]	1732 ± 377[b]	26.2 ± 2.8	2319 ± 1795	89.5 ± 19.5[d]	5.28 ± 1.16[d]

[a] Abbreviations as in Table VI; values expressed as mean ± standard deviation.
[b] Unpaired t test, $p < 0.01$.
[c] Unpaired t test, $p < 0.001$.
[d] Unpaired t test, $p < 0.5$.

Table XI
ESTIMATES OF CORRELATIONS (r) BETWEEN MER AND FER, ON THE ONE HAND, AND RELATIVE BODY WEIGHT AND LIPID PARAMETERS, ON THE OTHER, IN HYPERTENSIVE SUBJECTS[a]

Parameter	Sex	MER	FER
BI	M	0.4043	NS[b]
	W	NS	NS
TC	M	0.5682	−0.4423
	W	0.6996	NS
UC	M	0.5032	−0.5890
	W	0.5739	−0.4807
UC (%)	M	NS	−0.4459
	W	−0.3256	−0.4950
TG	M	0.4735	NS
	W	NS	NS
MER	M	1.0000	0.3237
	W	1.0000	0.4178

[a] Dobiášová *et al.* (unpublished results). Abbreviations as in Table VI.
[b] Numerical values not given if not significant.

found that the unesterified cholesterol fraction accounted for the total increase in plasma cholesterol in a greater proportion of men than women.

By examining the same subjects repeatedly in a follow-up study for 2–3 years, it could be established that FER decreased progressively in those patients who developed serious atherosclerotic complications, whereas the changes in cholesterol concentration were scattered without any apparent correlation (Fig. 4). The degree to which the disease could be controlled by pharmacological means, assessed by the response of diastolic blood pressure to drug administration, was reflected in the frequency of deviations from the standard deviation limits of the mean of the FER and the MER/BI index characteristic for the reference group: Those patients whose diastolic blood pressure could be kept below 13.3 kPa by appropriate medication exhibited much a lower incidence of atherogenous complications, and also a lower frequency of reduced values of FER and LCAT/BI. On the other hand, cholesterolemia was relatively constant without any relation to the effect of medication (Dobiášová *et al.*, 1979b).

Despite the fact that the quoted results are only the first evidence suggesting the participation of defective recirculation of cholesterol in hypertension, it seems that the frequency of deviations was rather proportional to the severity of the disease, and that they were expressed more significantly in men than in women.

Table XII

COMPILATION OF PUBLISHED RESULTS ON PLASMA LIPID CONCENTRATION AND ESTERIFICATION RATE: AN ATEMPT TO PROVIDE A BASIS FOR MUTUAL COMPARISON BY CONVERTING REPORTED ABSOLUTE DATA TO RELATIVE VALUES (WITH REFERENCE TO THE PERTAINING CONTROL DATA TAKEN AS 100%)

Subjects and treatment[a]	Number	TC	UC	TG	MER	FER	Reference
(C) Men, normolipemic, BI < 115	83	100	100	100	100	100	Dobiášová et al. (unpublished results)
(T) Men, normolipemic, BI = 115–125	12	110	109	162	115	103	
Men, normolipemic, BI > 125	12	116	116	157	117	102	
Men, athletes	12	75	81	71	90	112	
(C) Women, normolipemic, BI < 115	57	100	100	100	100	100	Dobiášová et al. (unpublished results)
(T) Women, normolipemic, BI > 115	13	105	105	159	126	118	
Women, athletes	7	87	85	66	104	122	
(C) Pregnancy, first trimester	27	100	100	100	100	100	Gillett and Silva (1978)
(T) Pregnancy, second trimester	25	132	132	141	123	93	
Pregnancy, third trimester	17	143	144	186	152	105	
Postpartum, 1 month	21	105	108	108	123	113	
(C) Diet, polyunsaturated/saturated ratio = 0.2	15	100	100	100	100	100	Miller et al. (1975)
(T) Diet, polyunsaturated/saturated ratio = 2.4	15	83	85	60	83	99	
(C) Hypertriglyceridemic subjects	39	100	100	100	100	100	Wallentin (1978a)
(T) Hypertriglyceridemic, TG-lowering diet (2 months)	39	87	87	76	82	93	
(C) Men, normolipemic	99	100	100	100	100	100	Dobiášová et al. (unpublished results)
(T) Men, hypertensive							
First stadium	12	98	102	137	98	96	
Second stadium	12	126	153	246	97	66	
Third stadium	27	127	143	151	91	64	
(C) Women, normolipemic	57	100	100	100	100	100	
(T) Women, hypertensive							
First stadium	14	119	116	218	117	100	
Second stadium	13	125	131	270	110	85	
Third stadium	13	111	120	208	95	78	

	n						Reference
(C) Men, normolipemic	99	100	100	100	100	100	Dobiášová and Vondra (1978)
(T) Men, diabetics	29	127	144	212	108	78	
(C) Women, normolipemic	57	100	100	100	100	100	Dobiášová and Vondra (1978)
(T) Women, diabetics	15	143	143	215	148	104	
(C) Male + female	11	100	100	—	100	100	Miller and Thompson (1973)
(T) Malabsorbers	16	74	82	—	67	80	
(C) Men, normolipemic	99	100	100	100	100	100	Dobiášová and Vondra (1978 and unpublished results)
(T) Men, 3 months after myocardial infarction, TC < 7.5	22	118	133	167	98	76	
Men, 3 months after myocardial infarction, TC > 7.5	16	140	156	281	103	67	
(C) Men, normolipemic, healthy	12	100	100	100	100	100	Soloff et al. (1973)
(T) Men, after myocardial infarction	17	125	140	250	101	72	
(C) Men, normolipemic, healthy	44	100	—	100	100	100	Wallentin and Moberg (1982)
(T) Men, coronary artery disease	38	114	—	147	102	90	
(C) Men, type IV hyperlipoproteinemia	29	100	—	100	100	100	
(T) Men, coronary artery disease	22	98	—	92	79	82	
(C) Men, healthy, normolipemic	99	100	100	100	100	100	Dobiášová et al. (1982b)
(T) Men, transplanted kidney	23	86	110	109	53	49	
(C) Women, normolipemic, healthy	57	100	100	100	100	100	
(T) Women, transplanted kidney	22	113	125	236	81	66	
(C) Men, normolipemic, no medication	7	100	100	100	100	100	Heller et al. (1981)
(T) Colestipol, 7 days	7	73	78	83	88	110	
Fenofibrate, 7 days	7	91	89	70	107	121	
Fenofibrate + colestipol, 14 days (2nd week)	7	62	57	76	107	184	
(C) Men, type II hyperlipoproteinemia, no medication	6	100	100	100	100	100	Dobiášová et al. (1982a)
(T) Etiroxate, 14 days	6	74	71	55	93	144	
Etiroxate, 16 weeks	6	97	85	67	164	195	

[a] C, control subjects; T, test subjects. Otherwise, abbreviations as in Table VI.

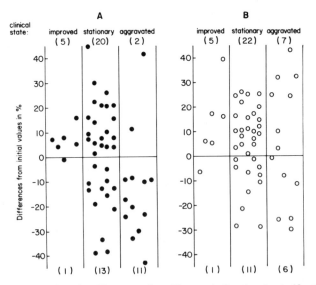

FIG. 4. Longitudinal study of hypertension. Changes in fractional esterification rate (A) and total cholesterol concentration (B) (expressed as percentage of initial values) in hypertensive patients 2-3 years after the first examination. The subdivision of patients into three groups is based on clinical symptomatology. (M. Dobiášová, unpublished data.)

2. Renal Diseases

The alarming discovery that cardiovascular diseases account for the mortality of as many as one-half of the dialysis patients (Lindner *et al.*, 1974), as well as for the frequently seen occurrence of atherosclerotic complications in patients who have undergone renal transplantation, has brought about an intensive search for lipid metabolism disturbances. Increased triglyceride levels were also reported as a common finding in hemodialysis patients, whereas cholesterol concentrations remained essentially normal (Bagdade, 1968). In a very detailed study, Norbeck (1981) confirmed the validity of this finding for patients suffering from chronic renal failure as well. This abnormality in the plasma lipid pattern was associated with relative changes in all main lipoprotein classes: in VLDL an increase in cholesterol and triglycerides; in abnormally large LDL an increase in triglycerides; in HDL a decrease in both cholesterol and triglycerides. The results of the intravenous fat tolerance test pointed to a reduced rate of triglyceride removal from the blood stream. Cholesteryl esters contained a reduced proportion of linoleic acid (52%, compared to 56% in the controls). Somewhat more conflicting results were reported for patients who had received a kidney graft: Casaretto *et al.*

(1974) reported that hyperlipemia appeared often after successful kidney transplantation, whereas Beaumont et al. (1975) found a normal lipid profile in 29 out of 37 patients. Bagdade and Albers (1977), examining a group of transplant patients with normal lipid levels, found a low content of HDL cholesterol; on the other hand, Curtis et al. (1978) were unable to detect any significant difference in HDL cholesterol between controls and treated subjects.

Since hypertriglyceridemia *per se* was not identified in extensive epidemiological studies as one of the significant atherogenous risk factors (Kannel et al., 1979), and the remaining concentration indicators varied considerably in renal diseases, it can be assumed that the underlying mechanism is instead a not easily detectable kinetic disorder in lipid metabolism. It must be particularly stressed that for such situations the measurement of LCAT activity has special merit as a highly efficient method of distinguishing between the characteristics of cholesterol metabolism in normal subjects and those found in patients suffering from renal diseases. Even the few papers published so far have cited convincing evidence that LCAT activity is markedly reduced in chronic uremia (Guarnieri et al., 1978), in renal diseases associated with plasma creatinine levels exceeding 800 μmol/liter (Morin and Piran, 1981), in hemodialysis patients (Jung et al., 1980), and in patients after kidney transplantation (Dobiášová et al., 1983a).

In the last-quoted study, 45 subjects with grafted kidney were examined; increased concentrations of cholesterol and triglycerides were found, respectively, in 25 and 31% of the tests carried out, whereas a subnormal molar esterification rate could be detected in 74%, and a reduced fractional esterification rate could be detected in as many as 92% of the cases. Examination at different time intervals after transplantation surgery revealed that the lowest rate of esterification, as well as the lowest concentration of cholesterol, was consistently found early after transplantation (within 10–15 days). The values rose later on, but even 6–24 months after a successful kidney transplantation the molar and fractional esterification rates did not attain in most of the patients the levels present in the healthy control subjects (Fig. 5). If the function of the grafted kidney failed and the patient had to return to the hemodialysis treatment program, all of the examined parameters remained permanently low. If the results are expressed as relative figures, related to the pertinent control values (Table XII), they also demonstrate an increased ratio of free-to-esterified cholesterol in most of the patients, though the increase is again more significant in males than in females.

With regard to the disorder in cholesterol metabolism, it is of interest that the reduced fractional esterification rate persisted even in those pa-

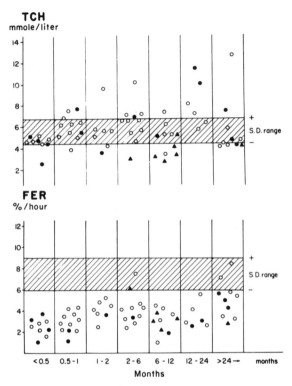

FIG. 5. Plasma total cholesterol (TCH) and fractional esterification rate (FER) in patients after kidney transplantation. (○) Plasma creatinine less than 300; (●) plasma creatinine greater than 300; (▲) patients returning to dialysis program (M. Dobiášová, unpublished data.)

tients whose clinical parameters indicated a normal renal function. According to our present understanding of the mechanism of action of LCAT, several causes can hypothetically be held responsible for the reduced activity of the enzyme: inhibition of enzyme secretion due to liver cell damage; an as yet unexplained function of the kidney in controlling the rate of cholesterol recirculation; and, finally, side effects of immunosuppressive drugs, or, possibly, inhibitory effects of phthalates released from plastic materials used in dialysis and surgical operation (Lagente *et al.*, 1979). It must be further admitted that surgically treated organisms may be more susceptible to the effects of stress factors (see Section VI,A,11) and occasionally of anesthetic drugs (Bell and Hubert, 1980), especially immediately after the operation. Because the presence of liver dysfunction could be recognized only in a minority of kidney trans-

·plantation cases (5/45), it follows that the impaired production of the enzyme cannot generally be held responsible for the effect, and consequently the kidney probably plays a much more substantial functional role in the regulation of endogenous cholesterol turnover than has been previously assumed. Conversely, it is not irrelevant that the most serious complication in LCAT deficiency is proteinuria associated with renal failure (Gjone, 1973).

3. Cardiovascular Diseases; Myocardial Infarction

Initial epidemiological studies dealing with the problem of cardiovascular diseases focused on serum total cholesterol, and demonstrated a powerful relationship between this lipid and the development of atherosclerosis. However, the predictive value of increased cholesterol concentration for the risk estimate is restricted to the age-group under 50 years, although even in this age-group the mean concentration of cholesterol determined among the victims of coronary heart disease (CHD) was only 244 mg/100 ml (Framingham study, Kannel et al., 1979).

Later, atherogenic potential was assessed in a lipoprotein typing system (Fredrickson and Levy, 1972) separately for each class of lipoproteins: chylomicronemia does not represent any atherogenic risk, whereas type III hyperlipoproteinemia, generally a rare disease characterized by a high concentration of intermediate-density lipoprotein enriched with apoprotein E, constitutes a definite risk. Low-density lipoproteins are endowed with the highest atherogenic potential; this fraction comprises most of the plasma cholesterol. In approximately 5% of cases of LDL hyperlipoproteinemia, the cause is a genetically determined malfunction of the peripheral receptors for LDL (Goldstein and Brown, 1975); in all other cases the disease is secondary, accompanying, for instance, hypothyroidism, adrenocortical disorders, nephrosis, and others, or is caused by other unidentified factors. Very low density lipoproteins may contribute to atherogenesis through secondary associations, since the increase in cholesterol accompanied by a concurrent elevation in triglycerides is frequent, but a specific atherogenic potential of VLDL or triglycerides has not been confirmed. Low-density lipoprotein hyperlipoproteinemia, or a combination of LDL and VLDL hyperlipoproteinemia, increases the likelihood of atherogenic risk; however, the predictive value for a given individual, as in the case of total cholesterol, is low. The discovery of an antiatherogenic potential of HDL, i.e., their binding capacity for cholesterol (Miller and Miller, 1975; Stein and Stein, 1976), increased the chance of narrowing the range of predictive confidence limits. However, the difference of 10 mg/100 ml in HDL cholesterol between men and women (45 mg/100 ml and 55 mg/100 ml, respectively) implies that the risk of CHD is twice as high in men as it is in women (Havel, 1979); consequently, the 5-10% error inherent in the method itself limits the validity of either pro- or antiatherogenic evaluation. Nevertheless, combined with an analysis of the lipoprotein pattern, the predictive value is much broader than a mere determination of plasma cholesterol, and can be applied to population groups over 50 years of age.

Lecithin : cholesterol acyltransferase is likely to be of considerable predictive value as an indicator of endogenous cholesterol kinetics, even though the participation of this enzyme in the presumed removal mechanism of cholesterol from peripheral tissues is still far from being entirely clear. Assuming that an increased concentration of cholesterol in HDL reflects a higher rate of extraction of peripheral cholesterol and its transfer to liver after esterification by means of LCAT, one would expect to find an activation of LCAT which is responsible for plasma homeostasis of the free cholesterol–esterified cholesterol system. In reality, however, LCAT is not only unrelated to the content of HDL (Stokke and Norum, 1971; Wallentin, 1978b), but even correlates inversely to HDL cholesterol (Soloff and Varma, 1978). This is also confirmed by the observation that HDL cholesterol decreases with rising relative body weight (Avogaro et al., 1978; Hulley et al., 1979) and triglyceridemia (Castelli et al., 1977), whereas the activity of LCAT increases under identical conditions (see previous sections). Moreover, HDL cholesterol need not be the exclusive source for LCAT, because of the rapid exchange of the pool of free cholesterol among lipoproteins. Further, it is not yet clear whether all the cholesterol that enters the plasma pool leaves it in the form of cholesteryl esters. Besides, hypertriglyceridemia and obesity (Havel et al., 1973) are accompanied by greater amounts of more voluminous HDL_2 particles, which are known to inhibit LCAT in vitro. Neither is the relationship between LCAT and VLDL concentration constant under all circumstances, as was shown in the preceding section. It appears, therefore, that attempting to explain the behavior of LCAT simply in terms of its relationship with lipoprotein levels would do nothing but obscure the full understanding of its function.

Thus if it can be assumed that LCAT is to a certain extent an independent factor capable of maintaining the balance of cholesterol recycling over a wide range of concentrations, as was repeatedly demonstrated and as can be seen also from the table listing the relative values pertaining to healthy subjects (Table XII), it appears that the identification of conditions under which LCAT is no longer able to assure a balanced UC/EC ratio and a normal recirculation rate of cholesterol could be considered useful for individual predictions. The notion of a relationship between LCAT and atherogenesis is obviously not new, but during the last decade the idea was repeatedly shattered by seemingly divergent findings reported in several independent studies. Soloff et al. (1973), examining 17 survivors of myocardial infarction several months after attack, demonstrated a significantly reduced fractional esterification rate and increased levels of cholesterol and triglycerides as compared to normolipemic control subjects. Ritland et al. (1975) questioned the general validity of these observations and argued that the reported changes in LCAT activity apply only to early postinfarction stages associated with reduced plasma protein levels. In their own study carried out on 10 victims of myocardial infarction, they found the LCAT activity reduced in the acute phase (after 3, 6, and 14 days), whereas immediately after the attack (22 hours) and then 7 weeks later the activity did not differ from that found in the group of normal control men. In the critical time period, the changes consisted in a reduced total plasma cholesterol level with a significantly increased UC/EC ratio and a gradual, though insignificant, elevation of triglycerides.

However, when the relative difference in FER is calculated by relating it to the appropriate control reference value, it becomes obvious that in patients recovering from myocardial infarction, the FER value is reduced by about 25% in all of the time intervals examined. Contrary to the control group, no correlation of enzyme activity was found with plasma cholesterol. Jansen et al. (1976) measured LCAT activity in 71 males who survived myocardial infarction and compared it with that of 30 controls of the same age-group. On the basis of plasma lipid values, the patients were subdivided into normolipemic, hyperlipemic, and hypercholesterolemic groups: in all three groups FER was reduced significantly and, as in the previous study, the proportion of unesterified cholesterol was increased. In 20 young (under 40 years of age) survivors of myocardial infarction, Wiklund and Gustafson (1979) found no difference in LCAT activity, FER, or the UC/EC ratio when the data were compared to those determined in cholesterol- and age-matched control groups. The only significant difference detected in the group of myocardial infarction patients was a higher relative proportion of arachidonic acid. However, in agreement with similar studies, no correlation was found with cholesterol in myocardial infarction patients, although LCAT activity and cholesterol did correlate significantly in the control group. Possible reasons for this particular finding, which differs from previously published results, might be the fact that the patients were examined after a longer time interval (1–6 years) following the cardial event, or that they were younger.

Bernstein and Bernstein (1978) reported on an investigation of the relationship between LCAT activity and coronary artery disease using coronary arteriography to classify the degree of coronary artherosclerosis: 116 patients were divided into four groups according to the extent of atheromas. Although a highly significant correlation was found between serum cholesterol and the severity of coronary artery atheroma, other parameters, such as free cholesterol, FER, and LCAT activity, did not correlate with the development of atheromatosis. In another group of 67 patients, of which at least 70% were characterized by stenosis or complete occlusion of one major artery, and 60% by a history of myocardial infarction (myocardial infarction was not mentioned in connection with the previously described group), FER was found significantly reduced when compared to the data pertaining to cholesterol-matched controls (Wallentin and Moberg, 1982). Similarly, in 33 patients with intermittent claudicaton, reduced fractional turnover was found in all cases, regardless of the plasma cholesterol level (Stříbrná et al., 1982).

In our own laboratory, we followed for a period of 5 years a group of 43 male patients recovering from myocardial infarction suffered 3 months earlier. Our first results (Dobiášová and Vondra, 1978) showed that as a group, the patients exhibited a significantly reduced FER and an increased UC/EC ratio. A more detailed analysis of the results revealed that the relationship between LCAT and the cholesterol level, known to be positively correlated among control individuals, was disturbed in the diseased subjects. If the patients were subdivided according to the presence or absence of additional risk factors (i.e., one subgroup of patients free from hypertension, obesity, and diabetes, and with cholesterol levels not exceeding 7.5 mmol/liter, and a second group with one or more risk factors present), the frequency of deviations (identified as ± the standard deviation limit) in the fractional esterification rate was about four times higher in the former group and six times higher in the latter (Fig. 6). The decrease

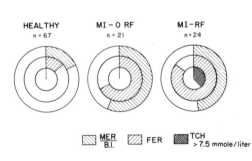

FIG. 6. Relative frequency of deviations from standard deviation limit of MER/BI and FER, and percentage of total cholesterol plasma level exceeding 7.5 mmol/liter in patients examined 3 months after myocardial infarction (MI). Patients were selected according to the presence (RF) or absence (0 RF) of additional risk factors such as obesity, diabetes, hypertension, or hyperlipoproteinemia. [Adapted from Dobiášová and Vondra (1982).]

(under the standard deviation limit) in the MER/BI index amounted in both groups to about 50%. Though in the majority of cases the fractional esterification rate was reduced, two patients displayed extremely high esterification rate values. Compared to the control level, the average values of FER were reduced by about 33 and 24%, respectively, depending on whether the cholesterol concentration exceeded 7.5 mmol/liter (Table XII). In those patients where the cardiovascular disease showed no further progression (13 subjects), the FER increased from its initial value by 17% on the average (during the 5-year period), whereas in other patients with aggravated symptoms it either did not change or further decreased.

These preliminary results of the quoted studies suggest that in CHD the LCAT function is somehow detached and is directly connected neither with the functional state of the liver nor with changes in the lipoprotein spectrum of the plasma. Reduction of the fractional esterification rate results basically from the fact that either the enzyme activity does not increase proportionally to the increase in cholesterol level, or its activity is inhibited at the normal cholesterol concentration.

4. Other Diseases

There are several reports on LCAT activity in diabetics (e.g., Yao *et al.*, 1981). In 13 nonobese, insulin-independent diabetics (9 men and 4 women, pooled data), LCAT activity was found reduced by about 20% from the control level, whereas the fractional esterification rate fell only insignificantly. The enzyme activity correlated positively with the cholesterol concentration both in the disease and in the control groups, even though the variability around the regression line was higher in the former than in the latter group. Concomitantly, an inverse correlation with the synthesis of cholesteryl linoleate was found in patients having diabetes. A reduced molar esterification rate is often related to the observed 2.5-fold increase in free fatty acids in diabetics; this may be analogous to the induction of lipoprotein lipase by means of heparin, with the consequent rising of free fatty acid levels and with a simultaneous drop in MER (Rutenberg *et al.*, 1973). Mattock *et al.* (1976) likewise reported on changes in fractional esterification rates in insulin-dependent and maturity-onset diabetes.

In our own study, we measured plasma lipids and LCAT activity in 29 male and 15 female diabetes patients (Dobiášová and Vondra, 1978). Although the molar esterification rate was significantly higher in diseased men than in control subjects, the fractional esterification rate was reduced due to the elevated cholesterol plasma levels (Table XII). The UC/EC ratio was significantly higher than in the controls, and was inversely related to the fractional esterification rate, whereas no significant correlation was found between LCAT activity and cholesterol. In women, on the other hand, LCAT activity was correlated positively to the cholesterol level with a high degree of significance, so that neither the fractional esterification rate nor the UC/EC ratio differed from the control group characterized by low cholesterol levels. This appears to reconfirm the conclusion on distinct differences in the regulation of endogenous turnover of cholesterol between men and women. With respect to the higher cholesterol levels in female diabetes patients, the data are in agreement with the results of the Framingham study (Kannel and McGee, 1979).

A significant finding was reported on the relationshp between LCAT and secondary hypolipoproteinemia (Miller and Thompson, 1973). Treatment with vitamin B_{12} of patients with pernicious anemia increased simultaneously the values for plasma cholesterol and LCAT activity by about 50%; only a slightly significant increase was noted, however, in the UC/EC ratio. Another type of secondary hypolipoproteinemia, that caused by intestinal malabsorption, was marked by reductions in both plasma cholesterol and LCAT activity (Table XII). In both of these instances of hypolipoproteinemia, a significant positive correlation, surprisingly, was found between LCAT activity and HDL cholesterol.

C. Effect of Hypolipemic Drugs

A vast number of experimental and clinical studies are concerned with testing the therapeutic efficiency of various hypolipemic drugs. The aim is to find a way of depressing pathologically elevated levels of cholesterol and triglycerides, or possibly of increasing the level of HDL cholesterol. However, only a few of the studies use these parameters to evaluate those effects that would make it possible to reach conclusions based on the recognition of changes in the kinetics of lipid metabolism. The urgency of the requirement to be able to assess, and correctly interpret, the dynamic changes induced by hypolipemic drugs is particularly evident in light of the finding that reduction of the lipid level by itself does not always have a beneficial effect (Coronary Drug Project Group, 1976).

As LCAT activity is normally roughly proportional to the amount of plasma cholesterol and to relative body weight, deviations from the normal esterification rate might suggest a possible defect. On the other hand, if the normalization of lipid levels were accompanied by a simultaneous normalization of the rate of exchange of cholesterol in the endogenous pool, the evaluation of the pharmacological effect would have a much more objective foundation. If all the new facts summarized in this article are taken into consideration, an ideally efficient hypolipemic pharmacological agent would be required to meet the following criteria: (1) it should reduce pathological levels of cholesterol (including a decrease in the ratio of LDL cholesterol to HDL cholesterol); (2) it should normalize the UC/EC ratio; (3) it should normalize FER; and (4) it should correct the production of cholesteryl esters and lysolecithin related to the body weight index. It is clear that determination of the molar esterification rate as a separate factor, without relating it to cholsterol or to relative body weight, has only restricted informative merit, in spite of the fact that the regulation of the molar esterification rate may be precisely the principal mechanism of cholesterol homeostasis. In practice, however, the interrelations of LCAT with other parameters make it possible to unify to some extent different, seemingly conflicting findings of "increased" or "decreased" MER in response to similar stimuli.

Three types of hypolipemic drugs have been examined so far in connection with their effect upon LCAT activity; they will be dealt with in turn.

1. Type Anion-Exchange Resin

The hypolipemic effect of anion-exchange resins, which bind bile acids in the intestinal lumen and interfere with cholesterol and bile acid absorption, has been known for a long time (Tennent et al., 1960; Haskins and Van Itallie, 1965). The effects of cholestyramine (Wallentin, 1978a; Miller, 1976) and colestipol (Clifton-Bligh et al., 1974; Heller et al., 1981), two similar bile acid-sequestering agents, were studied in both normo- and hyperlipemic subjects after administration of the drug for a time period ranging from several days to several months. In all published reports, therapy by such a resin resulted in a reduction of the plasma cholesterol concentration, and an increase in FER, at an unchanged UC/EC ratio. Lecithin : cholesterol acyltransferase activity (MER) was found reduced in one case, elevated in one case, and unchanged in two cases. However, when the results are analyzed in more detail, certain patterns do emerge. With the highest increase in FER, i.e., by 50%, MER was also increased (Clifton-Bligh et al., 1974); if FER rose by 20 or 30%, MER remained unchanged (Wallentin, 1978a; Miller, 1976), but if the increase in FER was only 10%, MER decreased (Heller et al., 1981). Since anion-exchange resins operate essentially by affecting the absorption of cholesterol, there are no grounds for assuming that this agent acts directly on the enzyme molecules; rather it acts through a regulatory effect on LCAT activity directly in the plasma.

2. Type Clifibrate

Clofibrate, which modifies the synthesis of cholesterol and steps up its mobilization from tissues, reduces the plasma level of triglycerides and, to a lesser extent, cholesterol (Grundy *et al.*, 1972). In a study in which he examined 20 hyperlipemic subjects, Wallentin (1978b) confirmed the reducing effect of this drug on triglycerides (decrease by 43%), as well as on free and esterified cholesterol (decrease by only 15%); in accordance with a previously reported finding, no alternation was found in the fractional esterification rate (D'Alessandro *et al.*, 1975). The molar esterification rate decreased in parallel with cholesterol. The fractional esterification rate was increased in patients with type IIb and type III hyperlipoproteinemia, in contrast to subjects having type IV hyperlipoproteinemia. The effect of fenofibrate (another derivative of clofibrate) administered for a period of 1 week to seven normolipemic subjects was described by Heller *et al.* (1981). The treatment did not affect MER, whereas FER increased due to a reduction in the cholesterol level. An interesting result was noted when fenofibrate and colestipol were combined. The cumulative effect of the two drugs resulted in a dramatic fall of the cholesterol level down to about 60%, and an increase in FER up to 180% (Table XII), without any change in MER. Simultaneously, the LDL cholesterol decreased while the HDL cholesterol increased.

3. Type Etiroxate

The effect of drugs derived from thyroid hormones on the total activation of cholesterol metabolism has been studied in detail (Starr, 1978; Abrams and Grundy, 1981). The action is manifested as an increased output of neutral steroids and an enhanced synthesis and absorption of cholesterol, as well as a reduction of plasma cholesterol. Since L-thyroxine potentiates the cardiostimulatory effect of catecholamines due to a synergetic effect, synthetic derivatives, such as D-thyroxine or etiroxate (DL-α-methylthyroxine ethyl ester hydrochloride) are preferred for hypolipemic therapy in euthyroid patients. We studied the effect of etiroxate on plasma lipid levels and LCAT activity during a 14-week test period, examining 15 patients with type II and type IV hyperlipoproteinemia who had been found refractory to other methods of treatment (Dobiášová *et al.*, 1982a). Contrary to other published results (Lageder and Irsigler, 1975), we were able to detect a transient decrease in plasma cholesterol during the first weeks of treatment, accompanied by a fall in MER (Table XII). At that time, however, FER was already significantly higher; it continued to rise with time, and by the end of the experiment attained a value almost twice as high as the initial level. The increase in MER during the later phases was attributable to LCAT activation, because the concentration of cholesterol, apart from the initial transient fall, remained practically unchanged throughout the entire remaining test period. However, the UC/EC ratio tended to decrease markedly. Compared to the control group, most of the FER values approached the ± standard deviation limit, with the cholesterol concentration almost doubled. On the other hand, the production of cholesteryl esters as related to the body weight index exceeded the standard deviation limit of the control value.

It may be asked whether the excess production of cholesteryl esters and lysolecithin in relation to body weight might not exert unfavorable effects, even though FER and the UC/EC ratio are normalized. As to the steep decrease in the cholesterol level at the beginning of the experiment, it is possible that it reflects just a transient redistribution of cholesterol between plasma and liver, something similar to what was found in a parallel experiment on rats given etiroxate (Dobiášová *et al.*, 1982b). It is not inconceivable that a similar phenomenon of a transient redistribution constitutes the underlying mechanism of action of other drugs as well (e.g., the synergetic effect of colestipol and fenofibrate mentioned earlier). Direct effects of thyroid therapy in hypothyroid patients were demonstrated by Thomas *et al.* (1981). The administration of a thyroid preparation resulted in an increase in LCAT activity by 37%, whereas total cholesterol fell by only 15%. However, antihyperthyroid therapy did not alter the LCAT activity.

On the basis of the described results, it seems possible that thyroid-like agents induce the direct activation (or production) of LCAT, whereas drugs of the anion-exchange resin and clofibrate types affect LCAT only indirectly, through their action on plasma lipoprotein patterns. However, a more conclusive statement must await additional experimental data.

VII. Closing Remarks

The preceding sections dealt more or less systematically with various aspects and open questions related to the physiological role of LCAT. In concluding this article, it seems appropriate to call attention to a few points which appear to be rather rarely taken into account in this connection. These are, in particular, the role of lysolecithin in cholesterol pathways, sex differences in the capacity to maintain the balance between the concentration of plasma cholesterol and its recycling, and, finally, the tendency to preserve a normal rate of cholesterol recirculation in relation to its plasma pool and the body mass.

Leaving aside the traditional notion that lysolecithin originating in the LCAT reaction is immediately and completely immobilized by albumin, one could associate lysolecithin with a more active function in the transfer of cholesterol. The process of interconversion of VLDL–IDL–LDL is bypassed in rat plasma, where phospholipids contain around 30% lysolecithin, and VLDL, like LDL, are preferentially directed to the liver. Since in rat plasma lysolecithin probably can take over the function of lipid transport vehicle, the presence of cholesteryl ester transfer protein is not required (Sections IV and V). If it is at all acceptable to make some comparisons between the rat model and humans, the question emerges of whether the hyper- or hypoproduction of lysolecithin in the LCAT reaction could affect in man as well the partition of plasma cholesterol (or LDL) between the liver and peripheral tissues. Assuming that lysolecithin stimulates LDL uptake in the liver, a decrease in lysophosphatidylcholine production or a disproportion between lysophosphatidylcholine and the plasma pool of cholesterol could possibly result in a reduced rate of LDL cholesterol removal from the plasma to the liver, with the consequent increase in its plasma concentration and possibly also a preferential uptake of LDL in peripheral receptors. Neither would the opposite situation be any more favorable. Though the excess lysophosphatidylcholine could accelerate cholesterol removal primarily from the plasma to the liver, it could also increase the uptake of lysophosphatidylcholine in tissues, including those of the slow pool, whereby the peripheral cell surface would be damaged due to the lytic effects of lysophosphatidylcholine. The consequence would then be a reduction in the number of peripheral receptors accompanied by a slowing down of LDL removal and an increase in their plasma levels. In addition to that, the effect of lysophosphatidylcholine on heart muscle activity must also be reconsidered.

Even though these conjectures are based on indirect evidence, they are not contradictory to the obvious tendency of a healthy organism to have the production rates of cholesteryl esters and lysolecithin correlated to both the plasma concentration of cholesterol and relative body mass, which actually indicates the size of the receptor pool. On the other hand, practically every risk disease mentioned in this article is characterized by an impaired relationship between LCAT activity and the size of its products on the one hand, and the plasma pool of cholesterol or relative body weight (or both) on the other. One may ask, however, how long an organism can cope with an overload of the products of the LCAT reaction in its struggle to balance the rate of recirculation of the plasma pool cholesterol, such as can be observed in that type of hypercholesterolemia (type IV hyperlipoproteinemia) in which a spontaneous LCAT activation is apparently possible, contrary to other primary or secondary hyperlipoproteinemias. As concerns the balance between the size of the plasma pool of cholesterol and its recirculation, it could be shown that female subjects, compared to males, are able to respond to metabolic changes with better efficiency. This finding may have some bearing on the known fact that women are more resistant than men to cardiovascular diseases. It appears that the activity of LCAT is subject to subtle regulation, possibly at the hormonal level, whereas lipoproteins, being substrats, modify the reaction rate only to a small extent.

It is obvious that a full clarification of the mechanism of endogenous regulation of LCAT requires more detailed information; nevertheless, data that are available now confirm the role of LCAT in the homeostasis of the intra- and extravascular pools of cholesterol, and assign LCAT an important position as an indicator of derangements in lipid metabolism.

References

Abrams, J. J., and Grundy, S. M. (1981). *J. Lipid Res.* **22**, 323–338.
Aftergood, L., and Alfin-Slater, R. B. (1967). *J. Lipid Res.* **8**, 126–130.
Akanuma, Y., and Glomset, J. A. (1968). *Biochem. Biophys. Res. Commun.* **32**, 639–643.
Akanuma, Y., Kuzuya, T., Hayashi, M., Ide, T., and Kuzuya, N. (1973). *Eur. J. Clin. Invest.* **3**, 136–141.
Akanuma, Y., Yokoyama, S., Imawari, M., and Itakura, H. (1978). *Scand. J. Clin. Lab. Invest.* **38** (Suppl. 150), 40–47.
Akino, T., Yamazaki, I., and Abe, M. (1972). *Tohoku J. Exp. Med.* **108**, 133–139.
Albers, J. J. (1978). *Scand. J. Clin. Lab. Invest.* **38**, 48–52.
Albers, J. J., Cabana V. G., and Stahl, Y. D. B. (1976). *Biochemistry* **15**, 1084–1087.
Albers, J. J., Lin, J. T., and Roberts, G. P. (1979). *Artery* **5**, 61–75.
Albers, J. J., Adolphson, J. L., and Chen, C.-H. (1981a). *J. Clin. Invest.* **67**, 141–148.
Albers, J. J., Chen, C.-H., and Adolphson, J. L. (1981b). *J. Lipid Res.* **22**, 1206–1213.
Albers, J. J., Gjone, E., Adolphson, J. L., Chen, C.-H., Teisberg, P., and Torsvik, H. (1981c). *Acta Med. Scand.* **20**, 455–459.
Albers, J. J., Bergelin, R. O., Adolphson, J. L., and Wahl, P. W. (1982). *Atherosclerosis* **43**, 369–379.
Alcindor, L. G., Melin, B., Benhamou, G., and Piot, M. C. (1977). *Clin. Chim. Acta* **81**, 177–182.
Alcindor, L. G., Dusser, A., Piot, M. C., Infante, R., and Polonovski, J. (1978). *Scand. J. Clin. Lab. Invest.* **38**, 12–15.
Amaducci, L., Antuono, P., Bartolini, L., De Medio, G. E., Inzitari, D., and Porcellati, G. (1978). *Neurochem. Res.* **3**, 725–731.
Anderson, D. W., Nichols, A. V., Forte, T. M., and Lindgren, F. T. (1977). *Biochim. Biophys. Acta* **493**, 55–68.
Aron, L., Jones, S., and Fielding, C. J. (1978). *J. Biol. Chem.* **253**, 7220–7226.
Assmann, G., and Fredrickson, D. S. (1974). *Proc. Int. Symp., Atheroscler., 3rd* pp. 640–651.

Assmann, G., Schmitz, G., Donath, N., and Lekim, D. (1978a). *Scand. J. Clin. Lab. Invest.* **38,** 16-20.
Assmann, G., Schmitz, G., and Heckers, H. (1978b). *Scand. J. Clin. Lab. Invest.* **38,** 98-102.
Avogaro, P., Cazzolato, G., Bittolo Bon, G., Quinci, G. B., and Chinello, M. (1978). *Atherosclerosis* **31,** 85-91.
Babický, A., Pařízek, J., Oštádalová, I., and Kolář, J. (1973). *Physiol. Bohemoslov.* **22,** 557-566.
Bagdade, J. D. (1968). *Am. J. Clin. Nutr.* **21,** 426-429.
Bagdade, J. D., and Albers, J. J. (1977). *New Engl. J. Med.* **296,** 1436.
Barter, P. J. (1974a). *J. Lipid Res.* **15,** 11-19.
Barter, P. J. (1974b). *J. Lipid Res.* **15,** 234-242.
Barter, P. J., and Jones, M. E. (1979). *Atherosclerosis* **34,** 67-74.
Barter, P. J., and Jones, M. E. (1980). *J. Lipid Res.* **21,** 238-249.
Barter, P. J., and Lally, J. I. (1978a). *Biochim. Biophys. Acta* **531,** 233-236.
Barter, P. J., and Lally, J. I. (1978b). *Atherosclerosis* **31,** 355-364.
Bartholomé, M., Niedmann, D., Wieland, H., and Seidel, D. (1981). *Biochim. Biophys. Acta* **664,** 327-334.
Beaumont, J. E., Galla, J. H., Luke, R. G., Rees, E. D., and Siegel, R. R. (1975). *Lancet* **15,** 599-601.
Bell, F. P. (1973). *Exp. Mol. Pathol.* **19,** 293-303.
Bell, F. P. (1975). *Biochim. Biophys. Acta* **398,** 18-27.
Bell, F. P. (1983). *Int. J. Biochem.* **15,** 133-136.
Bell, F. P., and Hubert, E. V. (1980). *Lipids* **15,** 811-814.
Bell, F. P., and Hubert, E. (1981). *Lipids* **16,** 815-819.
Bennett-Clark, S., and Norum, K. R. (1977). *J. Lipid Res.* **18,** 293-300.
Bernstein, M., and Bernstein, V. (1978). *Scand. J. Clin. Lab. Invest.* **38** (Suppl. 150), 124-128.
Bjornson, L. K., Gniewkowski, C., and Kayden, H. L. (1975). *J. Lipid Res.* **16,** 39-53.
Blanchette-Mackie, E. J., and Scow, R. O. (1976). *J. Lipid Res.* **17,** 57-67.
Blomhoff, J. P., Skrede, S., and Ritland, S. (1974). *Clin. Chim. Acta* **53,** 197-207.
Blomhoff, J. P., Holme, R., and Östrem, T. (1978). *Scand. J. Gastroenterol.* **13,** 693-702.
Bojensen, E. (1978). *Scand. J. Clin. Lab. Invest.* **38** (Suppl. 150), 26-31.
Bretherton, K. N., Day, A. J., and Skinner, S. L. (1976). *Atherosclerosis* **24,** 99-106.
Bron, A. F., Lloyd, J. K., Fosbrook, A. S., Winder, A. F., and Tripathi, R. C. (1975). *Lancet* **1,** 928-929.
Brunzell, J. D., Chait, A., and Bierman, E. L. (1978). *Metabolism* **27,** 1109-1127.
Carroll, K. K. (1964). *Can. J. Biochem.* **42,** 79-86.
Casaretto, A., Marchioro, T. L., Goldsmith, R., and Bagdade, J. D. (1974). *Lancet* **2,** 481-484.
Castelli, W. P., Doyle, J. T., Gordon, T., Hames, C. G., Hjortland, M. T., Hulley, S. B., Kagan, A., and Zukel, W. J. (1977). *Circulation* **55,** 767-772.
Chaje, T., and Fielding, C. J. (1978). *Proc. Natl. Acad. Sci. U.S.A.* **75,** 3445-3449.
Chajek, T., Aron, A., and Fielding, C. J. (1980). *Biochemistry* **19,** 3673-3677.
Chen, C.-H., and Albers, J. J. (1981). *Biochem. Med.* **25,** 215-226.
Chen, C.-H., and Albers, J. J. (1982a). *Biochim. Biophys. Res. Commun.* **107,** 1091-1097.
Chen, C.-H., and Albers, J. J. (1982b). *J. Lipid Res.* **23,** 680-691.
Chevalier, F., D'Holander, F., and Vaughan, M. (1971). *Biochim. Biophys. Acta* **248,** 524-529.

Chevet, D., Ramée, M. P., Le Pogam, P., Thomas, R., Garré, M., and Alcindor, L. G. (1978). *Nephron* **20**, 212–219.
Chobanian, A. V., Taal, A. R., and Brecher, P. I. (1979). *Biochemistry* **18**, 180–187.
Chong, K. S., Davidson, L., Huttash, R. G., and Lacko, A. (1981). *Arch. Biochem. Biophys.* **211**, 119–124.
Chung, J. A., Scanu, A. M., and Reman, F. (1973). *Biochim. Biophys. Acta* **296**, 116–123.
Chung, J. A., Abano, D. A., Fless, G. M., and Scanu, A. M. (1979). *J. Biol. Chem.* **254**, 7456–7464.
Cisternas, J. R. (1979). *Rev. Bras. Pesqu. Med. Biol.* **12**, 317–324.
Cisternas, J. R., Santa Rosa, C. A., Milstein Kuschnaroff, T., and Feres, A. C. (1981). *IRCS Med. Sci.* **9**, 1054–1055.
Clifton-Bligh, P., Miller, N. E., and Nestel, P. J. (1974). *Metabolism* **23**, 437–444.
Coronary Drug Project Research Group (1976). *J. Am. Med. Assoc.* **220**, 996.
Curry, M. D., Alaupovic, P., and Suenram, C. A. (1976). *Clin. Chem.* **22**, 315–322.
Curtis, J. J., Galla, J. H., Woodford, S. Y., Rees, E. D., and Luke, R. G. (1978). *Transplantation* **26**, 364–366.
Daerr, W. H., and Greten, H. (1982). *Biochim. Biophys. Acta* **710**, 128–133.
D'Alessandro, A., Zucconi, A., Bellini, F., Boncinelli, L., and Chiostri, R. (1975). *Lipids* **10**, 804–807.
David, A., Soloff, J. S. K., and Lacko, A. G. (1976). *Life Sci.* **18**, 701–706.
David, J. S. K., Fahmy, W. F., Reichle, R. M., and Richle, F. A. (1978). *Scand. J. Clin. Lab. Invest.* **38** (Suppl. 150), 142–146.
Day, E. E., and Levy, R. I. (1969). *J. Theor. Biol.* **23**, 387–399.
Demel, R. A., Kessel, W. S. M. G., and van Deenen, L. L. M. (1972a). *Biochim. Biophys. Acta* **266**, 26–40.
Demel, R. A., Bruckdorfer, K. R., and van Deenen, L. L. M. (1972b). *Biochim. Biophys. Acta* **255**, 311–320.
Dieplinger, H., and Kostner, G. (1980). *Clin. Chim. Acta* **106**, 319–324.
Dobiášová, M., and Faltová, E. (1974). *Physiol. Bohemoslov.* **23**, 479–480.
Dobiášová, M., and Faltová, E. (1976). *Int. Conf. Biochem. Probl. Lipids, 19th, Paris* Abstr. No 2A2, p. 31.
Dobiášová, M., and Linhart, J. (1969). *Lipids* **5**, 445–451.
Dobiášová, M., and Vondra, K. (1978). *Scand. J. Clin. Lab. Invest.* **38** (Suppl. 150), 129–133.
Dobiášová, M., and Vondra, K. (1982). *Czech. Med.* **5**, 16–28.
Dobiášová, M., Linhart, J., and Bíbr, B. (1971). *Physiol. Bohemoslov.* **20**, 489–497.
Dobiášová, M., Faltová, E., Marek, J., and Obenberger, J. (1975). *Atherosclerosis* **21**, 337–347.
Dobiášová, M., Kymla, J., and Faltová, E. (1976). *Atherosclerosis* **24**, 421–429.
Dobiášová, M., Vondra, K., and Karen, P. (1978a). *Čas. Lék. Česk.* **117**, 1530–1534.
Dobiášová, M., Kopecká, J., Živný, K. (1978b). *Čas. Lék. Česk.* **117**, 1394–1396.
Dobiášová, M., Kuhn, E., Djubeková, E., Vondra, K., and Brodan, V. (1979a). *Čas. Lék. Česk.* **118**, 727–729.
Dobiášová, M., Stříbrná, J., and Vondra, K. (1979b). *Čas. Lék. Česk.* **118**, 579–583.
Dobiášová, M., Vondra, K., Matoušek, V., and Válek, J. (1982a). *Atherosclerosis* **42**, 251–261.
Dobiášová, M., Vítek, V., and Kopecká, J. (1982b). *Physiol. Bohemoslov.* **31**, 323–327.
Dobiášová, M., Kristl, J., and Stříbrná, J. (1983). *Physiol. Bohemoslov.* **32** (in press).
Doi, Y., and Nishida, T. (1981). *Methods Enzymol.* **71**, 753–767.

Douset, N., and Douste-Blazy, L. (1975). *C.R. Séances Soc. Biol.* **169**, 440–443.
Douset, N., Douset, J. C., Soula, G., and Douste-Blazy, L. (1978). *Scand. J. Clin. Lab. Invest.* **38**, 21–25.
Duncan, L. E., Jr., Buck, K., and Lynch, A. (1965). *J. Atheroscler. Res.* **5**, 69–74.
Dustan, H. P. (1974). *Circulation* **50**, 871–879.
Edelstein, C., Kézdy, F. J., Scanu, A. M., and Shen, B. W. (1979). *J. Lipid Res.* **20**, 143–153.
Eder, H. A., and Steinberg, D. (1955). *J. Clin. Invest.* **34**, 932.
Eisenberg, S., and Rachmilewitz, D. (1973). *Biochim. Biophys. Acta* **326**, 378–390.
Erbland, J. F., and Marinetti, G. V. (1965). *Biochim. Biophys. Acta* **106**, 128–138.
Fabien, H. D., Davignon, J., and Marcel, Y. L. (1973). *Can. J. Biochem.* **51**, 550–555.
Felker, T. E., Fainaru, M., Hamilton, R. L., and Havel, R. J. (1977). *J. Lipid Res.* **18**, 465–473.
Fielding, C. J. (1974). *Scand. J. Clin. Lab. Invest.* **33** (Suppl. 137), 15–17.
Fielding, C. J., and Fielding, P. E. (1971). *FEBS Lett.* **15**, 355–358.
Fielding, C. J., and Fielding, P. E. (1981). *J. Biol. Chem.* **256**, 2102–2104.
Fielding, C. J., Shore, V. G., and Fielding, P. E. (1972a). *Biochem. Biophys. Res. Commun.* **46**, 1493–1498.
Fielding, C. J., Shore, V. G., and Fielding, P. E. (1972b). *Biochim. Biophys. Acta* **270**, 513–518.
Forte, T., and Nichols, A. V. (1972). *Adv. Lipid Res.* **10**, 1–38.
Fredrickson, D. S., and Levy, R. I. (1972). *In* "The Metabolic Basis of Inherited Disease" (J. B. Stanbury, J. B., Wyngaarden, and D. S. Fredrickson, eds.), p. 545. McGraw-Hill, New York.
Frohlich, J., Godolphin, J., Reeve, C. E., and Evelyn, K. A. (1978). *Scand. J. Clin. Lab. Invest.* **38** (Suppl. 150), 156–161.
Furukawa, Y., and Nishida, T. (1979). *J. Biol. Chem.* **254**, 7213–7219.
Giese, J. (1973). *Am. J. Med.* **55**, 315.
Gillett, M. P. T. (1978). *Scand. J. Clin. Lab. Invest.* **38**, (Suppl. 150), 32–39.
Gillett, M. P. T., and Silva, T. B. (1978). *Scand. J. Clin. Lab. Invest.* **38** (Suppl. 150), 118–123.
Gjone, E. (1973). *Acta Med. Scand.* **194**, 353–356.
Gjone, E., and Norum, K. R. (1968). *Acta Med. Scand.* **183**, 107–112.
Gjone, E., and Norum, K. R. (1970). *Acta Med. Scand.* **187**, 153–161.
Gjone, E., Blomhoff, J. P., and Wiencke, I. (1972a). *Scand. J. Gastroenterol.* **6**, 161–
Gjone, E., Nordby, A., Blomhoff, J. P., and Wiencke, I. (1972b). *Acta Med. Scand.* **191**, 481–484.
Gjone, E., Norum, K. R., and Glomset, J. A. (1978). *In* "The Metabolic Basis of Inherited Disease" (J. B. Stanbury, J. B. Wyngaarden, and D. S. Fredrickson, eds.), pp. 589–603. McGraw-Hill, New York.
Glomset, J. A. (1962). *Biochim. Biophys. Acta* **65**, 128–135.
Glomset, J. A. (1963). *Biochim. Biophys. Acta* **70**, 389–395.
Glomset, J. A. (1968). *J. Lipid Res.* **9**, 155–167.
Glomset, J. A. (1979). *Prog. Biochem. Pharmacol.* **15**, 41–66.
Glomset, J. A., and Norum, K. R. (1973). *Adv. Lipid Res.* **11**, 1–65.
Glomset, J. A., and Wright, J. L. (1964). *Biochim. Biophys. Acta* **89**, 266–276.
Glomset, J. A., Norum, K. R., and King, W. (1970). *J. Clin. Invest.* **49**, 1827–1837.
Glomset, J. A., Mitchell, C. D., King, W. C., Applegate, K. R., Forte, T., Norum, K. R., and Gjone, E. (1980). *Ann. N.Y. Acad. Sci.* **348**, 224–243.
Glomset, J. A., Norum, K. A., and Gjone, E. (1983). *In* "Metabolic Basis for Inherited

Disease" (J. B. Stanbury, J. B. Wyngaarden, D. S. Fredrickson, J. L. Goldstein, and M. S. Brown, eds.), 5th ed. McGraw-Hill, New York, in press.
Godin, D. V., and Frohlich, J. (1981). *Res. Commun. Chem. Pathol. Pharmacol.* **31,** 555–566.
Goldstein, J. L., and Brown, M. S. (1975). *J. Lab. Clin. Med.* **85,** 15–25.
Goodman, D. S. (1964). *J. Clin. Invest.* **43,** 2026–2036.
Groot, P. H. E., and van Tol, V. (1978). *Biochim. Biophys. Acta* **530,** 188–196.
Grundy, S. M., Ahrens, E. H., Jr., Salen, G., Schreibman, P. H., and Nestel, P. J. (1972). *J. Lipid Res.* **13,** 531–551.
Guarnieri, G. F., Moracchiello, M., Campanacci, L., Ursini, F., Ferri, L., Valente, M., and Gregolin, C. (1978). *Kidney Int.* **13,** S-26-S-30.
Gustow, E., Varma, K. G., and Soloff, L. A. (1978). *Scand. J. Clin. Lab. Invest.* **38** (Suppl. 150), 1–5.
Hamilton, R. L., Williams, M. C., Fielding, C. J., and Havel, R. J. (1976). *J. Clin. Invest.* **58,** 667–680.
Hamnström, B., Gjone, E., and Norum, K. R. (1969). *Br. Med. J.* **2,** 283–286.
Harvey, P. W., and Allen, K. G. D. (1981). *J. Nutr.* **111,** 1855–1858.
Haskins, S. A., and van Itallie, T. B. (1965). *J. Am. Med. Assoc.* **192,** 289.
Havel, R. J. (1979). *Circulation* **60,** 1–3.
Havel, R. J., Kane, J. P., and Kashyap, M. L. (1973). *J. Clin. Invest.* **52,** 32–38.
Helenius, A., and Simons, K. (1975). *Biochim. Biophys. Acta* **415,** 29–79.
Heller, F. R. (1979). *Artery* **5,** 110–116.
Heller, F. R., and Desager, J. P. (1978). *Artery* **4,** 231–238.
Heller, F. R., Desager, J. P., and Harvengt, C. (1981). *Metabolism* **30,** 67–71.
Ho, K. J., Cisternas, R., and Boyd, G. S. (1976). *Atherosclerosis* **23,** 145–153.
Hsia, J. C., Long, R. A., Hruska, F. E., and Gesser, H. D. (1972). *Biochim. Biophys. Acta* **290,** 22–31.
Hulley, S., Ashman, P., Kuller, L., Lasser, N., and Sherwin, R. (1979). *Lipids* **14,** 119–125.
Ihm, J., Harmony, J. A. K., Ellsworth, J., and Jackson, R. L. (1980). *Biochem. Biophys. Res. Commun.* **93,** 1114–1120.
Ihm, J., Ellsworth, J. L., Chataing, B., and Harmony, J. A. K. (1982). *J. Biol. Chem.* **257,** 4818–4827.
Iwamoto, A., Naito Ch., Teramoto T., Kato, H., Kako, M., Kariya, T., Shimizu, T., Oka, H., and Oda, T. (1978). *Acta Med. Scand.* **204,** 219–227.
Iwasaki, M., and Hamada, C. (1981). *Anal. Chim. Acta* **126,** 237–240.
Jackson, R. L., Sparrow, J. T., Baker, H. N., Morrisett, J. D., Taunton, O. D., and Gotto, A. M. (1974). *J. Biol. Chem.* **249,** 5308–5311.
Jackson, R. L., Cardin, A. D., Barnhart, R. L., and Johnson, J. D. (1980). *Biochim. Biophys. Acta* **619,** 408–413.
Jahani, M., and Lacko, A. G. (1981). *J. Lipid Res.* **22,** 1102–1110.
Jahani, M., and Lacko, A. G. (1982). *Biochim. Biophys. Acta* **713,** 504–511.
Janiak, M. L., Small, D. M., and Shipley, G. G. (1979). *J. Lipid Res.* **20,** 183–199.
Jansen, W., Murawski, U., and Blümchen, G. (1976). *Int. Conf. Biochem. Probl. Lipids, 19th, Paris* Abstr. No. 2A6, 33.
Jonas, A., and Matz, C. E. (1982). *Biochemistry* **21,** 6867–6872.
Jung, K., Neumann, R., Precht, K., Nugel, E., and Scholz, D. (1980). *Enzyme* **25,** 273–275.
Kannel, W. B., and McGee, D. L. (1979). *Circulation* **59,** 8–13.
Kannel, W. B., Castelli, W. P., and Gordon, T. (1979). *Ann. Intern. Med.* **90,** 85–91.

Kitabatake, K., Piran, U., Kamio, Y., Doi, Y., and Nishida, T. (1979). *Biochim. Biophys. Acta* **573**, 145–154.
Klein, P. D., and Dahl, R. M. (1961). *J. Biol. Chem.* **236**, 1658–1660.
Klopfenstein, W. E. (1969). *Biochim. Biophys. Acta* **187**, 272–274.
Kostner, G. M. (1978). *Scand. J. Clin. Lab. Invest.* **38**, (Suppl. 150), 66–71.
Kostner, G. M., and Alaupovic, P. (1972). *Biochemistry* **11**, 3419–3428.
Kuczyńska, K., and Sznajderman, M. (1980). *Pol. Arch. Med. Wewn.* **64**, 335–340.
Kudchodkar, B. J., and Sodhi, H. S. (1976). *Clin. Chim. Acta* **68**, 187–194.
Kuhn, E. (1968). *Ärztl. Praxis* **20** (67), 2913–2918.
Kunkel, H. G., and Bearn, A. G. (1954). *Proc. Soc. Exp. Biol. Med.* **86**, 887–891.
Lacko, A. G. (1976). *Clin. Biochem.* **9**, 212–215.
Lacko, A. G., Rutenberg, H. L., and Soloff, L. A. (1972). *Lipids* **7**, 426–432.
Lacko, A. G., Varma, K. G., Rutenberg, H. L., and Soloff, L. A. (1973). *Circulation* **48** (Suppl. 4), 113.
Lacko, A. G., Rutenberg, H. L., and Soloff, L. A. (1974). *Atherosclerosis* **19**, 297–305.
Lageder, H., and Irsigler, K. (1975). *Atherosclerosis* **22**, 473–484.
Lagente, M., de La Farge, F., and Valdiguie, P. (1979). *Lipids* **14**, 533–534.
Langer, W., Munder, P. G., Weltzien, H. U., and Westphal, O. (1972). *Tag. Ges. Immunol., 4th, Bern* Abstr. No. 81.
Larking, P. V., and Sutherland, W. H. F. (1977). *Atherosclerosis* **26**, 225–232.
Lau, B. W. C., and Klevay, L. (1981). *J. Nutr.* **111**, 1698–1703.
Leiss, O., Murawski, U., and Egge, H. (1978). *Scand. J. Clin. Lab. Invest.* **38** (Suppl. 150), 77–84.
Lewis, L. A., and Naito, H. K. (1978). *Clin. Chem.* **24**, 2081–2098.
Lichtenstein, A. H., Nicolosi, R. J., and Hayes, K. C. (1980). *Atherosclerosis* **37**, 603–616.
Lindner, A., Charra, E., Sherrard, D. J., Scribner, B. H. (1974). *N. Engl. J. Med.* **290**, 697–701.
Lopez, S. A., Vial, R., Balart, L., and Arroyave, G. (1974). *Atherosclerosis* **20**, 1–9.
Lucy, J. A. (1970). *Nature (London)* **227**, 815–817.
McConathy, W., and Lane, D. M. (1980). *Pediatr. Res.* **14**, 757–761.
Magnani, H. N. (1976). *Biochim. Biophys. Acta* **450**, 390–401.
Mannien, V., Mälkönen, M., Blomhoff, J. P., and Gjone, E. (1978). *Scand. J. Clin. Lab. Invest.* **38** (Suppl. 150), 147–150.
Marcel, Y. L., and Vezina, C. (1973a). *J. Biol. Chem.* **248**, 8254–8259.
Marcel, Y. L., and Vezina, C. (1973b). *Biochim. Biophys. Acta* **306**, 497–504.
Marcel, Y. L., Vezina, C., Emond, D., and Suzue, G. (1980). *Proc. Natl. Acad. Sci. U.S.A.* **77**, 2969–2973.
Marcel, Y. L., Vezina, C., Emond, D., Verdery, R. B., and Milne, R. W. (1981). *J. Lipid Res.* **22**, 1198–1205.
Marniemi, J., Dahlström, S., Kvist, M., Seppänen, A., and Hietanen, E. (1982). *Eur. J. Appl. Physiol.* **49**, 25–35.
Marsh, J. B. (1976). *J. Lipid. Res.* **17**, 85–90.
Marsh, J. B., and Kashub, E. (1970). *Fed. Proc. Fed. Am. Soc. Exp. Biol.* **29**, 673.
Mattock, M., Fuller, J., Pinney, S., Stringer, K., and Keen, H. (1976). *Eur. Soc. Clin. Invest.* **6**, 318.
Matz, C. E., and Jonas, A. (1982). *J. Biol. Chem.* **257**, 4541–4546.
Miller, C. J., and Miller, N. E. (1975). *Lancet* **1**, 16–19.
Miller, J. P. (1976). *Eur. J. Clin. Invest.* **6**, 477–479.
Miller, J. P. (1978). *Scand. J. Clin. Lab. Invest.* **38** (Suppl. 150), 138–141.
Miller, J. P., and Thompson, G. R. (1973). *Eur. J. Clin. Invest.* **3**, 401–406.

Miller, J. P., Chait, A., and Lewis, B. (1975). *Clin. Sci. Mol. Med.* **49,** 617–620.
Monger, E. A., and Nestel, P. J. (1967). *Clin. Chim. Acta* **15,** 269–273.
Morin, R. J., and Piran, U. (1981). *Clin. Chim. Acta* **111,** 211–218.
Morrisett, J. D., Jackson, R. L., and Gotto, A. M., Jr. (1977). *Biochim. Biophys. Acta* **472,** 93–133.
Nagasaki, T., and Akanuma, Y. (1977). *Clin. Chim. Acta* **75,** 371–375.
Nakagawa, M., and Nishida, T. (1973). *J. Biochem.* **74,** 1263–1266.
Nakagawa, M., and Uchiyama, M. (1974). *Biochem. Pharmacol.* **23,** 1641–1646.
Nakagawa, M., Takamura, M., and Kojima, S. (1977). *J. Biochem. (Tokyo)* **81,** 1011–1016.
Nakagawa, M., Fujimoto, Y., and Motojima, S. (1982a). *J. Biochem. (Tokyo)* **91,** 59–66.
Nakagawa, M., Motojima, S., Fujimoto, Y., Furusawa, K., Murata, K., and Kojima, S. (1982b). *Chem. Pharm. Bull.* **30,** 1884–1888.
Nakagawa, M., Kobayashi, H., Katsua, M., Takada, N., and Kojima, S. (1982c). *Chem. Pharm. Bull.* **30,** 214–218.
Nestel, P. J. (1970). *Adv. Lipid Res.* **8,** 1–39.
Nestel, P. J., and Monger, E. A. (1967). *J. Clin. Invest.* **46,** 967–974.
Nichols, A. V., and Gong, E. L. (1971). *Biochim. Biophys. Acta* **231,** 175–184.
Nichols, A. V., and Smith, L. (1965). *J. Lipid Res.* **6,** 206–210.
Nichols, A. V., Forte, T., Gong, E., Blanche, P., and Verdery, R. B. (1974). *Scand. J. Clin. Lab. Invest.* **33,** (Suppl. 137), 147–156.
Nicoll, A., Miller, N. E., and Lewis, B. (1980). *Adv. Lipid Res.* **17,** 54–96.
Nikkilä, E. A., Taskinen, M.-R., Rehunen, S., and Härkönen, M. (1978). *Metabolism* **27,** 1661–1671.
Noble, R. C., O'Kelly, J. C., and Moore, J. H. (1972). *Biochim. Biophys. Acta* **270,** 519–528.
Norbeck, H. E. (1981). *Acta Med. Scand. Suppl.* **649,** 7–42.
Nordby, G., and Norum, K. (1978). *Scand. J. Clin. Lab. Invest.* **38,** 111–114.
Nordby, G., Berg, T., Nilsson, M., and Norum, K. R. (1976). *Biochim. Biophys. Acta* **450,** 69–77.
Nordby, G., Hustvedt, B. E., and Norum, K. R. (1979). *Scand. J. Clin. Lab. Invest.* **39,** 235–240.
Norum, K. R. (1965). *Biochim. Biophys. Acta* **99,** 511–522.
Norum, K. R. (1974). *Scand. J. Clin. Lab. Invest.* **33,** 7–13.
Norum, K. R., and Gjone, E. (1967). *Scand. J. Clin. Lab. Invest.* **20,** 231–243.
Norum, K. R., Glomset, J. A., Nichols, A. V., Forte, T., Albers, J. J., King, W. C., Mitchell, C. D., Applegate, K. R., and Gong, E. L. (1975). *Scand. J. Clin. Lab. Invest.* **35,** 31–55.
Norum, K. R., Lilljeqvist, A. C., Helgerud, P., Normann, E. R., Mo, A., and Selbekk, B. (1979). *Eur. J. Clin. Invest.* **9,** 55–62.
Norum, K. R., Berg, T., and Drevon, C. A. (1982). *In* "Metabolic Risk Factors in Ischemic Cardiovascular Disease" (L. A. Carlson and B. Pernow, eds.), pp. 35–47. Raven, New York.
Osuga, T., and Portman, O. W. (1971). *Am. J. Physiol.* **220,** 735–741.
Owen, J. S., Ramalho, V., Costa, J. C. M., and Gillett, M. P. T. (1979). *Comp. Biochem. Physiol. B* **63,** 261–265.
Patsch, W., Sailer, S., and Braunsteiner, H. (1976). *J. Lipid Res.* **17,** 182–185.
Patsch, W., Lisch, H. J., Sailer, S., and Braunsteiner, H. (1978). *Eur. J. Clin. Invest.* **8,** 209–213.
Pattnaik, N. M., Montes, A., Hughes, L. B., and Zilversmit, D. B. (1978). *Biochim. Biophys. Acta* **530,** 428–438.
Phillips, G. B., and Wille, L. E. (1973). *Clin. Chim. Acta* **49,** 153–160.

Pinon, J. C., and Laudat, M. H. (1978). *Scand. J. Clin. Lab. Invest.* **38** (Suppl. 150), 85–90.
Pinon, J. C., Bridoux, A. M., and Laudat, M. H. (1980). *J. Lipid Res.* **21**, 406–414.
Piran, U., and Morin, R. J. (1979). *J. Lipid Res.* **20**, 1040–1043.
Piran, U., and Nishida, T. (1976). *J. Biochem.* **80**, 887–889.
Piran, U., and Nishida, T. (1979). *Lipids* **14**, 478–482.
Portman, O. W., and Alexander, M. (1969). *J. Lipid Res.* **10**, 158–165.
Portman, O. W., and Alexander, M. (1976). *Biochim. Biophys. Acta* **450**, 322–334.
Portman, O. W., and Illingworth, D. R. (1974). *Biochim. Biophys. Acta* **348**, 136–144.
Portman, O. W., and Sugano, M. (1964). *Arch. Biochem. Biophys.* **105**, 532–540.
Pownall, H. J., Van Winkle, W. B., Pao, Q., Rohde, M., and Gotto, A. M., Jr. (1982). *Biochim. Biophys. Acta* **713**, 494–503.
Proudfit, W. L., Shirey, E. K., and Sones, F. M. (1966). *Circulation* **33**, 901–910.
Quarfordt, S. H., and Goodman, D. S. (1969). *Biochim. Biophys. Acta* **176**, 863–872.
Ragland, J. B., Bertram, P. D., and Sabesin, S. M. (1978). *Biochem. Biophys. Res. Commun.* **80**, 81–88.
Reaven, G. M., and Bernstein, R. M. (1978). *Metabolism* **27**, 1047–1054.
Reichl, D., Simons, L. A., Myant, N. B., and Pflug, J. J. (1973). *Clin. Sci. Mol. Med.* **45**, 313–329.
Ritland, S., and Bergan, A. (1975). *Scand. J. Gastroenterol.* **10**, 17–24.
Ritland, S., Blomhoff, J. P., Enger, S. C., Skrede, S., and Gjone, E. (1975). *Scand. J. Clin. Lab. Invest.* **35**, 181–187.
Rose, H. G. (1978). *Scand. J. Clin. Lab. Invest.* **38**, 91–97.
Rose, H. G. (1981). *In* "High-Density Lipoproteins" (C. E. Day, ed.), pp. 213–80. Dekker, New York.
Rose, H. G., and Juliano, J. (1976). *J. Lab. Clin. Med.* **88**, 29–43.
Rose, H. G., and Juliano, J. (1977). *J. Lab. Clin. Med.* **89**, 524–532.
Rose, J. (1972). *Biochim. Biophys. Acta* **260**, 312–326.
Rosseneu, M., Soetewey, F., Middelhoff, G., Peeters, H., and Brown, W. V. (1976). *Biochim. Biophys. Acta* **411**, 68–80.
Rowen, R., and Martin, J. (1963). *Biochim. Biophys. Acta* **70**, 396–405.
Rubenstein, B., and Rubinstein, D. (1972). *J. Lipid. Res.* **13**, 317–324.
Russell, R. L., Soler-Argilaga, C. S., and Heimberg, M. (1976). *Life Sci.* **19**, 1347–1350.
Rutenberg, H. L., Lacko, A. G., and Soloff, L. A. (1973). *Biochim. Biophys. Acta* **326**, 419–427.
Saier, E. L., Nordstrand, E., Juves, M. W., and Hartsock, R. J. (1979). *Am. J. Clin. Pathol.* **71**, 83–87.
Scanu, A. M., Vitello, L., and Deganello, S. (1974). *Crit. Rev. Biochem.* **2**, 175–196.
Schaefer, E. J., Levy, R. I., Anderson, D. W., Brewer, H. B., Danner, R. N., and Blackwelder, W. C. (1978). *Lancet* **2**, 391.
Schmitz, G., Assmann, G., and Melnik, B. (1982). *Clin. Chim. Acta* **119**, 225–236.
Schumaker, V. N., and Adams, G. H. (1970). *J. Theor. Biol.* **26**, 89–91.
Segrest, J. P., Jackson, R. L., Morrisett, J. D., and Gotto, A. M., Jr. (1974). *FEBS Lett.* **38**, 247–253.
Sgoutas, D. S. (1972). *Biochemistry* **11**, 293–296.
Sigler, G. F., Soutar, A. K., Smith, L. C., Gotto, A. M., and Sparrow, J. T. (1976). *Proc. Natl. Acad. Sci. U.S.A.* **73**, 1422–1426.
Simko, V., and Kelley, R. E. (1979). *Am. J. Clin. Nutr.* **32**, 1376–1380.
Simon, J. B. (1979). *Yale J. Biol. Med.* **52**, 117–126.
Simon, J. B., and Boyer, J. L. (1970). *Biochim. Biophys. Acta* **218**, 549–551.
Simon, J. B., and Poon, R. W. M. (1978). *Gastroenterology* **75**, 470–473.

Simon, J. B., and Scheig, R. (1970). *New Engl. J. Med.* **283**, 153–161.
Smith, N. B., and Kuksis, A. (1980). *Can. J. Biochem.* **58**, 1286–1291.
Sniderman, A., Teng, B., Vezina, C., and Marcel, Y. L. (1978). *Atherosclerosis* **31**, 327–333.
Sobel, B. E., Corr, P. B., Robinson, A. K., Goldstein, R. A., Witkowski, F. X., and Klein, M. S. (1978). *J. Clin. Invest.* 546–553.
Sodhi, H. S., Kudchodkar, B. J., and Mason, D. T. (1980). *Adv. Lipid. Res.* **17**, 107–151.
Soler-Argilaga, C., Russell, R. L., Goh, E. H., and Heimberg, M. (1977). *Biochim. Biophys. Acta* **488**, 69–75.
Solera, M. L. (1978). *Lipids* **13**, 619–611.
Soloff, L. A., and Varma, K. G. (1978). *Scand. J. Clin. Lab. Invest.* **38**, 72–76.
Soloff, L. A., Rutenberg, H. L., and Lacko, A. G. (1973). *Am. Heart J.* **85**, 153–161.
Soutar, A. K., Pownall, H. J., Hu, A. S., and Smith, L. C. (1974). *Biochemistry* **13**, 2828–2836.
Soutar, A. K., Garner, C. W., Baker, H. N., Sparrow, J. T., Jackson, R. L., Gotto, A. M., and Smith, L. C. (1975). *Biochemistry* **14**, 3057–3064.
Soutar, A. K., Sigler, G. F., Smith, L. C., Gotto, A. M., Jr., and Sparrow, J. T. (1978). *Scand. J. Clin. Lab. Invest.* **38**, 53–58.
Starr, P. (1978). *Adv. Lipid Res.* **16**, 345–371.
Steim, J. M., Edner, O. J., and Bargoot, F. G. (1968). *Biochim. Biophys. Acta* **162**, 34–42.
Stein, O., and Stein, Y. (1976). *Biochim. Biophys. Acta* **431**, 363–368.
Stein, O., Goren, R., and Stein, Y. (1978). *Biochim. Biophys. Acta* **529**, 309–318.
Stein, O., Fainaru, M., and Stein, Y. (1979). *Biochim. Biophys. Acta* **574**, 495–504.
Steinberg, D. (1974). *Proc. Int. Symp., Atheroscler., 3rd* pp. 658–671.
Stokke, K. T. (1972). *Biochim. Biophys. Acta* **270**, 156–166.
Stokke, K. T. (1974). *Atherosclerosis* **19**, 393–406.
Stokke, K. T., and Norum, K. R. (1971). *Scand. J. Clin. Lab. Invest.* **27**, 21–27.
Stožický, F., Slabý, P., and Voleníková, L. (1982). *Acta Paediatr. Scand.* **71**, 239–241.
Stříbrná, J., Dobiášová, M., Vlček, J., Hejnal, J., and Kováč, J. (1982). *VASA* (in press).
Subbaiah, P. V., and Bagdade, J. D. (1978). *Life Sci.* **22**, 1971–1978.
Subbaiah, P. V., and Bagdade, J. D. (1979). *Biochim. Biophys. Acta* **573**, 212–217.
Subbaiah, P. V., Albers, J. J., Chen, C.-H., and Bagdade, J. D. (1980). *J. Biol. Chem.* **255**, 9275–9280.
Subbaiah, P. V., Chen, C.-H., Albers, J. J., and Bagdade, J. D. (1982). *Atherosclerosis* **45**, 181–190.
Sugano, M., and Portman, O. W. (1964). *Arch. Biochem. Biophys.* **107**, 341–351.
Sugano, M., Chinen, I., and Wada, M. (1967). *J. Biochem.* **61**, 320–327.
Sutherland, W. H. F., Temple, W. A., Nye, E. R., and Herbison, P. G. (1979). *Atherosclerosis* **34**, 319–327.
Suzue, G., Vezina, C., and Marcel, Y. L. (1980). *Can. J. Biochem.* **58**, 539–541.
Takatori, T., and Privett, O. S. (1974). *Lipids* **9**, 1018–1023.
Takatori, T., and Privett, O. S. (1978). *Endocrinology* **103**, 748–751.
Tall, A. R., and Small, D. M. (1980). *Adv. Lipid Res.* **17**, 2–44.
Tennent, D. M., Siegel, H., Zanetti, M. E., Kuron, G. W., Ott, W. H., and Wolf, F. J. (1960). *J. Lipid Res.* **1**, 469–476.
Thanabalasingham, S., Thompson, G. R., Trayner, I., Myant, N. B., and Soutar, A. K. (1980). *Eur. J. Clin. Invest.* **10**, 45–48.
Thomas, M., Charpentier, G., Duron, F., Aubert, P., Dussert, A., Alcindor, L. G., Pepin, D., Bereziat, G., and Robert, A. (1981). *Rev. Med. Interne II,* 83–89.
Utermann, G., Schoenborn, W., Langer, K. H., and Dieker, P. (1972). *Hum. Genet.* **16**, 295.
Utermann, G., Menzel, H. J., and Langer, K. H. (1974). *FEBS Lett.* **45**, 29–32.

Utermann, G., Menzel, H. J., Adler, G., Dieker, P., and Weber, W. (1980). *Eur. J. Biochem.* **107**, 225–241.
Varma, K. G., and Soloff, L. A. (1976). *Biochem. J.* **155**, 583–588.
Varma, K. G., Nowotny, A. H., and Soloff, L. A. (1977). *Biochim. Biophys. Acta* **486**, 378–384.
Varma, K. G., Soloff, L. A., and Frohlich, J. (1978). *Scand. J. Lab. Clin. Invest.* **38** (Suppl. 150), 6–11.
Vaurik, M., Priddle, W. W., and Liu, S. F. (1974). *J. Am. Gerontol. Soc.* **22**, 56–
Verdery, R. B. (1981). *Biochem. Biophys. Res. Commun.* **98**, 494–500.
Verdery, R. B., and Gatt, S. (1981). *Methods Enzymol.* **72**, 375–384.
Wallentin, L. (1977). *Atherosclerosis* **26**, 233–248.
Wallentin, L. (1978a). *Eur. J. Clin. Invest.* **8**, 383–389.
Wallentin, L. (1978b). *Atherosclerosis* **31**, 41–52.
Wallentin, L., and Moberg, B. (1982). *Atherosclerosis* **41**, 155–165.
Wallentin, L., and Vikrot, O. (1975). *Scand. J. Clin. Lab. Invest.* **35**, 661–667.
Wallentin, L., and Vikrot, O. (1976). *Scand. J. Clin. Lab. Invest.* **36**, 473–479.
Warth, M. R., Arky, R. A., and Knopp, R. H. (1975). *J. Clin. Endocrinol. Metab.* **41**, 649–655.
Webster, G. R. (1965). *Biochim. Biophys. Acta* **98**, 512–519.
Wiklund, O., and Gustafson, A. (1979). *Atherosclerosis* **33**, 1–8.
Wille, L. E., Heiberg, A., and Gjone, E. (1978). *Scand. J. Clin. Lab. Invest.* **38**, 59–65.
Williams, D. R., Barter, P. J., and Mackinnon, A. M. (1981). *Digestion* **21**, 273–278.
Wilson, D. B., Ellsworth, J. L., and Jackson, R. L. (1980). *Biochim. Biphys. Acta* **620**, 550–561.
Yamamoto, M., Tanaka, Y., and Sugano, M. (1979a). *Comp. Biochem. Physiol. B* **62**, 185–193.
Yamamoto, M., Tanaka, Y., and Sugano, M. (1979b). *Comp. Biochem. Physiol. B* **63**, 441–449.
Yao, J. K., and Dyck, P. J. (1977). *Clin. Chem.* **23**, 447–453.
Yao, J. K., Chang, S. C. S., Ryan, R. J., and Dyck, J. (1980). *Biochem. Biophys. Res. Commun.* **95**, 738–744.
Yao, J. K., Palumbo, P. J., and Dyck, P. J. (1981). *Artery* **9**, 262–274.
Yashiro, M., and Kimura, S. (1980). *J. Nutr. Sci. Vitaminol.* **26**, 59–69.
Yokoyama, S., Takajama, M., and Akanuma, Y. (1977). *J. Biochem.* **81**, 1227–1230.
Yokoyama, S., Murase, T., and Akanuma, Y. (1978). *Biochim. Biophys. Acta* **530**, 258–266.
Yokoyama, S., Fukushima, D., Kupferberg, J. P., Kézdy, F. J., and Kaiser, E. T. (1980). *J. Biol. Chem.* **255**, 7333–7339.
Zierenberg, O., Odenthal, J., and Betzing, H. (1979). *Atherosclerosis* **34**, 259–276.
Zilversmit, D. B., Hughes, L. B., and Balmer, J. (1975). *Biochim. Biophys. Acta* **409**, 393–398.
Zvezdina, N. D., Prokasova, N. V., Vaver, V. A., Bergelson, L. D., and Turpaev, T. M. (1978). *Biochem. Pharmacol.* **27**, 2793–2801.

Nicotinic Acid and Its Derivatives: A Short Survey

W. HOTZ

Research and Development Division
Merz and Company, GmbH
Frankfurt-Am-Main, Federal Republic of Germany

I. Introduction 195
II. Pharmacology 196
    A. Pharmacokinetics 196
    B. Pharmacodynamics 203
III. Mechanism of Action 207
IV. Toxicology and Side Effects 210
V. Summary 212
    References 213

I. Introduction

Nicotinic acid (pyridine-3-carboxylic acid, niacin) belongs to the group of water-soluble B vitamins. The human organism has a daily nicotinic acid requirement of approximately 15 mg (Dietary Allowances Committee of the National Research Council). A certain degree of substitution is possible by tryptophan in the diet, since nicotinic acid is produced in its route of degradation, but as a rule the above requirement of 15 mg only fails to be reached when a diet is predominantly free of nicotinic acid (e.g., maize). At the beginning of this century it was discovered that pellagra, a disease with symptoms affecting the skin, gastrointestinal tract, and the central nervous system, is caused by a nicotinic acid deficiency. In 1913, Funk became the first person to isolate nicotinic acid from foodstuffs.

After its rapid amidation in the liver, nicotinic acid buildup to NAD (nicotinamide adenine dinucleotide) or NADP (nicotinamide adenine dinucleotide phosphate) takes place via a number of steps (Preiss and Handler, 1957). These two compounds are physiologically active forms of nicotinic acid and act as hydrogen-transferring coenzymes of most dehydrogenases.

The pharmacological effects of nicotinic acid are only manifested at high doses. Altschul *et al.* (1955) were the first to demonstrate the lowering action of nicotinic acid, after oral administration at a dose of 3 g/day, on the plasma cholesterol concentration, and in 1962 Carlson and Orö reported the same for the concentration of nonesterified fatty acids and triglycerides in the plasma. This lowering of the plasma lipid concentration is independent of the action of the substance as a vitamin and is not caused by nicotinamide. Other special pharmacological properties are vasodilation and activation of fibrinolysis. A detailed summary of the pharmacological and metabolic characteristics of the substance was published by Gey and Carlson (1971).

The commonly occurring side effects, such as flushes and pruritus, due in particular to the high dose of the substance needed to produce the desired pharmacological effect, for a time brought into question the use of nicotinic acid as a lipid reducer. However, the development of chemically retarding forms, largely by esterification with various polyhydric alcohols, has led to a marked reduction both in the dose needed and in the incidence of the side effects. In addition, in some cases it has been possible to incorporate via the ester component molecular variants (e.g., clofibric acid) with pharmacologically synergistic activity, and thus to improve the spectrum of the lipid-reducing effect. Moreover, the properties of reducing certain atherogenic fractions of the plasma lipids and elevating the fraction of vasoprotective lipids have resulted in new spheres of therapeutic application of nicotinic acid and its derivatives, particularly for the prevention and possibly also for the remission of atherosclerotic processes.

Although the exact mechanism of action of nicotinic acid is still unknown, the effect of the substance on the prostaglandin system discovered a few years ago seems to permit a connection between the various pharmacological effects and the side effects (Carlson, 1978; Vincent and Zijlstra, 1978).

II. Pharmacology

A. Pharmacokinetics

After oral administration, nicotinic acid is rapidly and completely absorbed (Carlson et al., 1968; Svedmyr, and Harthon, 1970). The maximum plasma concentration is reached after some 30–60 minutes; after an oral administration of 1 g of nicotinic acid the plasma concentration in man increases from the normal 0.04 μg/ml to 9.4–38.5 μg/ml (Carlson et al., 1968). In whole blood the content can be 20–100 times higher than in the serum (Frank et al., 1963; Gaut and Solomon, 1970), since metabolites less capable of diffusion, such as NAD, concentrate in the erythrocytes and plateletes.

The plasma half-life of nicotinic acid after pharmacological doses is determined largely by the rapid renal excretion and, at about 1 hour, is relatively short (Petrack et al., 1966). According to Miller et al. (1960), in humans 88% of an oral dose of 3 g is recovered in the urine. In tissue investigations using radiolabeling, no free nicotinic acid could be detected either in the liver or in the fat tissue of rats already 6 hours after subcutaneous (sc) injection of 250 mg/kg (Carlson and Nye, 1966).

In connection with this rapid biotransformation, high doses are needed if constant plasma concentrations are to be sustained over a prolonged period of time. According to various studies, plasma concentrations of only 0.5–2 µg/ml of free nicotinic acid are sufficient to achieve maximum pharmacological effects (Carlson and Orö, 1962; Ekström-Jodal et al., 1970; Svedmyr and Harthon, 1970), provided that they are maintained over a prolonged period of time. Apparently the constancy of a pharmacologically active concentration is more important for the lipid-lowering action than particularly high plasma levels (Svedmyr and Harthon, 1970).

The derivatives of nicotinic acid, mainly esters with polyhydric alcohols, give rise to prolonged constant plasma levels of the acid as a result of gradual hydrolysis and release of the free acid, and thus in particular reduce the incidence of flush. The rebound phenomenon, a strong increase in the plasma lipids on discontinuation of the therapy, is also reduced. Moreover, some of the derivatization components apparently exhibit synergistic pharmacological properties (Benzce, 1975). Table I gives a survey of the various nicotinic acid derivatives. In the case of Complamin (xanthinol nicotinate), the nicotinic acid is only the anion of the xanthinol base. In spite of this, the pharmacological effects observed after the administration of equimolar quantities of the individual components are slightly different and less pronounced (Brenner and Brenner, 1972).

Nicotinic acid is slowly released from its esters with polyhydric alcohols such as niceritrol (pentaerythritol tetranicotinate) (Brattsand and Harthon, 1971; Harthon and Sigroth, 1974), hexanicit (meso-inositol tetranicotinate) (Harthon and Brattsand, 1979), bradilan (tetranicotinoyl fructose) (Salmi and Frey, 1974), sorbinicate (nicotinic acid hexaester of sorbitol) (Avogaro et al., 1977, 1978; Subissi et al., 1980), and nicotinic acid linked to polymeric carbohydrates (Puglisi et al., 1976, 1979). A maximum plasma concentration is achieved only after 2–4 hours. Apparently these derivatives are absorbed in the gastrointestinal tract in the unhydrolyzed form. Cleavage into the nicotinic acid and the alcohol components takes place mainly in the blood and in various tissues by esterases only after absorption (Subissi et al., 1980; Svedmyr et al., 1969). At comparable doses, the plasma concentration is about half that of after the

Table I
NICOTINIC ACID AND ITS DERIVATIVES

Chemical structure and formula[a]	INN (proprietary name)	Human dosage (g/die)	Molecular weight
$C_6H_5NO_2$	Nicotinic acid (Niconacid®)	3–15 (Carlson, 1978)	123.1
$C_{13}H_{22}N_5O_4 \cdot C_6H_4NO_2$	Xanthinol-nicotinate (Complamin®)	0.5–1.0 (Cultrera et al., 1971)	434.5
$C_{29}H_{24}N_4O_8$	Niceritrol (PETN) (Pericyt®)	3.0 (Olson et al., 1974)	556.5
$C_{42}H_{30}N_6O_{12}$	Mesoinositol-tetranicotinate (MIHN) (Hexanicit®)	1.0 (Harthorn and Brattsand, 1979)	810.7
$C_{30}H_{24}N_4O_{10}$	Nicofuranose (Bradilan®)	1.5 (Benaim and Dewar, 1975) 0.5 (Salmi and Frey, 1974)	600.5
$C_{42}H_{32}N_6O_{12}$	Sorbinicate	1.6 (Avogaro et al., 1978)	812.7

(continued)

Nicotinic Acid and Derivatives

Table I (*continued*)

Chemical structure and formula	INN (proprietary name)	Human dosage (g/die)	Molecular weight
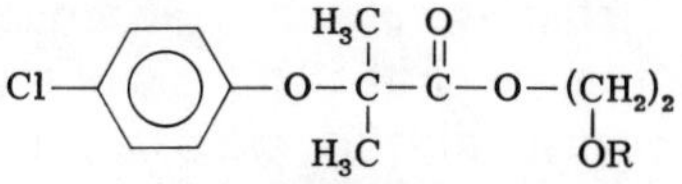 Cl—⟨C₆H₄⟩—O—C(CH₃)₂—C(=O)—O—(CH₂)₂—OR $C_{18}H_{18}ClNO_5$	Etofibrate (Lipo Merz®)	0.5–0.9 (Mertz et al., 1982; Spöttl and Froschauer, 1976)	363.65
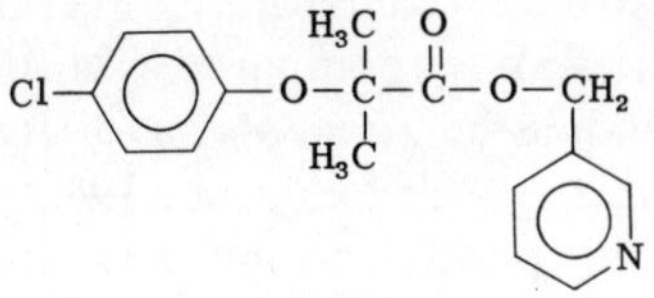 Cl—⟨C₆H₄⟩—O—C(CH₃)₂—C(=O)—O—CH₂—⟨pyridyl⟩ $C_{16}H_{16}ClNO_3$	Nicofibrate (Arterium®)	0.8–1.2 (Marmo et al., 1971)	342.2
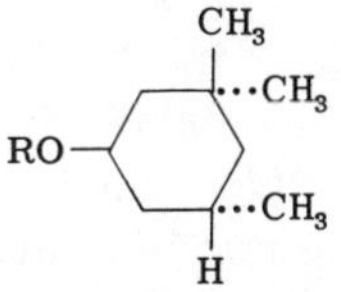 RO—⟨cyclohexyl with 3 CH₃⟩—H $C_{15}H_{21}NO_2$	Ciclonicate (Cortofludan®)	0.5–1.5 (Rimondi et al., 1980)	247.9
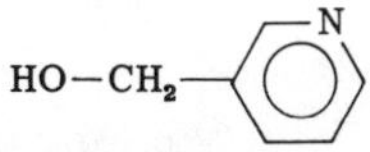 HO—CH₂—⟨pyridyl-N⟩ C_6H_7NO	β-Pyridylcarbinol (Ronicol®)	1.5–2.0 (Zöllner and Gudenzi, 1966; Zöllner and Wolfram, 1970)	109.1

[a] $R = -C(=O)-⟨pyridyl-N⟩$

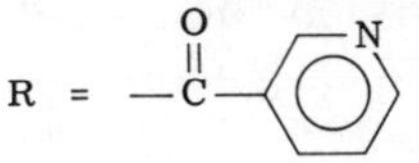

ingestion of free nicotinic acid (Harthon and Brattsand, 1979). In accordance with this difference in pharmacokinetics, after the administration of niceritrol, for example, only about 30% of the administered dose is excreted within the first 24 hours in metabolite form (Harthon and Sigroth, 1974). The elimination half-life is thus correspondingly prolonged. As a result of these special pharmacokinetics, a relatively low dose is sufficient to produce the complete pharmacological effect (Brattsand, 1976; Kruse et al., 1979; Puglisi et al., 1976, 1979).

Etofibrate is the double ester of nicotinic and clofibric acids with ethylene glycol; during biohydrolysis the corresponding half-esters of ethylene glycol appear predominantly as the primary metabolites (Garrett and Gardner, 1982; Kummer et al., 1979). At cellular pH the half-ester of nicotinic acid in particular exhibits an elevated esterase stability (Oelschlager et al., 1980), as a result of which the half-life of the nicotinic acid is increased. Obviously, because of this difference in pharmacokinetics, the metabolization too is different. For example, N-methylnicotinamide, a qualitatively important metabolite of nicotinic acid in the rat, is not found after the administration of etofibrate. Overall, this results in an increased pharmacological activity compared with the effects of equimolar quantities of the individual components or of free nicotinic acid (Ortega et al., 1980; Priego et al., 1979; Schatton, 1982).

The very similar compound obtained from β-pyridylcarbinol and clofibric acid (nicofibrate) likewise does not seem to be a pure prodrug. Increased activity compared with equimolar quantities of the individual components was measured in rats, dogs, cats, and chickens (Marmo et al., 1971). In the case of ciclonicate the steric hindrance of the methyl groups probably inhibits the attack by esterases and thus delays hydrolysis (Rimondi et al., 1980; Turba et al., 1980). β-Pyridylcarbinol acts as a precursor of nicotinic acid. It is rapidly oxidized to nicotinic acid in the liver, and the latter substance is then responsible for the pharmacological effects (Zöllner and Wolfram, 1970).

Metabolism

The biotransformation of nicotinic acid depends largely on the ingested dose. Under normal conditions the principal metabolites in human urine are N-methylnicotinamide and N-methyl-2- or 4-pyridone carboxamide (Chang and Johnson, 1961b; Fumagalli, 1971). The 4-pyridone derivative evidently originates mainly from nucleotide catabolism (Chang and Johnson, 1959, 1961a). After the administration of pharmacologically active doses it is the fractions of unmetabolized nicotinic and nicotinuric acid excreted in the urine that are particularly elevated (Table II). In addition, the number of metabolites that can be detected increases. The excretion of nonbiotransformed nicotinic acid, however, only increases when the nicotinuric acid production capacity in the liver and kidneys is exceeded (Petrack et al., 1966).

Table II gives a synopsis of the quantitative distribution of the individual metabolites in man and in rats after the administration of various

Table II
METABOLIC PATTERN OF NICOTINIC ACID AFTER THE ADMINISTRATION OF PHARMACOLOGICAL DOSES IN HUMANS AND RATS[a]

Species	Dose (mg/kg)	Nicotinic acid	Nicotinamide	Nicotinuric acid	N-Methyl-nicotinamide	N-Methyl-2-pyridone-5-carboxamide	N-Methyl-4-pyridone-5-carboxamide	Nicotin-amide-N-oxide
Humans[b] (70 kg)	~15	—	2	13	17	26	5	5
	~43	16	—	25	18	28	5	1
Rats[c]	5	6.3	1.0	26.6	14.8	0.25	0.25	0.5
	500	60.9	1.7	19.5	2.8	0.10	0.10	1.8

[a] As percentage of the administered dose.
[b] Modified from Mrochek et al. (1976).
[c] From Lee et al. (1972).

pharmacological doses. In man the quantity of unmetabolized nicotinic acid varies, in particular, depending on the dose. The concentrations of the other metabolites remain about the same. Figure 1 gives a summary of the metabolism of nicotinic acid. After rapid amidation in the liver, the nicotinamide that is produced is N-methylated with the participation of ATP and Mg^{2+}, and oxidized to the 2- or 4-pyridone derivative by aldehyde dehydrogenase. In addition, *N*-ribosyl-2-pyridone-5-carboxamide

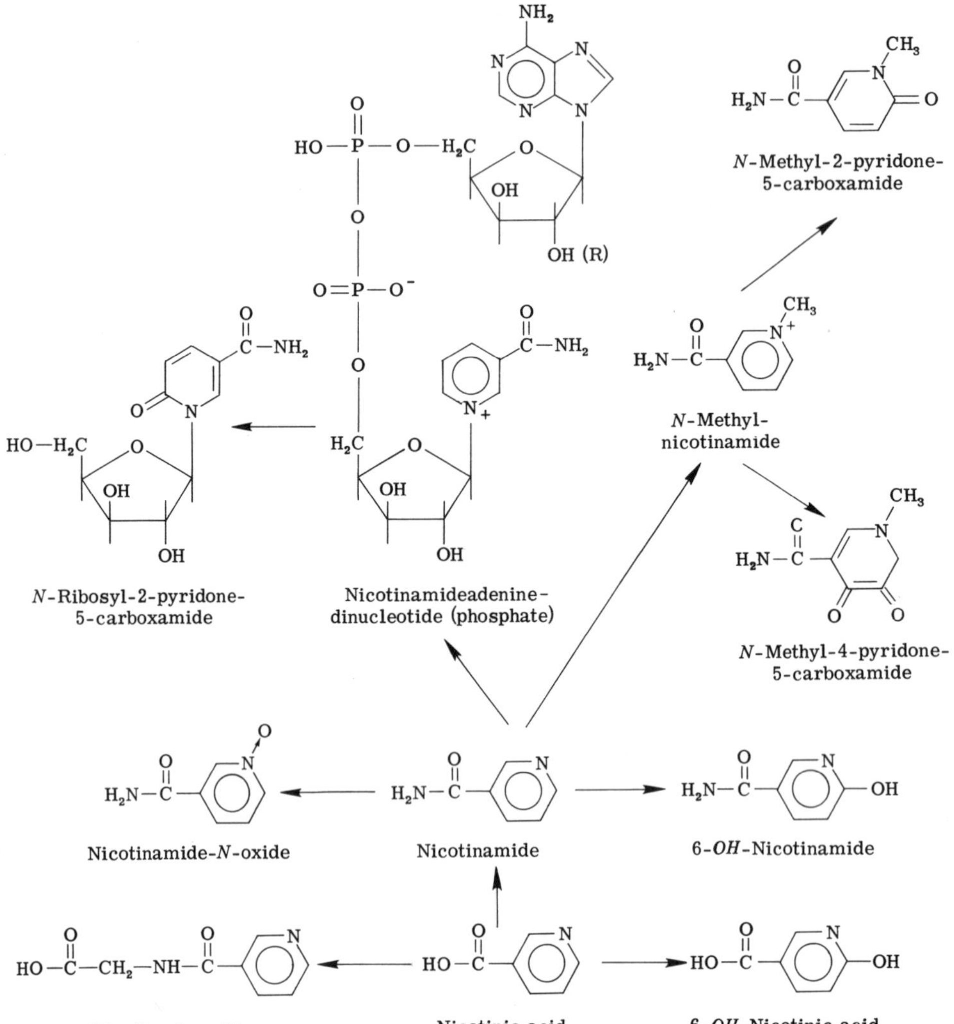

FIG. 1. A summary of the metabolism of nicotinic acid. See text for details.

is produced from the nucleotide metabolism as a degradation product (Mrochek et al., 1976).

In chickens and rats, traces of, respectively, β-nicotinylornithine (Chang and Johnson, 1959) and β-nicotinyl-D-glucuronide (Eyes et al., 1955) can also be found. In mice nicotinamide-N-oxide is a particularly important metabolite (Chaykin et al., 1965). Interspecies variation is also found in the case of the 2-pyridone derivative, which is a quantitatively important metabolite in man but hardly occurs in dogs, cats, sheep, and rats (Brown and Price, 1965; Chang and Johnson, 1962).

B. PHARMACODYNAMICS

1. Effects on Lipid Metabolism

The action of nicotinic acid on lipid metabolism was first described in 1955 by Altschul et al. (1955). At an oral nicotinic acid dose of 3 g/day, these authors observed a significant reduction of the plasma cholesterol concentration in man.

Cholesterol can either be taken in with food or synthesized in the liver. A reduced absorption of dietary cholesterol under the influence of nicotinic acid can be largely excluded (Kritchevsky et al., 1971; Kudchodkar et al., 1978; Miettinen, 1971), and animal experiments and in vitro investigations (Gamble and Wright, 1961; Miettinen, 1971; Priego et al., 1979), as well as clinical studies on hyperlipemic patients (Grundy et al., 1981; Moutafis and Myant, 1971; Parsons, 1971), indicate that an inhibition of the endogenous cholesterol synthesis in the liver is taking place. Moreover, after the administration of nicotinic acid to rats intensified catabolic effects, such as increased cholesterol oxidation (Kritchevsky et al., 1971) and increased excretion of cholesterol in the feces (Miettinen, 1971), are observed. In clinical studies, Charman et al. (1972) measured a mean cholesterol reduction of 26% in 160 hyperlipemic patients, while in the Coronary Drug Project (1975), a broadly based 5-year study carried out in the United States and involving some 8000 patients, a mean cholesterol reduction of some 10% was determined under the influence of nicotinic acid.

In 1962 Carlson and Orö discovered that the oral administration of 200 mg of nicotinic acid to fasting subjects likewise effects a rapid reduction of the plasma concentration of nonesterified fatty acids (NEFA). This is evidently due to an inhibition of the lipolysis of fat tissue (Carlson and Orö, 1962; Carlson et al., 1963), only nicotinic acid or its glycine conjugate nicotinuric acid exerting an antilipolytic influence; nicotinamide is inactive (Carlson and Orö, 1962). In consequence of the increased lipolysis during the night, over the entire 24-hour period the NEFA concentration is seemingly not reduced by nicotinic acid. However, this may be due to lower plasma concentrations and can be prevented by regular infusions of nicotinic acid (Schlierf and Hess, 1977). On the other hand, during the night there is no clear correlation between nicotinic acid concentration and lipolysis inhibition (Kruse *et al.*, 1979). The reasons for this are largely unclear. It is possible that a diurnal rhythm of the participating enzyme is responsible for the phenomenon, as is known to be the case for the activity of HMG-CoA reductase, the key enzyme in cholesterol biosynthesis (Edwards and Gould, 1974).

As a result of the smaller influx of NEFA to the liver from fat tissue under the influence of nicotinic acid, hepatic triglyceride synthesis can be reduced, and thus the very low density lipoprotein (VLDL) production rate can be slowed down, the latter being one of the triglyceride-richest lipoprotein fractions in the plasma (Carlson and Orö, 1962; Kissebah *et al.*, 1974; Magide *et al.*, 1975; Rudermann *et al.*, 1968). Grundy *et al.* (1981) were correspondingly able to measure a smaller proportion of VLDL triglycerides with an otherwise unchanged quantity of VLDL. A reduced availability of NEFA can also reduce chylomicron production in the intestine (Marsh, 1971).

Within the framework of lipoprotein catabolism, low density lipoproteins (LDL) are produced from VLDL synthesized in the liver. The plasma concentration of LDL is almost exclusively determined by the degradation of VLDL (Carlson, 1978; Eisenberg and Levy, 1975), and some 80% of the cholesterol detected in the serum is present in LDL. This precursor–product relationship between VLDL and LDL is thus the reason why under the influence of nicotinic acid a reduced plasma triglyceride concentration is observed first, followed after a few days by a fall in cholesterol concentration (Carlson *et al.*, 1968; Eisenberg *et al.*, 1973; Gitlin *et al.*, 1958).

In contrast to this, in the course of nicotinic acid therapy the concentration of high density lipoproteins (HDL) in the plasma increases (Blum *et al.*, 1977). According to Nikkilä (1978), an increase in HDL seems in general to correlate with an increased activity of lipoprotein lipase, the most important enzyme in lipoprotein catabolism. Under the influence of nicotinic acid, NEFA, whose production is also increased as a result of elevated lipoprotein lipase activity, are reabsorbed into the fat tissue (Carlson, 1978).

In general, a protective action is attributed to HDL in the development of atherosclerosis (Barr *et al.*, 1951), since they can carry cholesterol away from the vascular wall. However, with elevated lipoprotein lipase activity and the associated increased mobilization of cholesterol from the tissues, lithogenicity can also be increased (Schwartz et al., 1978), the cholesterol crystallizing out in the bile. According to more recent findings, an elevated chylomicron turnover, such as is caused by an increased activity of lipoprotein lipase under the influence of nicotinic acid (Nikkilä, 1971; Otway et al., 1971; Shepherd et al., 1979), apparently makes available more proteins for HDL synthesis. The turnover of apolipoproteins AI and AII is particularly increased (Assmann, 1980), and these make up the bulk of the protein component in HDL. Blum et al. (1977) were thus able to detect a higher apoprotein AI/AII coefficient in HLP after treatment with nicotinic acid. Apolipoprotein AI apparently stimulates the enzyme lecithin:cholesterol acyltransferase (LCAT), which removes tissue cholesterol by esterification (Fielding et al., 1972; Glomset et al., 1966).

Thus, Shephard et al. (1979) measured a strong increase of the HDL_2/HDL_3 coefficient under nicotinic acid therapy, especially due to a reduction of the HDL_3 concentration and an increase of the HDL_2 concentration, whereas Packard et al. (1980) observed this effect only in patients with type IIa hyperlipoproteinemia and did not find any change in the HDL concentration in type IIb patients. Together with these changes in the apoproteins, different compositions were found within the lipoprotein fractions corresponding to a changed lipoprotein catabolism caused by the nicotinic acid (Shepherd et al., 1979).

Overall, the influence of nicotinic acid on the concentration of nonesterified fatty acids in the plasma seems to be less reliable than the lowering of the plasma cholesterol. For this reason nicotinic acid tends as a rule to be used for the treatment of elevated cholesterol concentrations, although in clinical studies reductions of NEFA by more than 60% of the initial values have also been reported (Grundy et al., 1981; Parsons, 1971).

2. Effects on Atherosclerotic Processes

Elevated cholesterol and possibly also elevated triglyceride concentrations in the serum can be among the risk factors relevant for the appearance of atherosclerosis (McGill, 1979; Perrin et al., 1968; Roberts et al., 1973). In particular, the cholesterol contained in LDL is regarded as an atherogenic factor in the narrower sense of the term, whereas an increased HDL cholesterol concentration is thought to have a certain protective action (Barr et al., 1951; Gordon et al., 1977; Miller, 1980). The reduction of the plasma lipid concentration by nicotinic acid and its derivatives can thus have a beneficial influence on the development of atherosclerotic processes. Lipid infiltration into the aorta could be reduced in rabbits by prophylactic administration of nicotinic acid, niceritrol, xanthinol nicotinate, and β-pyridylcarbinol (Brattsand, 1975;

Brattsand and Lundholm, 1971; Parwaresch et al., 1978). Parwaresch et al. (1978) also observed a smaller number of arterial lesions by planimetric calculations in experimentally atherosclerotic rabbits. Likewise, under the influence of niceritrol and β-pyridylcarbinol the deposits of free and esterified cholesterol in the intima of minipigs are reduced (Lundholm et al., 1978).

An HDL concentration increased by nicotinic acid is evidently responsible for the mobilization and removal of cholesterol from the vascular walls (Glomset et al., 1966). This can be attributed to the fact that protein components from the HDL fraction (apoA-I) stimulate LCAT, as a result of which increased free cholesterol is esterified with lecithin (Glomset et al., 1966). Since these effects could also be observed without a simultaneous lowering of the plasma cholesterol concentration (Brattsand, 1976), a direct effect on the metabolism of the vascular walls is likewise assumed. This may be attributed to an inhibition of the proliferation of myocytes, a process normally involved in atherosclerotic plaque formation (Brattsand, 1976). On the other hand, in atherosclerosis the NAD content of the vascular wall can be reduced (Kirk, 1968), whereas nicotinic acid can increase the tissue NAD content in animal experiments (Harthon et al., 1971).

According to more recent work, when rabbits with experimentally induced atherosclerosis are fed a cholesterol diet, the formation of prostacyclin (PGI_2) by the endothelium of the vascular wall is reduced (D'Angelo et al., 1978; Dembinska-Kiec et al., 1977). Nicotinic acid stimulates the formation of PGI_2 (Puglisi, 1982) and can thus inhibit the progression of the atherosclerotic processes. Paoletti (1982) was able to show that under the influence of the nicotinic acid derivative etofibrate as well as the prostacyclin concentration in the arterial endothelium is significantly increased, an effect that is maintained over a fairly long period of time. It is concluded that some antithrombotic and antisclerotic effects of β-pyridylcarbinol in patients also are due to the release of PGI_2 (Grodzinska et al., 1982).

However, less research has been done on whether, in addition to the preventive activity of nicotinic acid and its derivatives on atherosclerotic processes, a remission of already existing atherosclerotic lesions can be achieved. Weitzel et al. (1962) reported that nicotinic acid reduced the cholesterol content in the aortas of older, spontaneously atherosclerotic hens, while Brattsand was unable to observe any remission of atherosclerotic alterations under the influence of niceritrol in animal experiments (Brattsand, 1976). A therapeutic effect on atherogenic processes is likewise exerted by the ability of nicotinic acid to inhibit platelet aggregation and to activate fibrinolysis (Lakin et al., 1980; de Nicola et al.,

1963; Shestakov et al., 1977). In particular, the risk of vascular occlusion is increased by an increased platelet aggregation tendency as a consequence of reduced PGI_2 production (D'Angelo et al., 1978; Gryglewski et al., 1978).

Nicotinic acid and its derivatives such as xanthinol nicotinate apparently intervene selectively in prostaglandin metabolism (Lakin et al., 1980) and inhibit the aggregation tendency both by a reduction of the synthesis of thromboxane A_2 (Ortega et al., 1980; Vincent and Zijlstra, 1978) and by a stimulation of prostacyclin (Paoletti, 1982; Puglisi, 1982). General inhibitors such as acetylsalicylcic acid, on the other hand, also bring about a reduction of antiaggregating factors such as prostacyclin by inhibiting the cyclooxygenase system (Lakin et al., 1980; Ortega et al., 1980).

In comparison with nicotinic acid, under the influence of etofibrate the inhibition of platelet aggregation is actually intensified (Ortega et al., 1980). A reduction of the fibrinogen concentration has been observed both after the administration of nicofuranose to elderly atherosclerotic patients (Benaim and Dewar, 1975) and after the administration of xanthinol nicotinate and etofibrate (Cultrera et al., 1971; Spöttl and Froschauer, 1976). Beneficial effects were thereby observed in the rheological parameters of the blood.

It may well be that all these effects were responsible for the 29% reduction in nonfatal myocardial infarctions after the administration of nicotinic acid within the framework of the Coronary Drug Project (1975). Likewise, in the Stockholm study the risk factors for ischemic heart disease were significantly reduced (Rosenhamer and Carlson, 1980) under nicotinic acid therapy.

III. Mechanism of Action

Nicotinic acid and its derivatives reduce the concentration of NEFA in the plasma *in vitro* and *in vivo* largely by inhibiting the lipolysis of fat tissue, whereas nicotinamide is inactive in this respect (Carlson and Orö, 1962). As a consequence of the antilipolytic activity it is possible, for example, to measure a lower content of free glycerol in the plasma (Carlson et al., 1963; Yu-Yan-Yeh, 1975), and owing to the reduced supply of NEFA the production of ketone bodies in the liver is similarly reduced (Yu-Yan-Yeh, 1975). Butcher et al. (1968) were the first to show that the cAMP concentration in the fat tissue of rats is reduced under the influence of nicotinic acid, which has subsequently also been confirmed for many other species (Aktories et al., 1980a,b; Burns et al., 1972; Fain, 1973). This is evidently due mainly to an inhibition of adenylate cyclase (Aktories et al., 1980b; Altschul et al., 1955; Kather and Simon, 1979; Moreno et al., 1979). The stimulation of phosphodiesterase in fat cells, likewise to be held responsible for the reduction of the cAMP concentration and observed by Krishna et al. (1966), has not, however, been confirmed by other researchers (Chmelar and Chmelarova, 1971). This mechanism is seemingly of importance in platelets (Abdulah, 1969) and may produce the previously described reduction of the release of thromboxane A_2 (Vincent and Zijlstra, 1978).

The direct consequence of the reduced cAMP concentration in the fat tissue is a reduction of the activity of the hormone-sensitive triglyceride lipase. This is associated with a reduced enzymatic cleavage of triglycerides into fatty acids, whose plasma concentration thus decreases to a considerable extent. The effect of the NAD is still unclear. Moreno et al. (1979) were able to observe an inhibitory influence of NAD on adenylate cyclase, but this has not been confirmed by other authors (Aktories et al., 1980a,b). The increased release of prostaglandins (PG) by nicotinic acid and its derivatives may be involved in the inhibition of lipolysis (Burns et al., 1972). Nicotinic acid gives rise in particular to an increased release of PGE_1, which in turn can reduce the concentration of cAMP in isolated hamster fat cells (Aktories et al., 1980a,b).

In addition to the influence on adenylate cyclase, the activity of lipoprotein lipase is increased (Priego et al., 1979; Shafir and Biale, 1971), as a result of which the triglycerides of lipoproteins and chylomicrons are hydrolyzed to an increased extent and the NEFA thus produced are absorbed into the fat tissue (Carlson, 1978). An increased activity of glucose-6-phosphate dehydrogenase is a further effect on isolated enzymes that has been observed in rats (Priego et al., 1979).

The influence of nicotinic acid on serum cholesterol has a bearing on endogenous biosynthesis. Thus, Kudchodkar et al. (1978) recorded a reduced incorporation of radioactive acetate in mevalonate. This inhibitory effect apparently takes place mainly at the stage of HMG-CoA reductase (Mertz et al., 1982; Priego et al., 1979) which, as the key enzyme of cholesterol biosynthesis, irreversibly catalyzes the conversion of β-hydroxy-β-methylglutaryl-CoA into mevalonic acid, and whose activity is reduced by nicotinic acid and, even more strongly, by derivatives such as etofibrate (Priego et al., 1979). Endogenous bile acids and cholesterol itself are also allosteric inhibitors of this enzyme, although this negative feedback mechanism seems insufficient for the regulation of the plasma cholesterol concentration in omnivores as opposed to carnivores (Endo et al., 1977).

Over and above the action on the HMG-CoA reductase, an inhibitory action of nicotinic acid at the stage of synthesis between mevalonic acid and squalene has been observed as a further effect (Hamilton et al., 1971; Miller et al., 1960). In addition, catabolic effects such as increased cholesterol oxidation in the rat (Kritchevsky et al., 1971; Lengsfeld and Gey, 1971) and increased excretion in the feces from type II hypercholesterolemic patients in the form of neutral steroids (Miettinen, 1971) have been observed. The elevated excretion of neutral steroids in the feces seems to be a measure of the mobilization of tissue cholesterol and is apparently only transient (Kudchodkar et al., 1978; Sodhi et al., 1973). Grundy et al. (1981) were consequently unable to observe an increased excretion of this type.

Normally a reduction of the plasma NEFA concentration is manifested very rapidly after the administration of nicotinic acid, whereas the cholesterol concentration only decreases after an interval of a few days (Carlson et al., 1968). It is therefore assumed (Carlson, 1978; Grundy et al., 1981; Kudchodkar et al., 1978) that as a result of the reduced NEFA supply the liver synthesizes fewer lipoproteins. This effect of nicotinic acid is particularly pronounced in fasting patients (Carlson, 1978; Carlson et al., 1963). When the diet is rich in carbohydrates, which stimulates the liver to produce NEFA, nicotinic acid seemingly has no influence on the lipoprotein concentration (Kudchodkar et al., 1978), since the NEFA necessary for the synthesis are provided by the diet. However, good therapeutic results have also been reported by other groups on alimentary lipemia (Fröberg et al., 1971). On the whole the primary mechanism of action of nicotinic acid seems to be the antilipolytic action on fat tissue (Kudchodkar et al., 1978).

A clear elevation of the cholic acid/chenodeoxycholic acid coefficient as a result of the action of nicotinic acid was observed by Einarson et al. (1977) in patients with type II hyperlipoproteinemia. From this the authors concluded that nicotinic acid has an adverse effect on the conversion of cholesterol into bile acids. The flush reaction occurring frequently after the administration of nicotinic acid can be reduced by various inhibitors of prostaglandin synthesis, such as indomethacin (Eklund et al., 1979; Kaijser et al., 1979), acetylsalicylic acid (Åberg, 1973; Anderson et al., 1977), or benorylate (Rothe et al., 1977), without any extensive impairment of lipolysis (Carlson, 1978; Steinberg et al., 1963). It is therefore very likely that prostaglandins are in some way responsible for the nicotinic acid-induced flush, though histaminergic, serotoninergic, and cholinergic mechanisms also seem to be involved (Rothe et al., 1977).

It is very probable that the strong vasodilating actions of PGE_1 and

PGI_2 (prostacyclin) play a decisive role in the flush attacks. Elevated PGE_1 and PGI_2 concentrations caused by nicotinic acid have been measured in rat heart (Kaijser and Wennmalm, 1978), isolated fat cells (Kather and Simon, 1979), and guinea pig ear (Anderson et al., 1977), and in plethysmographic investigations on human forearm (Kaijser et al., 1979). According to more recent studies, the increased production of prostaglandins under the influence of nicotinic acid seems to affect chiefly PGI_2 and to have PGI_2-like activity (Eklund et al., 1979; Weithmann and Granzer, 1981). In any event, PGI_2 apparently has the strongest vasodilating activity of all the prostaglandins (Szczeklik et al., 1979), and thus the greatest potential for the precipitation of the flush symptoms. The connection between the release of prostaglandins and the inhibition of lipolysis due to nicotinic acid is, however, still largely unclear. An inhibition of lipolysis by PGE_1 has been described by most authors (Aktories et al., 1980a,b; Carlson, 1978; Steinberg et al., 1963), although the opposite effect, stimulation of adenylate cyclase in fat cells and an associated increase of lipolysis, has also been reported (Kather and Simon, 1979).

The prostaglandins are obviously decisively involved in the inhibition of platelet aggregation caused by nicotinic acid (Kirstein, 1980; Vincent and Zijlstra, 1978; Weithmann and Granzer, 1981). Thus, PGE_1 and PGI_2 exert aggregation-inhibiting and disaggregating effects, and are both used clinically for this purpose, although PGI_2 seems to be about 30 times more effective than PGE_1 (Carlson and Erikson, 1973; Gorman et al., 1979; Gryglewski et al., 1976; Szczeklik et al., 1979). Moreover, the production of thromboxane A_2 in platelets is reduced by nicotinic acid (Gryglewski et al., 1978; Tateson et al., 1977), and thus the tendency toward platelet aggregation is restricted. In addition, PGI_2, the strongest known antiaggregation compound (Gryglewski et al., 1976; Tateson et al., 1977), is likewise able to promote the breakdown of thrombi that have already been formed (Szczeklik et al., 1979).

Little information is available on the mechanism of action of nicotinic acid on fibrinolysis (Tesi, 1975; Turazza et al., 1973). The direct release of an activator is assumed to play a part (Weiner, 1979); prostacyclin itself seemingly has no effect on the fibrinolytic process (Szczeklik and Gryglewski, 1979).

IV. Toxicology and Side Effects

Because of rapid elimination of nicotinic acid, the doses necessary to produce pharmacological effects are, at an average of 1–3 g/day, several times the amount required for the substance to act as an antipellagra vitamin. There are even reports of the administration of up to 15 g of nicotinic acid per day (Carlson, 1978). However, since nicotinic acid, as a vitamin of the B group, is relatively well soluble in water, it has a low toxicity. The LD_{50} values are correspondingly high (5–7 g/kg po for mice and rats) (Unna, 1939).

The principal side effects of nicotinic acid listed here must be attributable to the high doses necessary to achieve a therapeutic effect, and can as a rule be prevented or alleviated by the use of nicotinic acid derivatives. The most common side effect is flush, a reddening of the skin mainly affecting the upper body and face, and associated with a sensation of heat. The worst cases of flush are produced after iv injection of nicotinic acid (Weiner et al., 1958), and develop in virtually every patient within 1–2 hours even at doses of 50–100 mg, which are insufficient to produce any lipid-lowering effect (Parsons, 1971). There is no exact dose–action relationship; apparently it is not so much the absolute dose administered that is responsible for the symptom as the continuous rise of the nicotinic acid level in the plasma (Eklund et al., 1979; Svedmyr et al., 1969). This also seems to be why flush occurs much less frequently with nicotinic acid derivatives with which, as retard compounds, a delayed release is possible. After repeated administration of nicotinic acid the flush intensity abates (Svedmyr et al., 1969; Zöllner and Gudenzi, 1966). Exhaustion of the previously mentioned mediator substances from the tissue is obviously taking place (Anderson et al., 1977; Credner, 1977).

Another relatively common side effect is a deterioration of oral glucose tolerance (Balasse and Neef, 1973; Gaut et al., 1971; Mosher, 1970). An elevated blood glucose level is evidently the consequence of increased glycogen mobilization in the liver and is a compensatory effect caused by the reduction of the NEFA concentration within the framework of the glucose–fatty acid cycle (Ammon et al., 1971; Benito et al., 1978). The result is an increased peripheral glucose utilization (Balasse and Neef, 1973). The strong increase of NEFA (rebound) after the end of nicotinic acid therapy can thus seemingly likewise be attributed to this regulation of the glucose–fatty acid cycle (Balasse and Neef, 1973; Benito et al., 1978). It may be, however, that the rebound phenomenon is also caused by an increased plasma concentration of somatotropin under the influence of nicotinic acid (Fain et al., 1966; Irie et al., 1967).

Hyperuricemia is a less common side effect of nicotinic acid therapy. Nicotinic acid seems to intensify purine synthesis (Becker et al., 1973), although it can also reduce renal uric acid clearance (Gershon and Fox, 1974). An elevation of transaminases and alkaline phosphatase has in many cases led to a discontinuation of nicotinic acid therapy (Olson *et al.*, 1974). Gastrointestinal irritation and liver dysfunction also occur sporadically (Mosher, 1970), as does bilirubinemia (Dietmann and Stork, 1976). All these side effects, however, are transient or cease on termination of the treatment (Mosher, 1970; Parsons, 1971).

On the basis of its high LD_{50}, nicotinic acid can be regarded as relatively nontoxic. The side effects are more unpleasant than they are dangerous. However, it should be used with caution in patients with diabetes mellitus or gout (Parsons, 1960). The stated nicotinic acid derivatives extensively reduce the side effects on the vessels and the gastrointestinal tract by lowering the plasma concentration of free nicotinic acid, thus permitting the use of lower doses. This is further supported by the intrinsic pharmacodynamic effects of many derivatization components. According to more recent investigations (Gad, 1982), the nicotinic acid derivative etofibrate actually seems to improve the glucose tolerance in non-insulin-dependent diabetics. In antihyperlipidemic therapy with nicotinic acid the new derivatives should always be given preference.

V. Summary

Nicotinic acid is used widely in the treatment of hyperlipoproteinemias and reduces both the cholesterol and the triglyceride concentrations in the plasma. However, very high doses are needed to achieve these therapeutic effects, which is why side effects are common and are particularly known to hinder the compliance of patients. Nicotinic acid derivatives have been developed to overcome this problem. As a result of chemical and galenic retardation, these derivatives lessen the side effects and the necessary doses are considerably reduced in comparison with pure nicotinic acid. However, most of these derivatives are not pure prodrugs, and exert their own synergistic pharmacokinetic and pharmacodynamic effects.

In general, when nicotinic acid and its derivatives are used in therapy a desirable modification of the composition of the plasma lipids in the sense of an antiatherosclerotic activity can be expected (reductions of VLDL and LDL and an increase of HDL). Good possibilities for using nicotinic acid in the prevention and remission of atherosclerotic processes arise in connection with the activation of fibrinolysis and the reduction of the tendency toward platelet aggregation.

In the light of recent studies on the mechanisms of action of nicotinic acid, an influence on the prostaglandin system has been found, as a result of which an interconnection between its various effects and side effects appears to be possible.

References

Abdulah, Y.-H. (1969). *J. Atheroscler. Res.* **9**, 171.
Åberg, G. (1973). *IRCS Gen. Pharmacol.* **10**, 13.
Aktories, K., Schultz, G., and Jacobs, H. (1980a). *Naunyn-Schmiedeberg's Arch. Pharmacol.* **312**, 167.
Aktories, K., Jacobs, K. H., and Schultz, G. (1980b). *FEBS Lett.* **115**, 11.
Altschul, R., Hoffer, A., and Stephen, J. D. (1955). *Arch. Biochem.* **54**, 558.
Ammon, H. P. T., Estler, C. J., and Heim, F. (1971). *In* "Metabolic Effects of Nicotinic Acid and its Derivatives" (K. F. Gay and L. A. Carlson, eds.), p. 799. Huber, Bern.
Anderson, R. G., Åberg, G., Brattsand, R., Ericsson, E., and Lundholm, L. (1977). *Acta Pharmacol. Toxicol.* **41**, 1.
Assmann, G. (1980). *Klin. Wochenschr.* **58**, 749.
Avogaro, P., Bittolo-Bon, G., Pais, M., and Taroni, G. C. (1977). *Pharmacol. Res. Commun.* **9**, 599.
Avogaro, P., Bittolo-Bon, G., and Cazzolato, G. (1978). *Pharmacol. Res. Commun.* **10**, 127.
Balasse, E. O., and Neef, A. (1973). *Metabolism* **22**, 1193.
Barr, D. B., Russ, E. M., and Eder, H. A. (1951). *Am. J. Med.* **11**, 480.
Becker, M. A., Raivio, K. O., and Meyer, L. J. (1973). *Clin. Res.* **21**, 616.
Benaim, M. E., and Dewar, H. A. (1975). *J. Int. Med. Res.* **3**, 423.
Benito, M., Moreno, F. J., Medina, J. M., and Major, F. (1978). *Arch. Biochem. Biphys.* **188**, 21.
Benzce, W. L. (1975). *Handb. Exp. Pharmacol.* **41**, 349.
Blum, C. B., Levy, R. I., Eisenberg, S., Hall, M., Goebel, R. H., and Bermann, M. (1977). *J. Clin. Invest.* **60**, 795.
Brattsand, R. (1975). *Acta Pharmacol. Toxicol.* **36**, 39.
Brattsand, R. (1976). Linköping University Medical Dissertations No. 34, p. 57.
Brattsand, R., and Harthon, L. (1971). *Arzneim. Forsch.* **21**, 1335.
Brattsand, R., and Lundholm, L. (1971). *Atherosclerosis* **14**, 91.
Brenner, G., and Brenner, H. (1972). *Arzneim. Forsch.* **22**, 754.
Brown, R. R., and Price, J. M. (1965). *J. Biol. Chem.* **219**, 985.
Burns, T. W., Langley, P. E., and Robinson, G. A. (1972). *Adv. Cyclic Nucleotide Res.* **1**, 63.
Butcher, R. W., Baird, E. E., and Sutherland, E. W. (1968). *J. Biol. Chem.* **243**, 1705.
Carlson, L. A. (1978). *Adv. Exp. Med. Biol.* **109**, 225.
Carlson, L. A., and Erikson, I. (1973). *Lancet* **I**, 155.
Carlson, L. A., and Nye, E. R. (1966). *Acta Med. Scand.* **179**, 453.
Carlson, L. A., and Orö, L. (1962). *Acta Med. Scand.* **172**, 641.
Carlson, L. A., Havel, R. J., Ekelund, L. G., and Holmgren, A. (1963). *Metabolism* **12**, 837.
Carlson, L. A., Orö, L., and Ostman, J. (1968). *Acta Med. Scand.* **183**, 457.
Chang, M. L. W., and Johnson, B. C. (1959). *J. Biol. Chem.* **234**, 1817.
Chang, M. L. W., and Johnson, B. C. (1961a). *J. Biol. Chem.* **236**, 799.
Chang, M. L. W., and Johnson, B. C. (1961b). *J. Biol. Chem.* **236**, 2096.
Chang, M. L. W., and Johnson, B. C. (1962). *J. Nutr.* **76**, 512.
Charman, R. C., Matthews, L. B., and Braeuler, C. (1972). *Angiology* **23**, 29.
Chaykin, S., Dagani, M., Johnson, L., and Marqueta, S. (1965). *J. Biol. Chem.* **240**, 932.

Chmelař, M., and Chmelařová, M. (1971). *In* "Metabolic Effects of Nicotinic Acid and Its Derivatives" (K. F. Gey and L. A. Carlson, eds.), p. 263. Huber, Bern.
Coronary Drug Project Research Group (1975). *J. Am. Med. Assoc.* **231**, 360.
Credner, K. (1977). Inaugural dissertation, Freie Universität, Berlin.
Cultrera, G. Giarola, P., Gibelli, A., Baldoni, E., Cuttin, S., Galetti, G., and Serenthà, P. (1971). *Arzneim. Forsch.* **21**, 954.
D'Angelo, V., Villa, S., Mysliewiec, M., Donati, M. B., and de Gaetano, G. (1978). *Thromb. Diath. Haemorrh.* **39**, 535.
Dembinska-Kiec, A., Gryglewska, T., Zmuda, A., and Gryglewski, R. J. (1977). *Prostaglandins* **14**, 1025.
Dietmann, K., and Stork, H. (1976). *Med. Klin.* **71**, 1047.
Edwards, P. A., and Gould, R. G. (1974). *J. Biol. Chem.* **249**, 2891.
Einarson, K., Hellström, K., and Leijd, B. (1977). *J. Lab. Clin. Med.* **90**, 613.
Eisenberg, S., and Levy, R. J. (1975). *Adv. Lipid Res.* **13**, 1.
Eisenberg, S., Bilheimer D. W., Levy, R. I., and Lindgren, F. T. (1973). *Biochim. Biophys. Acta* **326**, 361.
Eklund, B., Kaijser, L., Nowak, J., and Wennmalm, A. (1979). *Prostaglandins* **17**, 821.
Ekström-Jodal, B., Harthon, L., Haggendahl, E., Malmberg, R., and Svedmyr, N. (1970). *Pharmacol. Clin.* **2**, 86.
Endo, A., Tsujita, Y., Kuroda, M., and Tanzawa, K. (1977). *Eur. J. Biochem.* **77**, 31.
Eyes, J., van, Toustar, O., and Darby, W. J. (1955). *J. Biol. Chem.* **217**, 287.
Fain, J. N. (1973). *Pharmacol. Rev.* **25**, 67.
Fain, J. N., Galton, D. J., and Kovacey, V. P. (1966). *Mol. Pharmacol.* **2**, 237.
Fielding, C. J., Shore, V., and Fielding, P. E. (1972). *Biochem. Biophys. Res. Commun.* **46**, 1493.
Frank, O., Baker, H., and Sobotha, H. (1963). *Nature (London)* **199**, 490.
Fröberg, S. O., Boberg, J., Carlson, L. A., and Ericsson, M. (1971). *In* "Metabolic Effects of Nicotinic Acid and Its Derivatives" (K. F. Gey and L. A. Carlson, eds.), p. 167. Huber, Bern.
Fumagalli, R. (1971). *In* "Metabolic Effects of Nicotinic Acid and its Derivatives" (K. F. Gey and L. A. Carlson, eds.), p. 33. Huber, Bern.
Funk, C. (1913). *J. Physiol. (London)* **46**, 173.
Gad, A. S. (1982). Inaugural dissertation, Cairo University.
Gamble, W., and Wright, L. D. (1961). *Proc. Soc. Exp. Biol.* **107**, 160.
Garrett, E. R., and Gardner, M. R. (1982). *J. Pharm. Sci.* **71**, 14.
Gaut, Q. N., and Solomon, H. M. (1970). *Biochim. Biophys. Acta* **201**, 316.
Gaut, Q. N., Solomon, H. M., and Miller, O. N. (1971). *In* "Metabolic Effects of Nicotinic Acid and Its Derivatives" (K. F. Gey and L. A. Carlson, eds.), p. 923. Huber, Bern.
Gershon, S. L., and Fox, I. H. (1974). *J. Lab. Clin. Med.* **84**, 179.
Gey, K. F., and Carlson, L. A., eds. (1971). "Metabolic Effects of Nicotinic Acid and Its Derivatives." Huber, Bern.
Giltin, D., Cornwell, D. G., Nikasato, D., Oncley, J. L., Hughes, W. L., and Janeway, C. A. (1958). *J. Clin. Invest.* **37**, 172.
Glomset, J. A., Janssen, E. T., Kennedy, R., and Robbins, J. (1966). *J. Lipid Res.* **7**, 638.
Grodzinska, L., Basista, M., Dembinska-Kiec, A., and Kostka-Trabka, E. (1982). Personal communication.
Gordon, T., Castelli, W. P., Hjortland, M. C., and Drawber, W. R. (1977). *Am. J. Med.* **62**, 707.
Gorman, R. R., Hamilton, R. D., and Hopkins, N. K. (1979). *In* "Prostacyclin" (J. R. Vane and S. Bergström, eds.), p. 85. Raven, New York.

Grundy, S. M., Mok, H. Y. I., Zech, L., and Berman, M. (1981). *J. Lipid Res.* **22**, 24.
Gryglewski, R. J., Bunting, S., Moncada, S., Flower, R. J., and Vane, J. R. (1976). *Prostaglandins* **12**, 685.
Gryglewski, R. J., Dembinska-Kiec, A., Zmuda, A., and Gryglewska, T. (1978). *Atherosclerosis* **31**, 385.
Hamilton, J. G., Sullivan, A. C., Gutierrez, M., and Miller, O. N. (1971). *Fed. Proc. Fed. Am. Soc. Exp. Biol.* **30**, 519.
Harthon, L., and Brattsand, R. (1979). *Arzneim. Forsch.* **29**, 1859.
Harthon, L., and Sigroth, K. (1974). *Arzneim. Forsch.* **24**, 1688.
Harthon, L., Brattsand, R., and Lundholm, L. (1971). In "Metabolic Effects of Nicotinic Acid and Its Derivatives" (K. F. Gey and C. A. Carlson, eds.), p. 115. Huber, Bern.
Irie, M., Sakma, M., Tsuskima, T., Skizume, K., and Nahao, K. (1967). *Proc. Soc. Exp. Biol. Med.* **126**, 708.
Kaijser, B., Eklund, B., Olson, A. G., and Carlson, L. A. (1979). *Med. Biol.* **57**, 114.
Kaijser, L., and Wennmalm, A. (1978). *Acta Physiol. Scand.* **102**, 246.
Kather, H., and Simon, B. (1979). *Res. Commun. Chem. Pathol. Pharmacol.* **23**, 81.
Kirk, J. E. (1968). In "The Biological Basis of Medicine" (E. Bittar, ed.), Vol. I, p. 493. Academic Press, New York.
Kirstein, A. (1980). *Scand. J. Haematol.* **34**, 105.
Kissebah, A. H., Adams, P. W., Harrigan, P., and Wynn, V. (1974). *Eur. J. Clin. Invest.* **4**, 163.
Krishna, G., Weiss, B., Davies, J. I., and Hynie, S. (1966). *Fed. Proc. Fed. Am. Soc. Exp. Biol.* **25**, 719.
Kritchevsky, D., Kitagawa, M., Tepper, S. A., and Davidson, L. M. (1971). In "Metabolic Effects of Nicotinic Acid and Its Derivatives" (K. F. Gey and L. A. Carlson, eds.), p. 585. Huber, Bern.
Kruse, W., Kruse, W., Raetzer, H., Heuck, C. C. et al. (1979). *Eur. J. Clin. Pharmacol.* **16**, 11.
Kudchodkar, B. J., Harbhajan, Ph.D., Sodki, S. et al. (1978). *Clin. Pharmacol. Ther.* **24**, 354.
Kummer, M., Schatton, W., Linde, H., and Oelschläger, H. (1979). *Pharm. Z.* **124**, 1312.
Lakin, K. M., Baluda, V. P., Lykoyanowa, I. I., Romanovskaya, V. N., and Baluda, M. V. (1980). *Farmakol. Toksikol.* **43**, 581.
Lee, Y. C. (1972). *Biochim. Biophys. Acta* **264**, 59.
Lengsfeld, H., and Gey, K. F. (1971). In "Metabolic Effects of Nicotinic Acid and Its Derivatives" (K. F. Gey and L. A. Carlson, eds.), p. 597. Huber, Bern.
Lundholm, L., Jacobsson, L., Brattsand, R., and Magnusson, O. (1978). *Atherosclerosis* **29**, 217.
McGill, H. C., Jr. (1979). *Am. J. Clin. Nutr. Suppl.* **32**, 2664.
Magide, A. A., Myant, N. B., and Reichl, D. (1975). *Atherosclerosis* **21**, 205.
Marmo, E., Imperatore, A., Caputi, A., and Cataldi, S. (1971). *Farmaco Prat.* **26**, 557.
Marsh, I. B. (1971). In "Plasma Lipoproteins" (R. M. S. Smellie, ed.), p. 89. Academic Press, New York.
Mertz, D. P., Göhmann, E., Suermann, I., Eisele, R., and Schatton, W. (1982). *Med. Welt* **33**, 405.
Miettinen, T. A. (1971). In "Metabolic Effects of Nicotinic Acid and Its Derivatives" (K. F. Gey and L. A. Carlson, eds.), p. 677. Huber, Bern.
Miller, N. E. (1980). *Atherosclerosis* **5**, 500.
Miller, O. N., Hamilton, I. G., and Goldsmith, G. A. (1960). *Circulation* **18**, 489.

Moreno, F. J., Shepherd, R. E., and Fain, J. E. (1979). *Naunyn-Schmiedeberg's Arch. Pharmacol.* **306**, 179.
Mosher, L. R. (1970). *Am. J. Psychiat.* **126**, 1290.
Moutafis, C. D., and Myant, N. B. (1971). *In* "Metabolic Effects of Nicotinic Acid and Its Derivatives" (K. F. Gey and L. A. Carlson, eds.), p. 659. Huber, Bern.
Mrochek, I. E., Jolley, R. L., Young, D. S., and Turner, W. G. (1976). *Clin. Chem.* **22**, 1821.
Nicola, P., de, Frandoli, G., and Gibelli, A. (1963). *Gen. Gerontol.* **11**, 955.
Nikkilä, E. A. (1971). *In* "Metabolic Effects of Nicotinic Acid and Its Derivatives" (K. F. Gey and L. A. Carlson, eds.), p. 487. Huber, Bern.
Nikkilä, E. A. (1978). *In* "High Density Lipoproteins and Atherosclerosis" (A. M. Gotto, ed.). Elsevier, Amsterdam.
Oelschläger, H., Rothley, D., Ewert, M., and Nachev, P. (1980). *Arzneim. Forsch.* **30**, 984.
Olson, A. G., Orö, L., and Rössner, S. (1974). *Atherosclerosis* **19**, 61.
Ortega, M. P., Sunkel, C., Armijo, M., and Priego, J. G. (1980). *Thromb. Res.* **19**, 409.
Otway, S., Robinson, D. S., Rogers, M. P., and Wing, D. R. (1971). *In* "Metabolic Effects of Nicotinic Acid and Its Derivatives" (K. F. Gey and L. A. Carlson, eds.), p. 497. Huber, Bern.
Packard, C. J., Steward, J. M., Third, J. L. H. C., Morgan, H. G., Lawrie, T. D. V., and Shepherd, J. (1980). *Biochim. Biophys. Acta* **618**, 53.
Paoletti, R. (1983). *Med. Klin. Prax.* **78**, 31.
Parsons, W. B., Jr. (1960). *Am. J. Clin. Nutr.* **8**, 471.
Parsons, W. B. (1971). *In* "Treatment of Hyperlipemic States" (H. R. Casdorph, ed.), p. 335. Thomas, Springfield, Illinois.
Parwaresch, M. R., Haake, H., and Mäder, Ch. (1978). *Atherosclerosis* **31**, 395.
Perrin, A., Loire, R., Couchat, N. C., Bourdillon, N. C., and Gras, J. (1968). *Rev. Lyon Med.* **17**, 727.
Petrack, B., Greengard, P., and Kalinsky, H. (1966). *J. Biol. Chem.* **241**, 2367.
Preiss, J., and Handler, P. (1957). *J. Am. Chem. Soc.* **79**, 4246.
Priego, J. G., Piña, M., Armijo, M., Sunkel, C., and Maroto, M. L. (1979). *Arch. Farm. Toxicol.* **5**, 3.
Puglisi, L. (1982). *Int. Etofibrate Symp., Munich, March 20.*
Puglisi, L., Caruso, V., and Paoletti, R. (1976). *Pharmacol. Res. Commun.* **8**, 379.
Puglisi, L., Maggi, F., Colli, S., and Paoletti, R. (1979). *Pharmacol. Res. Commun.* **11**, 775.
Rimondi, S., Descovich, G. C., and Lenzi, S. (1980). *Int. Symp. Drugs Affect. Lipid Metab., 7th, Milan, May 28* Abstr., p. 180.
Roberts, W. C., Ferrans, V. J., Levi, R. I., and Fredrickson, D. S. (1973). *Am. J. Cardiol.* **31**, 557.
Rosenhamer, G., and Carlson, L. A. (1980). *Atherosclerosis* **37**, 129.
Rothe, O., Thormälen, D., and Ochlich, P. (1977). *Arzneim. Forsch.* **27**, 2347.
Rudermann, N. B., Richards, K. C., Valles de Bourges, V., and Jones, A. L. (1968). *J. Lipid Res.* **9**, 613.
Salmi, H. A., and Frey, H. (1974). *Curr. Ther. Res.* **16**, 669.
Schatton, W. (1982). *Int. Etofibrate Symp., Munich, March 20.*
Schlierf, G., and Hess, G. (1977). *Artery* **3**, 174.
Schwartz, C. G., Halloran, L. G., Vlahcewics, Z. R., Gregory, D. H., and Swell, L. C. (1978). *Science* **200**, 62.
Shafir, E., and Biale, Y. (1971). *In* "Metabolic Effects of Nicotinic Acid and Its Derivatives" (K. F. Gey and L. A. Carlson, eds.), p. 515. Huber, Bern.
Shepherd, J., Packard, C. J., Patch, J. R., and Antonio, M. (1979). *J. Clin. Invest.* **63**, 858.

Shestakov, V. A., Ilyin, V. N., Danilova, L. M., Alexandrova, N. P., and Rodionov, S. V. (1977). *Prob. Gematol. Pereliv. Krovi.* **22**, 29.
Sodhi, H. S., Kudchodkar, B. J., and Harlick, L. (1973). *Atherosclerosis* **17**, 1.
Spöttl, F., and Froschauer, J. (1976). *Atherosclerosis* **25**, 293.
Steinberg, D., Vaughan, M., Nestel, P., Strand, O., and Bergström, S. (1963). *Biochem. Pharmacol.* **12**, 764.
Subissi, A., Biagi, M., and Murmann, W. (1980). *Arzneim. Forsch.* **30**, 1278.
Svedmyr, N., and Harthon, L. (1970). *Acta Pharmacol. (Copenhagen)* **28**, 66.
Svedmyr, N., Harthon, L., and Lundholm, L. (1969). *Clin. Pharmacol. Ther.* **10**, 559.
Szczeklik, A., and Gryglewski, R. J. (1979). *In* "Prostacyclin" (J. R. Vane and S. Bergström, eds.), p. 393. Raven, New York.
Szczeklik, A., Skawinski, S., Gluszko, P., and Nisankowski, R. (1979). *Lancet* **I**, 26.
Tateson, J. E., Moncada, S., and Vane, J. R. (1977). *Prostaglandins* **13**, 389.
Tesi, M. (1975). *In* "Progress in Chemical Fibrinolysis and Thrombolysis" (J. F. Davidson, M. M. Samana, and P. C. Desnoyers, eds.), Vol. 1, p. 255. Raven, New York.
Turazza, G., Spreafico, P. L., and Frandoli, G. (1973). *Arzneim. Forsch.* **23**, 654.
Turba, C., Faini, D., and Pagella, P. G. (1980). *Int. Symp. Drugs Affect. Lipid Metab., 7th, Milan, May 28* Abstr., p. 82.
Unna, K. (1939). *J. Pharmacol. Exp. Ther.* **65**, 95.
Vincent, J. E., and Zijlstra, F. J. (1978). *Prostaglandins* **15**, 629.
Weiner, M., Redish, W., and Steele, J. M. (1958). *Proc. Soc. Exp. Biol.* **98**, 755.
Weiner, M. (1979). *Drug Metab. Rev.* **9**, 99.
Weithmann, K. U., and Granzer, E. (1981). *Artery* **8**, 475.
Weitzel, G., Wahl, P., and Buddecke, E. (1962). *Hoppe-Seylers Z. Physiol. Chem.* **327**, 109.
Yu-Yan-Yeh (1975). *Life Sci.* **18**, 33.
Zöllner, N., and Gudenzi, M. (1966). *Med. Klin.* **61**, 2036.
Zöllner, N., and Wolfram, G. (1970). *Hautarzt* **21**, 443.

Heparin and Atherosclerosis

HYMAN ENGELBERG

Cedars–Sinai Medical Center
Los Angeles, California

I. Introduction 219
   A. Nature of Heparin Activity 219
   B. Etiology of Atherosclerosis 220
II. Heparin Actions in Systems Involved in Atherogenesis 221
   A. Endothelium 221
   B. Coagulation 222
   C. Triglyceride Transport 222
   D. Platelets 224
   E. Complement 226
   F. Macrophages (Reticuloendothelial System) 227
   G. Fibrinolysis 228
III. Effects of Heparin at Different Sites 229
   A. Endothelial Surface 230
   B. Bloodstream 235
IV. Results of Heparin Therapy 245
   References 246

I. Introduction

A. NATURE OF HEPARIN ACTIVITY

Heparins are naturally occurring substances found throughout the animal kingdom and probably in the plant kingdom (Engelberg, 1963). There are a variety of heparins found in different species. They have been classified as acidic mucopolysaccharides, glycosaminoglycans, and linear anionic polyelectrolytes (Jaques, 1980). They are highly sulfated compounds with a strong negative surface charge and complex carbohydrate chains which provide for specific binding of proteins and dyes. This structure provides an ion-exchange vehicle. Heparin–protein complexes are rather loose at physiologic pH and ionic strength (Godal, 1960), and the bond is readily dissociable with changing conditions. Similar labile redistribution when the binding forces are primarily electrostatic has been described with synthetic heparin-like polysaccharides (Mora and Young, 1959). When heparin complexes with a protein the high charge density of heparin produces conformational changes in the charged regions of the protein, thus affecting its biologic activity (Blackwell et al., 1977). The nature of the conformational changes in proteins due to interaction with the heparin charge varies depending on the constellation of amino acids (Stone, 1977). The physiologic activity of the protein is either enhanced or inhibited, and the original activity is restored when the heparin–protein bond is broken. The reaction between heparin and a protein represents a kinetic equilibrium which is readily altered if another protein or cation is added. The heparin will then redistribute itself between the proteins according to their concentrations and binding activities (Jaques, 1967). This suggests that heparin may function as part of the homeostatic mechanism regulating physiologic processes that involve a series of protein interactions.

Although the main clinical use of heparin thus far has been for anticoagulation, heparin has a wide range of biologic activities. In this presentation we will consider how these actions of heparin might affect the development and complications of atherosclerosis. Since this disease is the leading cause of death in industrialized societies, it is unfortunate that the evidence about heparin and atherogenesis has been ignored (Braunwald, 1980).

B. Etiology of Atherosclerosis

This enormous subject can only be briefly summarized here. The currently most acceptable hypothesis is that atherogenesis involves the following (Ross et al., 1976): (1) initial endothelial injury leading to a breakdown of the endothelial barrier; (2) adherence of platelets to the exposed subendothelial tissue and release of platelet constituents, including the platelet mitogenic factor, into that tissue; (3) migration of medial smooth muscle cells (SMC) into the intima and proliferation there; (4) formation of connective tissue matrix, including collagen, elastin, and mucopolysaccharides, by these cells; and (5) intra- and extracellular accumulation of lipid from the plasma into the smooth muscle cell lesion and in macrophages. Following the initial arterial wall injury, the subsequent steps in the reaction to that injury frequently overlap. Thrombus formation at atherosclerotic sites often occurs, contributing a major clinical role. Microthrombi may also be involved in the atherosclerotic process. Thus human atherogenesis involves the functions and reactions of the endothelial and smooth muscle cells of the arterial wall, platelets, factors contributing to elevated plasma lipid levels, and aberrations of coagulation which predipose the victim to thrombus formation. Hemodynamic forces also play a role in the localization and acceleration of atherosclerotic lesions.

II. Heparin Actions in Systems Involved in Atherogenesis

A. ENDOTHELIUM

The endothelium is the subject of intense investigation, and our knowledge of this important tissue has expanded rapidly since the relation of heparin to its integrity and function was last summarized (Engelberg, 1978a,b). It was shown years ago that areas of metachromatic staining characteristic of heparin and heparin-like substances were present on the internal vascular surface of animals and man (Zugibe, 1962; Velican, 1967). This metachromatic material had antithrombin cofactor and antithromboplastin-generating activity (Marin and White, 1961), both of which are properties of heparin. Endothelial cells in tissue culture synthesize and secrete sulfated mucopolysaccharides related to heparin (Buonassi, 1973), the bulk of which is apparently heparan (Buonassi, 1975), which is also a surface component of other cell lines (Kraemer, 1971). Endothelial cell surface heparan has a more heparin-like structure than the heparan of other cell types (Gamse *et al.*, 1978). The question of the presence of heparin *per se* together with heparan at the endothelial surface may be moot from the functional standpoint, as X-ray diffraction methods showed that 20% of the heparan molecule is identical to heparin (Atkins and Neiduszynski, 1977). This heparin moiety of heparan is probably the site of the heparin-like activity of heparan.

It is possible that heparin itself is normally present and functioning at the endothelial cell luminal surface. Glomerular capillary walls contain anionic sites which influence glomerular permeability to plasma proteins. Perfusion of rat kidneys with protamine, which binds to and neutralizes heparin, resulted in a marked reduction of these anionic sites and leakage of ferritin particles through the glomerulus. Reperfusion with heparin restored the anionic sites and normalized glomerular function (Kelley and Cavallo, 1979). The anionic sites on the luminal surface of capillary endothelial cells are apparently heparan or heparin, or both (Simionescu *et al.*, 1981). Cultured bovine aortic endothelial cells produced an inhibitor of smooth muscle cells which was apparently heparin and not heparan (Castellot *et al.*, 1981). Thrombin in the blood binds to high-affinity binding sites on endothelial cell surfaces which catalyze the inactivation of the bound thrombin by antithrombin (Lollar and Owen, 1980). This is heparin activity. Heparan and dermatan sulfate from human aortas also show anticoagulant activity, but it is less than 2% of the activity of heparin by the antifactor Xa or activated partial thromboplastin time (APTT) tests (Teien et al., 1976). Antithrombin cofactor activity has been detected on endothelial cell surfaces *in vitro* (Owen and Esmon, 1981). When polyethylene was coated with heparin, heparan, or dermatan sulfate, only the heparin-bound surface completely neutralized surface-bound thrombin upon exposure to antithrombin-containing blood plasma (Larsson et al., 1980). These various findings imply that heparin is a normal functioning constituent of vascular endothelial cell surfaces, and that complete proof will be forthcoming when better methods of identification are developed. Proteoglycans from the subendothelial area of bovine aorta also have potent anti-platelet-aggregating and antithrombotic activity which could act as a self-protective mechanism after mild endothelial injury (Klein et al., 1980).

B. Coagulation

The results of many investigations have shown that the anticoagulant activity of heparin is due to its marked acceleration of the inactivation of serum proteases involved in the coagulation cascade by the plasma protein antithrombin III (Rosenberg, 1977). Heparin apparently acts as a catalyst in this reaction, facilitating the formation of a thrombin–antithrombin complex (Griffith et al., 1979; Machovich et al., 1975; Pomerantz and Owen, 1978; Smith, 1977). In the case of thrombin inactivation, only a minute amount of heparin, .005–.01 units per milliliter of blood, is required (Machovich, 1975; Smith and Sundboom, 1981). The net result of this action of heparin is to prevent the generation of thrombin or to facilitate its almost immediate inactivation if it is formed. Another action of heparin in preventing thrombin formation is its neutralization of released tissue thromboplastin (Biggs et al., 1953). It has also been reported that heparin, at 0.1 units per milliliter, inhibited the activation of factor X via the intrinsic pathway even in the absence of antithrombin III (Ofosu et al., 1980).

C. Triglyceride Transport

The mechanisms involved in the removal of ingested fat from the blood were not understood until it was accidentally observed that heparin abolished postalimentary lipemia in dogs (Hahn, 1943). This finding was soon confirmed in other animals and in man (Weld, 1944). The presence of a lipemia-clearing factor in the blood after the injection of heparin was then demonstrated (Anderson and Fawcett, 1950). Subsequent studies in rabbits and man showed that after the injection of heparin there was a rapid disappearance of triglyceride-rich lipoproteins from the plasma (Graham et al., 1951), a loss of neutral fat from the blood (Brown, 1952; Grossman, 1952; Van Eck et al., 1952), an increase in α- or high-density lipoproteins (Boyle et al., 1952, Nikkila, 1953), and a shift in cholesterol from the β- to the α-lipoproteins (Nikkila, 1953). These significant changes in serum lipids were explained by the observation that postheparin lipemia-clearing factor caused the lipolysis of serum triglycerides into their constituent fatty acids and glycerol (Nichols et al., 1952), which then left the vascular compartment via the carrier protein, albumin (Gordon et al., 1953; Robinson and French, 1953). The active lipolytic enzyme was extracted from rat heart and named lipoprotein lipase (Korn, 1954). It was also found in some human plasmas without the prior injection of heparin (Engelberg, 1955), and its identity with postheparin lipoprotein lipase was shown (Engelberg, 1956). Although other lipolytic enzymes appear in the blood after heparin, many investigations have established that lipoprotein lipase enzymatic activity is the major physiologic pathway for the removal from the blood of both postalimentary and hepatically synthesized triglycerides, transported respectively as chylomicrons and very low density lipoproteins. It is generally accepted that the enzyme normally functions at the endothelial surface where the bulk of triglyceride lipolysis takes place. However, a small amount of endogenous plasma lipolytic activity was present in 75% of normal people when investigated using better methods for demonstration of the enzyme (Engelberg, 1964). As subsequently shown (Engelberg, 1969), the use of platelet-rich plasma would increase the incidence above 75%. In a few individuals this circulating lipoprotein lipase activity was substantial enough to account for the removal of the average daily fat intake from the bloodstream (Engelberg, 1958c; Muir, 1968).

Whether lipoprotein lipase can function without heparin remains an unsettled question. Highly purified enzyme obtained from postheparin rat plasma effects the hydrolysis of lipoprotein triglycerides (Fielding et al., 1974). There was no evidence for a direct effect of heparin in the activity of bovine milk lipoprotein lipase using rat very low density lipoprotein as a substrate (Bengtsson and Olivecrona, 1981). These workers believe that heparin binding is involved in the attachment of the enzyme to the capillary endothelium rather than in the lipolytic action, and that heparin increases the stability and solubility of lipoprotein lipase. Heparin had only a slight enhancing effect on the lipolytic activity of purified bovine milk lipoprotein lipase (Iverius and Lindahl, 1972). The problem is that minute amounts of endogenous heparin may have been present. It has been pointed out (Iverius and Lindahl, 1972) that even treatment with heparinase did not completely remove heparin from the purified enzyme, and so "it could not be stated that lipoprotein lipase can function in the complete absence of heparin." On the other hand, there is much evidence that endogenous heparin normally plays a role in the lipoprotein lipase mechanism. Heparin-neutralizing drugs increase the lipemia of rats and dogs (Spitzer, 1953), elevate serum lipoproteins in rabbits and rats (Hewitt et al., 1952; Bragdon and Havel, 1954), and increase chylomicronemia (Gruner et al., 1953) and serum triglycerides (Engelberg, 1965) in man. Lipoprotein lipase is inhibited by heparin-binding agents (Korn, 1954; Engelberg, 1956, 1964, 1969). Bacterial heparinase reduces the activity of chicken adipose tissue lipoprotein lipase (Korn, 1957). Tricalcium phosphate gel, which adsorbs heparin, inactivates endogenous plasma lipemia-clearing activity (lipoprotein lipase) (Engelberg, 1957). Circulating heparin-binding immunoglobulins are associated with markedly elevated serum triglyceride levels (Glueck et al., 1972). Lipoprotein lipase is apparently inactivated as triglyceride lipolysis occurs, and there is evidence that the inactivation is a result of breaking the heparin–enzyme bond (Engelberg, 1958d). This bond is probably ionic and reversible (Egelrud, 1973). Heparin stabilizes the activity of the enzyme and binds it to its substrate (Korn and Quigley, 1957; Patten and Hollenberg, 1969), the negatively charged groups of heparin probably complexing with the positively charged amino groups of the lipoproteins (Iverius, 1972). Exogenous heparin may also increase the net synthesis of lipoprotein lipase by displacing the enzyme into the surrounding medium from the cell surface (Chajek et al., 1978). The action of heparin in facilitating the lipolysis of lipoprotein triglycerides via lipoprotein lipase resembles its role in catalyzing the thrombin–antithrombin reaction.

D. PLATELETS

Platelet aggregation, platelet adhesion to injured vascular walls, and the release of platelet products are fundamental reactions in normal hemostasis. However, with recurrent endothelial injury, the same processes are important contributors to athero- and thrombogenesis. Here the effect of heparin is complex with some contradictory findings and many gaps in our understanding of the subject. Heparin, in as low a concentration as 0.06 units per milliliter, potentiates ADP or epinephrine-induced platelet aggregation in citrated platelet-rich plasma *in vitro* (Eika, 1973; Thomson et al., 1973; Michalski et al., 1977). This was associated with a shape change in the platelets (McLean and Hause, 1982). The effect of heparin on collagen-induced platelet aggregation is less clear (Tiffany and Penner,

1979). Even minute doses of heparin enhanced platelet aggregation, but less so in the presence of antithrombin (Salzman et al., 1980). Antithrombin had previously been found to neutralize the platelet-aggregating effect of heparin (Shanberge et al., 1976; Lindon et al., 1978). Antithrombin, of course, is present in vivo in blood plasma and so would minimize the more marked platelet aggregation seen with heparin in vitro. Thus, in vitro tests may be misleading about events in the body (O'Brien, 1975; Chen and Wu, 1980). Nevertheless, mild thrombocytopenia does occur with continued heparin therapy, but this is apparently usually asymptomatic and harmless, in contrast to the infrequent but severe thrombocytopenia which results from an immunological reaction (Chong et al., 1982).

Evidence as to the platelet release reaction in the presence of heparin is also confusing. Both decreased and increased release of thromboxane A_2 by platelets in vitro in the presence of heparin at a concentration of 1.25 units per milliliter has been reported (Mohammad et al., 1981; Huang and LeBlanc, 1981). Thrombin-induced platelet aggregation and the subsequent release of platelet products is inhibited by heparin (Eika, 1971; Michalski et al., 1977; Tiffany and Penner, 1979; Workman et al., 1977). There is considerable evidence that irreversible platelet aggregation and the platelet release reaction usually are dependent on the generation of thrombin from the coagulation factors in the plasma around the platelet surface (Akbar and Ardlie, 1978; Walsh, 1976). In vivo, platelet aggregates are unstable and disaggregate unless fibrin forms around them (Pan et al., 1978; Wautier and Caen, 1979). Both the rate and extent of platelet aggregation initiated by collagen or epinephrine in vitro were reduced in blood samples taken after the injection of heparin (Besterman and Gillett, 1973).

The relevance of in vitro studies on the effect of heparin on prostacyclin (PGI_2) and prostaglandin E_1 (PGE_1) inhibition of platelet aggregation to in vivo events is also unclear. Heparin, at concentrations of 1–3 units per milliliter, reversed the inhibition of platelet aggregation by PGI_2 (MacIntyre et al., 1981) and PGE_1 (Saba and Saba, 1981) in vitro. However, heparin does not inhibit increased PGI_2 production by the vascular wall unless it is the result of thrombin generation (Buchanan et al., 1979; Ronsaglioni et al., 1980). Heparin did not block the in vitro coronary artery vasodilating effect of PGI_2 (Eldor et al., 1980). In fact, heparin itself, at 8 $\mu g/ml$, relaxed coronary strips in vitro (Eldor et al., 1980). In 18 patients with coronary artery disease, the intravenous injection of 5000 units of heparin caused an increase in the coronary sinus blood content of PGI_2 and an increase in coronary blood flow, a beneficial effect that was blocked by pretreatment with aspirin (Wallis et al., 1982). Thus, despite the in vitro observations previously noted, it appears that heparin does not directly neutralize the biologic activity of PGI_2 (Eldor et al., 1980). Indeed, since both PGI_2 and heparin are antithrombotic, it would make little physiologic sense for one to inhibit the other.

Finally, it is probable that *in vivo* only aggregated platelets adherent to the vascular wall undergo the release reaction, and not those which are aggregated but not adherent (Witte et al., 1978). Heparin minimized the adherence of platelets to endothelial cell junctions (Samuels and Webster, 1952; McGovern, 1955) and to injured arterial walls and artificial shunts (Essien et al., 1978; Gregorius and Rand, 1976; Groves et al., 1982; Lageg-ren et al., 1975; Larsson et al., 1980), although there is contradictory evidence (Salzman et al., 1980). Heparin reduced platelet adhesion to collagen (Busch et al., 1979b). Some findings have indicated that heparin has a generalized inhibitory action on platelet–collagen interactions (Davies and Mengs, 1982) which is not mediated via antithrombin III. Postheparin plasma corrected postoperative increased platelet adhesiveness (Ham, 1969). Repeated small doses of heparin normalized the increased platelet adhesiveness to glass present in some patients with coronary heart disease (McDonald and Edgill, 1961). Thrombin increased platelet adhesiveness to vascular walls (Essien et al., 1978; Nordoy et al., 1978). An antithrombin agent (hirudin) markedly decreased the adherence of thrombin-aggregated platelets to endothelial and smooth muscle cells (Czervionki et al., 1978). Minute quantities of heparin, together with antithrombin, prevented the increased platelet adhesiveness resulting from thrombin generation. The effect of heparin on platelet adhesiveness is probably much more relevant to atherogenesis and thrombosis than its *in vitro* actions on platelet aggregation. The relation of heparin to released platelet products will be discussed later.

E. COMPLEMENT

The complement system is an effector pathway of the inflammatory response which normally is present in the bloodstream in an inactive form. It is activated specifically by infections and immune complexes, or it may be activated nonspecifically, as by aggregated macromolecules (Ruddy et al., 1972). Atheroma lipids may activate complement (Hammerschmidt et al., 1981a). The consequences of complement activation include increased vascular permeability, activation of neutrophiles, and alterations in cell membranes that can lead to lysis and cell death (Ruddy et al., 1972). Activation of the complement system involves the sequential interaction of the serum proteins composing the system. Since complement activation can potentially produce profound effects on cell membranes, the system has inhibitors. C1 esterase inhibitor is a serum protein which inhibits the activation of the first component of complement. Heparin, at very low concentrations, greatly potentiates this inhibiting activity (Rent et al., 1976). It has been proposed that heparin simultaneously binds to C1 and C1 esterase inhibitor, bringing the molecules into close proximity and thus stabilizing and kinetically favoring their interaction (Caughman et al., 1982). The similarity to the catalytic action of heparin in the thrombin–antithrombin reaction is clear. The C1 macromolecule also may directly be inactivated by heparin (Minick, 1980). Purified C1q has two high-affinity binding sites for heparin, and heparin inhibited the ability of C1q to recombine with other C1 components to form hemolytically active C1 (Alameda et al., 1983). Complement activation ultimately involves amplification of the third component of complement (C3) via the formation of an amplification convertase. Heparin inhibits the generation of this amplification convertase (Weiber et al., 1978), and this is independent of antithrombin-binding activity (Kazatchkine et al., 1981). Heparin also inhibits complement activity at a late stage, just prior to cell lysis (Baker et al., 1975). The classic pathway of complement activation is progressively inhibited by increasing concentrations of heparin, whereas its effect in the alternative pathway is more complex (Logue, 1978). Heparin augmented the action of control proteins in inhibiting one type of complement activation via the alternative pathway (Kazatchkine et al., 1979). Certain complement proteins bound strongly to heparin, and the authors suggested that this indicated a multifaceted regulatory role for heparin in the complement system (McKay and Laurell, 1980; McKay et al., 1981).

F. Macrophages (Reticuloendothelial System)

The reticuloendothelial (macrophage) system is a major host defense system (Saba, 1970). It monitors the vascular compartment and actively removes bacteria, immune complexes, injured platelets, effete red blood cells, denatured proteins, fibrin aggregates, circulating thromboplastin, tumor cells, and a variety of colloids. Thus, apart from forming foam cells, macrophages remove some agents that injure endothelium or contribute to hypercoagulability. Plasma fibronectin, also known as opsonic α-2-SB glycoprotein or cold-insoluble globulin, plays a role in macrophage phagocytic activity (Blumenstock et al., 1978; Molnar et al., 1977; Saba et al., 1966). The ingestion of nonbacterial particulate matter and of some bacterial strains by Kupffer cells is augmented by preliminary coating by fibronectin. This action of fibronectin requires the presence of heparin in test systems using rat liver slices (Allen et al., 1973; Molnar et al., 1977) or isolated macrophages (Blumenstock et al., 1980; Guderovich et al., 1980;

Van de Water et al., 1981). Heparin was not obligatory in the take-up of gelatin-coated beads by macrophage monolayers but, when present, augmented the extent of phagocytosis by 25–45% (Doran et al., 1981). Human plasma fibronectin binds to heparin (Ruoslahti and Engvall, 1980), as does cellular fibronectin from chick embryo fibroblasts (Yamada et al., 1980). There is evidence that heparin enhances reticuloendothelial function in vivo (Kaplan and Saba, 1981; Lahnborg et al., 1976), and that fibronectin participates (Kaplan and Saba, 1981). The authors suggested that endogenous heparin was a cofactor for fibronectin-mediated macrophage phagocytosis, that it might be depleted or inhibited in response to intravascular coagulation, and that this heparin depletion would explain the macrophage depression observed during intravascular coagulation. The change in the heparin content of tissues after birth suggests that heparin functions in defense mechanisms (Nader et al., 1982). Heparin restored the antibody-forming capacity in cortisol-suppressed mice apparently by restoration of the impaired macrophage function. The authors summarized considerable evidence that heparin beneficially influences the regulatory functions of macrophages (Jokay et al., 1980). The formation of pinocytic vesicles in mouse macrophages was stimulated by microgram quantities of heparin (Cohn and Parks, 1967). A slight temporary impairment of reticuloendothelial function in rats after physiologic doses of heparin also has been reported (Bergheim et al., 1978).

G. Fibrinolysis

Fibrinolysin, or plasmin, activity is a physiologic process which is involved in thrombus dissolution and may be related to atherogenesis. Older studies showed contradictory results about the effect of heparin on this process, reporting that heparin potentiated, did not potentiate, or inhibited fibrinolysis. It was also shown that heparin accelerated the neutralization of plasmin by antithrombin, but the concentration of heparin required for this effect was 20–30 times that required for full anticoagulation (Collen et al., 1978). Only 3–11% of in vivo-formed plasmin was neutralized (Machovich et al., 1975). Thus, it is apparent that the inactivation of fibrinolysis is not a physiologically significant action of heparin. Therapeutic doses of heparin normalized prolonged postoperative clot lysis times (Comp et al., 1979). Endothelium is rich in plasminogen activator (Todd, 1972). Heparin-enhanced release of plasminogen activator occurred in pigs at heparin concentrations of 0.05 units/ml of perfusion fluid (Markwardt and Klocking, 1977). Plasminogen activator release by heparin also has been observed in rabbits (Vairel et al., 1982). The total inhibition of endothelial cell production of plasminogen activator by thrombin (Loskutoff and Levin, 1982) is also relevant, as heparin would mitigate any thrombin effect. It has been found that fibrinolysis is increased by heparin added to human blood *in vitro*, as well as following its injection *in vivo* (Vinazzer et al., 1982). Heparin may also enhance fibrinolysis of non-cross-linked fibrin clots via a process that does not involve plasmin (Malakhova et al., 1978).

III. Effects of Heparin at Different Sites

Years ago studies suggested that injected heparin had an affinity for endothelium. Uptake of heparin at endothelial cell junctions was shown by various investigators using tissue-staining techniques (Florey et al., 1959; Ohta et al., 1962; Samuels and Webster, 1952). Tissue culture studies using human and other animal endothelial cells showed metachromatic granules on the cell surface when heparin was in the culture medium (Lazzarini-Robertson, 1963). Heparin coats cells *in vitro* (Gasic and Baydack, 1962). More recent work demonstrated that heparin concentrated on rat endothelium after its injection by any route (Hiebert and Jaques, 1976a). When aortic endothelium of dogs, rabbits, or rats was exposed to dilute heparin solutions *in vitro*, it took up heparin avidly (Hierbert and Jaques, 1976b). The endothelial heparin was more concentrated than the heparin in the medium and was not removed by repeated washings. Studies using labeled heparin showed a high binding affinity for endothelial cells even when large amounts of other glycosaminoglycans were present (Glimelius et al., 1978). This binding affinity corresponded to about 10^6 heparin molecules per cell. Other findings suggested that the heparin was bound to the cell surface and was not rapidly or extensively degraded since it retained the ability to bind antithrombin. It is not known precisely how long heparin persists on the endothelial surface, but apparently it is for much longer than the interval during which anticoagulant activity can be demonstrated in the blood. Early work in rats showed that increased tissue metachromosia was present for 3–4 weeks after daily subcutaneous injections of heparin (Asplund et al., 1939). Exogenous heparin persists for about 3 weeks in mouse peritoneal macrophages (Mims, 1969). It was stored in corrective tissue for several weeks after injection (Rigdon and Wilson, 1941). Exogenous heparin taken up by liver sinusoidal and Kupffer cells was apparently not altered (Hiebert, 1981), although the evidence on this point is conflicting (Dawes and Pepper, 1979).

A. ENDOTHELIAL SURFACE

1. Restoration of Normal Electronegativity

Observations made years ago showed that a metachromatic substance released from mast cells, i.e., heparin, appeared to play a role in the repair of injured vascular and peritoneal endothelium (McGovern, 1955, 1956). Following experimental injury in rats, the reformation of the intercellular cement and the endothelial surface film involved ground substance and the metachromatic material. Normal intima has a negative electric potential which becomes positive after injury (McGovern, 1955). There is an inverse correlation between the negative charge at the endothelial surface and vascular injury *in vitro* (Sawyer et al., 1973). The injection of heparin restores normal electronegativity, apparently as a result of the attachment of heparin to the injured endothelial surface (Srinivasan et al., 1968). Surface charge affects cellular permeability. The normal negative charge on the surface of glomerular endothelial cells is a charge-based barrier to the passage of macromolecules (Kelley and Cavallo, 1979). The anionic sites distributed over the surface of vascular endothelial cells are capable of lateral migration. The cell surface can be temporarily depleted of these negatively charged sites under the influence of cationic substances (Pelikan et al., 1979; Skutelsky and Danon, 1976). The interaction of endothelium with cationic ferritin particles resulted in rapid aggregation of most anionic sites on the luminal endothelial surface, with subsequent internalization of the particles. This left most of the endothelial surface devoid of anionic sites for a time (Skutelsky and Danon, 1976). As mentioned earlier in this article, the injection of heparin restored normal glomerular filtration after capillary endothelial anionic sites were removed by protamine (Kelley and Cavallo, 1979). These various observations indicate that the normal endothelial surface electronegativity is due to functioning anionic sites which can be depleted, and that exogenous heparin corrects that depletion and rapidly restores normal function.

2. Prevention of Endothelial Injury

Heparin binds and inactivates many of the agents that injure endothelium. Among these are histamine, serotonin, lysozymes, bradykinin, angiotensin, bacterial endotoxins, and many toxins (Engelberg, 1963, 1978b; Jaques 1967, 1980). It has been known for some time that histamine injures endothelium (Majuro and Palade, 1961; Majuro et al., 1969). Histamine, serotonin, bradykinin, and angiotensin enlarge endothelial intercellular gaps, thus increasing the influx of circulating lipoproteins and other macromolecules into the arterial wall where they may become trapped. There is evidence that *de novo* histamine formation is a mediator of the increased endothelial permeability which occurs early in the atherosclerotic process (Owens and Hollis, 1979), and that histamine plays a role in lipid deposition in the plaque (Hollander et al., 1979). Antihistamine drugs have antiatherosclerotic effects in cholesteral-fed rabbits (Harman 1962; Hollander et al., 1974), and antagonists of vasoactive amines suppressed the deposition of circulating immune complexes in vascular walls (Cochrane, 1971; Knicker and Cochrane, 1968). Heparin suppressed the increased vascular permeability induced by histamine, bradykinin, and prostaglandin E_1 (Carr, 1979). Aggregated white blood cells adhere to cultured endothelial cells (Jacob, 1980). Activated complement causes white blood cell aggregation *in vivo,* and this is associated with extravascular extravasation of plasma proteins, suggesting endothelial damage (Hammerschmidt et al., 1981b). Neutrophiles are essential mediators of tissue damage in many forms of immune complex-induced injury in blood vessel walls. Heparin neutralizes lysosomal cationic proteins released by leukocytes (Saba et al., 1968) and so may decrease the resulting endothelial injury. Large doses of heparin decreased the cytotoxic effects of eosinophil major basic protein (Frigas et al., 1980). The endothelial desquamation that occurs after citrate injection (Hladovec, 1979) is inhibited by extremely small quantities of heparin (J. Hladovec, personal communication). Thrombin injures endothelial cells in culture (Galdahl and Evenson, 1981; Lough and Moore, 1975). Injected heparin localized on the endothelial surface would minimize its formation and catalyze its rapid inactivation by antithrombin. This would prevent the formation of fibrin, which itself disorganizes endothelium (Kadish, 1979). Fibrin also is an important component of atherosclerotic plaques which may enhance low-density lipoprotein deposition and which stimulate the proliferation of smooth muscle cells (Ishida and Tanaka, 1982).

Viruses injure endothelium and, when associated with hypercholesterolemia, induce an atherosclerosis in experimental animals which closely resembles the disease in man (Minick et al., 1979). Several common viruses infect and replicate in human endothelial cell cultures causing varying degrees of injury (Friedman et al., 1981). Herpesvirus infection altered intracellular lipid metabolism and increased the lipid content of chicken arterial smooth muscle cells (Fabricant et al., 1981). Receptors that promote the deposition of immune complexes in vascular tissue were induced in human endothelial cells by herpes simplex virus type 1 infection (Cines et al., 1982). There is a marked increase in platelet adherence to virally injured human vascular endothelial cells (Curwen et al., 1980). Heparin inhibits the infectivity of many of the herpesviruses (but not herpes simplex virus type 2) (Choi et al., 1978; Vaheri, 1964) in cell cultures, possibly by interfering with viral adsorption to the cell surface (Choi et al., 1978). Bacterial adherence to cell surfaces is also decreased by heparin or heparin-like compounds. Protamine pretreatment of rabbit urinary bladders results in a ninefold increase in bacterial adherence which is a prerequisite to bacterial infection (Parsons et al., 1981). Small amounts of heparin restored the normal activity of the protective bladder mucin layer (Hanno et al., 1981).

Low-density lipoproteins injure human endothelial cells in tissue culture (Evensen, 1979; Henriksen et al., 1979a), and very low density lipoproteins alter the function and viability of bovine aortic endothelial cells (Gianturco et al., 1980). Lipoproteins increase platelet adherence to arterial tissue and decrease the inhibitory action of endothelial cells on platelet function (Armstrong et al., 1980; Nordoy et al., 1978). Hypercholesterolemia inhibited the recovery of prostacyclin (PGI_2) production in reendothelialized areas of previously injured rabbit aorta (Eldor et al., 1982). Minute amounts of heparin in the culture medium inhibited the uptake of serum lipoproteins by human and animal endothelial cells in tissue culture (Lazzarini-Robertson, 1961), and heparin was the mucopolysaccharide which most effectively inhibited the uptake of low-density lipoproteins by rabbit aortic tissue (Day et al., 1974–5). This action of heparin at the endothelial surface would ameliorate endothelial injury by serum lipids and also would decrease lipid infiltration into the vascular wall.

There is no doubt that immune vascular injury enhances atherogenesis in experimental animals (Friedman et al., 1975; Minick, 1980). However, there is as yet no proof that this actually occurs in human atherosclerotic disease, although there is much circumstantial evidence to that effect. Circulating immune complexes are more frequently present in patients with acute and chronic vascular disease than in normal individuals (Furst et al., 1978; Gallagher et al., 1982). Antibodies against homologous arterial tissue have been found in the sera of over 50% of patients with arteriosclerosis (Gero et al., 1980). However, as the authors noted, these findings could be a consequence of the disease, not a primary cause. However, even if they developed later in the disease process, such antibodies could then play a harmful role. Human atherosclerotic plaques contain both soluble and tissue-bound immunoglobulins, and the terminal component of activated complement, C3 (Hollander et al., 1979). The investigators could not determine whether these findings are due to a secondary localization of antigen–antibody complexes which had been formed elsewhere, or whether they represented an autoimmune event directed against arterial tissue. They did believe that the antigen–antibody activated complement gave rise to vascular lesions. The beneficial modification of allergic reactions by heparin has been known for years (Dolowitz and Daugherty, 1965; Engelberg, 1963; Jaques, 1967). It binds and inactivates histamine, it increases circulating histaminase activity (Mahadoo et al., 1981), which would also decrease histamine effects, and it limits the activation of complement. It has been found that small doses of heparin correct the T-cell deficiency present in many patients with migraine (Thonnard-Neumann, 1981), demonstrating that heparin may be beneficial in some abnormalities of the immune system.

3. Prevention of Thrombin-Mediated Actions

Together with plasma antithrombin, endothelial heparin prevents the local generation of thrombin. Injured endothelial cells and the exposed subendothelial tissue and media have high thromboplastic activity (Astrup et al., 1959), as does human atheroma material (Lyford et al., 1967), and so may directly initiate coagulation via the generation of thrombin. Heparin blocked thrombus formation in injured rat carotid arteries (Philp et al., 1978). There is evidence that thrombin must bind to heparin before the accelerated thrombin–antithrombin reaction can take place (Griffith et al., 1979; Machovich et al., 1975; Smith, 1977). Preformed human thrombin is taken up specifically by endothelial cells, and this is not prevented by heparin (Aubrey et al., 1979). There are high-affinity binding sites for thrombin throughout the vascular bed which probably are the heparin-like anionic sites on the endothelial cell surface (Lollar and Owen, 1980), and the investigators suggested that the binding of active thrombin to these sites is required for the rapid inactivation of thrombin *in vivo*.

Thrombin, of course, is fundamentally involved in normal hemostasis. However, as with platelets, under certain conditions many of its actions can contribute to pathologic events. Normal physiology becomes pathobiology. As previously mentioned, thrombin injures cultured endothelial cells. It inhibits plasminogen activator production by endothelial cells. It is a potent mitogenic agent (Chen and Buchanan, 1975). It increases platelet adhesion to the vessel wall (Essien et al., 1978) and to endothelial cells in tissue culture (Czervionki et al., 1978). As noted before, thrombin probably functions in irreversible platelet aggregation and the platelet release reaction. In the presence of thrombin even high concentrations of prostacyclin only partially prevented platelet adherence to cultured endothelial cells (Fry et al., 1980). There are two binding sites for thrombin on platelets, and heparin inhibited high-affinity binding (Workman et al., 1978). Platelet growth factor is quickly released by small amounts of thrombin, as are platelet factor 4 and β-thromboglobulin, which are also stored in the platelet α-granules (Witte et al., 1978). There also may be a mechanism of thrombus formation in which thrombin generation and fibrin formation may precede platelet accumulation on the injured vessel wall (Packham and Mustard, 1980). There is good evidence in man that once an atheromatous plaque has formed it can grow by accretion of thrombus material. Microthrombi have been identified on the endothelial surface and were incorporated into the thickened intima in areas of developing atheromatous plaque (Velican and Velican, 1980). This occurred less often in younger subjects. Increased endothelial heparin activity following heparin injection would enhance thrombin inactivation and so minimize these harmful effects.

4. Inhibition of Other Platelet Effects and of Smooth Muscle Cell Proliferation

Apart from decreasing thrombin-mediated platelet adhesion to the vascular wall and the release of platelet products, there are other protective actions of heparin in relation to harmful platelet effects. Human platelets contain an enzyme capable of degrading the heparan present on the endothelial cell surface. Heparin inhibits this degradation although heparin is itself attacked and neutralized by the enzyme (Wasteson et al., 1977; Oldberg et al., 1980). Heparin prevents the binding of released platelet factor 4 to endothelial cells (Busch et al., 1979a). This factor also enters endothelial cells (Goldberg et al., 1980). Endothelial-bound platelet factor 4 might adversely affect the nonthrombogenic role of the endothelium.

It is accepted that smooth muscle cells from the arterial media are the cells that proliferate in the neointima of atheromatous plaques, and that platelet-derived mitogenic growth factor (PDGF) plays an important role in SMC proliferation (Ross and Glomset, 1976). It has been estimated that two-thirds of the growth-potentiating activity of normal human serum on human arterial SMC can be attributed to factors released from platelets (Witte et al., 1978). Heparin, at low concentrations, inhibited the SMC growth-stimulating effect of PDGF in vitro although it did not bind to PDGF (Hoover et al., 1980). Since these investigators found that radiolabeled heparin readily bound to the surface of SMC, they believe that an interaction of heparin with the SMC surface was involved in the inhibition of PDGF. This action of heparin was not mediated by thrombin inactivation. In vivo studies from the same laboratory showed that the heparin antiproliferative effect amounted to about a 75% reduction in plaque volume (Guyton et al., 1980). There may also be more than one platelet mitogenic factor. A platelet basic protein which stimulated growth of mouse cells was inhibited by very small amounts of heparin, in this instance apparently by direct complexing with heparin (Paul et al., 1980). There are no published studies thus far on the effect of heparin on the macrophage growth factor (Glenn and Ross, 1981; Martin et al., 1981).

Subsequent investigation showed that bovine and rat aortic endothelial cells produce both positive and negative effectors of SMC growth (Castellot et al., 1981; Karnofsky, 1981). The inhibitory activity had many properties characteristic of heparin. Additional evidence that it was heparin came from *in vitro* data. Exogenous heparin at 10 ng/ml inhibited SMC growth whereas other glycosaminglycans, including heparan, had no inhibitory activity at 10 μg/ml. These findings indicate that exogenous heparin reinforces the physiologic inhibiting influences of endothelial heparin on medical SMC overproliferation. Other studies (Castellot et al., 1982) indicate that a platelet endoglycosidase (Oosta et al., 1982) may release heparin-like components from aortic endothelial cells. The authors suggested that alterations of the normal relationships between platelet endoglycosidases, endothelial cell heparin-like molecules, and smooth muscle cells could, in part, be responsible for initiation of the atherosclerotic process.

B. BLOODSTREAM

1. Correction of Hypercoagulability

Clinicians have long been aware that some patients require much more heparin than usual to achieve therapeutic anticoagulant levels. For years there have been reports indicating an enhanced tendency to coagulation in many patients with atherosclerotic disease. Although documentation of a hypercoagulable state was difficult in the past, newer methods have afforded increasing evidence that such a condition exists and may play a role in thrombogenesis, and perhaps in atherogenesis as well. We will describe the evidence for this.

Resistance to heparin was increased by abnormal plasma globulins (Glueck et al., 1972; Holger-Madsen, 1962; Wolf and Williams, 1976) and by α-1-acid glycoprotein (Anderson and Godal, 1977). The latter is elevated in acute inflammatory and in some degenerative diseases, and may interfere with heparin activity by a steric hindrance of the heparin–thrombin interaction (Anderson and Godal, 1979). Increased serum heparin-neutralizing activity, probably due to released platelet factor 4, was found in patients with severe coronary heart disease (Cella and Russo, 1977; O'Brien, 1974). High levels of heparin-neutralizing activity are present in and released from aggregated platelets of men with coronary disease (Lin et al., 1977). These aggregated platelets disaggregate much more slowly, possibly as a result of increased thrombin generation secondary to heparin inactivation by the released platelet factors. These factors markedly impair heparin anticoagulant efficacy (Handin et al., 1981). Slow platelet disaggregation also is present in patients with peripheral arterial disease (Davis, 1973). Markedly higher blood levels of fibrinopeptide A, a very sensitive index of thrombic activity in the blood, were found in patients with ischemic heart disease (Neri-Serneri et al., 1981). Even minor respiratory illnesses may be associated with an accelerated heparin thrombin clotting time (O'Brien et al., 1980), a test of coagulation that probably reflects platelet activation. Increased levels of factor VIII shorten clotting times (Edson et al., 1967; Glynn, 1979) and the activated partial thromboplastin time, and raise the heparin requirement for attaining therapeutic anticoagulation (Glynn, 1979). Increased levels of factor VIII were present in adults who had transient ischemic attacks or completed strokes before the age of 55 years (Mettinger, 1982). Mean levels of factor VII, factor VIIIc, and of fibrinogen were significantly higher in men who subsequently died of cardiovascular disease, findings compatible with the presence of hypercoagulable states (Meade et al., 1980). The children of men who had died of ischemic heart disease before the age of 45 had lower antithrombin and higher fibrinogen levels, indicating an altered hemostatic balance (Hansen and Ibsen, 1982). Lipoproteins interfere with the activation of antithrombin III by heparin (Bleyl et al., 1975), and hyperlipemia produces a hypercoagulable state in rats (Gautheron and Renaud, 1972). Estrogens decrease the activated factor X inhibitory activity of plasma without changing the level of antithrombin III, and so produce a hypercoagulable state which is completely reversed by trace amounts of heparin (Wessler et al., 1976; Gitel et al., 1978). Activation of the coagulation system increased the sizes of experimental thrombi in rats, whereas platelet activators had no such effect (MacIomhair and Lavelli, 1978). Heparin given twice a week corrected the increased coagulability (Akman and Ulutin, 1966) and platelet adhesiveness (McDonald and Edgill, 1961) found in some patients with atherosclerotic disease. The 31% reduction in the mortality rate of seriously ill noncardiac patients in an intensive care unit who received low-dose heparin therapy probably resulted from the prevention of thromboembolic episodes (Halkin et al., 1982).

Diabetic patients are particularly prone to develop atherosclerosis, and many articles have reported increased platelet aggregation in diabetics. There is also hypercoagulability. Insulin-dependent diabetics had evidence of increased activation of the coagulation system with thrombin generation (Ek et al., 1982). Accelerated blood coagulation (Egebert, 1963) and increased factor VIII levels (Mayne et al., 1970) are present in many diabetic patients. Decreased fibrinogen survival, probably resulting from an abnormal plasma environment, was found in patients with adult-onset diabetes (Jones and Peterson, 1979). Heparin normalized the fibrinogen survival time whereas the antiplatelet drugs, aspirin plus dipyridamole, did not. Aspirin plus dipyridamole, or sulfinpyrazone, does not correct hypercoagulability (Peck, 1979). There is enhanced activation of the contact phase of the intrinsic coagulation system in diabetics (Petrassi et al., 1982).

There is a relation between viscosity and the extent of coronary disease determined by angiography (Lowe et al., 1980). Increased blood viscosity, possibly secondary to a high hematocrit value, is associated with an increased risk of subsequent coronary disease (Sortie et al., 1981). Heparin decreases blood viscosity (Erdi et al., 1976; Ruggiergo et al., 1982).

In addition, heparin neutralizes many of the cationic proteins released by leukocytes and from other sources. These proteins influence blood coagulation and fibrin precipitation and so have potential significance in thrombosis and atherogenesis (Muller-Berghauf et al., 1976). Thus, for the various reasons which have been presented, heparin is the most efficacious approach for the correction of hypercoagulability. Interestingly this has often been accomplished with intermittent smaller doses, considerably less than those required for full anticoagulation.

2. Enhancement of Fibrinolysis

The beneficial effects of heparin upon the fibrinolytic process were discussed earlier in this article. Evidence supporting the role of fibrinolysis in atherogenesis has been fully presented (Kwaan, 1979). This includes the demonstration of enhanced experimental atheroma formation when fibrinolytic inhibitors were administered, decreased circulating fibrinolytic activity in patients with atherosclerotic disease, and the inhibition of fibrinolytic activity by atherosclerotic risk factors such as increasing age, cigarette smoking, hypercholesterolemia, hypertriglyceridemia, and diabetes mellitus.

3. Inhibition of Complement Activation

The inhibition of complement activation by heparin was presented earlier. Can complement contribute to atherogenesis? It has been proposed that activated serum complement is a key factor involved in endothelial injury (Geertinger and Sorensen, 1973). The terminal component of complement activation, C3, has been identified in atherosclerotic plaques (Hollander et al., 1979), but, as mentioned before, that does not establish causality. As far as this author is aware, there are no published reports of the effect of activated complement on endothelial cells. It can contribute to myocardial damage. Inhibition of C3 by cobra venom decreased the extent of myocardial necrosis after coronary occlusion in experimental animals, illustrating the important role of complement in the inflammatory response even when activated nonspecifically (Maroko et al., 1978). Rabbits with an inherited defect of the complement system (C6 deficiency) which impairs complement activation developed significantly less atherosclerosis on an atherogenic diet than control rabbits (Geertinger and Sorensen, 1974-1975). It is possible that decreased production of complement proteins by the liver is at least partly responsible for the lack of atherosclerotic disease in chronic alcoholics (Geertinger, 1980).

4. Increase in High-Density Lipoproteins

A relative or absolute reduction in the alpha or high-density lipoproteins in patients with known coronary atherosclerotic disease was found long ago (Barr et al., 1951; Nikkila, 1953), but its significance was not fully appreciated at that time. Since then many epidemiologic studies have provided evidence that low levels of these lipoproteins are an independent and important risk factor for atherosclerosis, and that above average levels are a protective factor contributing to increased longevity (Glueck et al., 1977). High-density lipoproteins inhibited the injury to human endothelial cells caused by low-density lipoproteins (Henricksen et al., 1979b). As mentioned before, when heparin is injected one of the lipid changes is an absolute increase in the α- or high-density lipoproteins (Boyle et al., 1952; Nikkila, 1953). Chemical analysis confirmed that a higher percentage of the total cholesterol was in the α-lipoprotein fraction after heparin administration. It was noted (Nikkila, 1953) that the redistribution of lipids elicited by heparin injection was toward the pattern found in normal subjects lacking known atherosclerotic disease. Heparin therapy raised the initially depressed high-density lipoprotein levels in patients with peripheral vascular disease (Masana et al., 1981). The relation of high-density lipoproteins to atherosclerosis, and the increase in the concentration of these particles in the blood after heparin-induced triglyceride lipolysis, fits well with the positive correlation that has been demonstrated between high-density lipoprotein cholesterol and the catabolism of triglyceride-rich lipoproteins (Sauar et al., 1980). Hypertriglyceridemia is frequently associated with abnormalities of the metabolism of the major high-density lipoproteins (Rao et al., 1980). Studies of human adipose tissue lipoprotein lipase (Nikkila et al., 1978) and of postheparin plasma lipases (Taskinen et al., 1980) also support the concept that a high rate of catabolism of triglyceride-rich chylomicrons and very low density lipoproteins via lipoprotein lipase is an important factor contributing to elevated high-density lipoprotein levels.

5. Displacement of Lipoprotein Lipase Activity

It is generally believed that normally the bulk of triglyceride lipolysis occurs at the vascular endothelial surface. It has been proposed that the interaction of endothelial lipoprotein lipase with chylomicrons and triglyceride-rich lipoproteins leaves cholesterol-rich remnants (Redgrave, 1970) and β-lipoprotein in high concentrations at the endothelium (Zilversmit, 1973). Rat aortic medial smooth muscle cells showed an enhanced uptake of these remnant particles (Bierman et al., 1973). Injected heparin mobilizes lipoprotein lipase and displaces its activity into the circulating blood. A much greater portion of triglyceride lipolysis then takes place in the blood plasma leaving a lower concentration of the cholesterol-rich remnant particles at the endothelial surface available for uptake by endothelial cells. If this theory has validity, it would add another way in which injected heparin might retard atherogenesis.

6. Reduction of Serum Lipemia

The mechanism involved in the accelerated removal of triglycerides from the blood following the injection of heparin was discussed in the beginning of this article. In general, after a 20,000-unit, subcutaneous dose of heparin, with the resultant decrease in triglyceride-bearing lipoproteins, it takes about 48–72 hours for the serum lipids to revert to their initial levels if the diet has remained unaltered (Engelberg, 1958b). Accompanying the reduction in serum triglycerides, there is an increased excretion of cholesterol end products in the feces (Engelberg, 1961c). These changes in the circulating lipids have various beneficial effects.

To begin with, the reduction in the average level of circulating atherogenic lipoproteins after heparin injection results in a decreased insudation of these particles into the arterial walls. Although the magnitude of this response is hard to quantitate, decreased infiltration of cholesterol-bearing lipoproteins would not only slow the atherosclerotic process but might allow mechanisms of regression or repair to operate more efficiently. Maintained lipemia continues to injure endothelium and also adversely affects regeneration of injured endothelial cells (Reidy and Bowery, 1978).

Lipemia also has harmful effects within the blood itself which would be decreased at lower lipid levels. Increased agglutination and aggregation of erythrocytes with resultant capillary stasis has been observed after high-fat meals (Cullen and Swank, 1954; Williams et al., 1957). The infusion of fat emulsions accelerated coagulation, as shown by thrombin generation tests, an action completely prevented by low doses of heparin (Brockner et al., 1965). Human plasma lipoproteins stimulated the activation of prothrombin by factor Xa and shortened the partial thromboplastin time (Vejayagopal and Ardlie, 1978). Fats increased the resistance of blood clots to lysis (Lee et al., 1962). Platelets are more sensitive to thrombin-induced and ADP-induced aggregation when serum is more lipemic (Nordoy et al., 1968; Renaud et al., 1970). Both saturated and unsaturated fats increase platelet adhesiveness in man, with saturated fats doing so to a greater degree (Moolten et al., 1963). This action of fats is more pronounced in patients with coronary disease (Horlick, 1961). The correction of hyperlipemia in coronary patients normalized their previously shortened platelet survival times (Steele and Rainwater, 1978).

Another harmful effect of increased lipemia is upon tissue oxygen supply. This is an important area which is relevant both to atherogenesis and to myocardial function, but it has been ignored. Before discussing it further it is necessary briefly to consider how the oxygen demand of the arterial wall is met. Oxygen is supplied to the arterial wall from the vascular lumen and from the vasa vasorum. The latter nourishes the outer two-thirds of the vessel wall. The intima and inner media are entirely dependent on the diffusion of oxygen from the lumen. In larger animals and in man the avascular wall thickness of large arteries is approximately 1 mm, close to the limiting distance over which oxygen diffusion will adequately supply tissue needs (Getz et al., 1969). Oxygen tension is lowest in the media with a sharp rise in the tissue just below the normal endothelium (Ninikowski et al., 1973). Thus, the media and its smooth muscle cells are very sensitive to even slight decreases in tissue oxygen tension. In man, due to intimal thickening, the distance over which oxygen must diffuse to reach the media increases with age. In addition, early fatty lesions of atherosclerosis markedly interfere with oxygen diffusion through the intima (Heughan et al., 1973). These anatomic facts implicate the oxygen economy of the arterial wall in the pathogenesis of atherosclerosis.

Does lipemia affect tissue oxygenation? Many years ago a decreased rate of oxygen uptake by the red blood cells of hypercholesterolemic rabbits was noted (Martin and Hueper, 1942). Abnormal ballistocardiograms improved after one injection of heparin, and it was suggested that this resulted from an increased supply of oxygen to the myocardium (Engelberg, 1952). This led to studies of total oxygen consumption after heparin administration in patients with coronary heart disease; these studies showed an increase in those individuals (one-half of the total group) whose control levels were below normal (Engelberg, 1958a). The increase in oxygen consumption did not occur at the height of the anticoagulant effect, but coincided with maximal lipid clearing. The induction of angina at peak serum lipemia after high-fat meals in ischemic heart disease patients was observed (Kuo and Joyner, 1955). This postlipemic angina was quickly relieved by intravenous heparin (Kuo and Joyner, 1957). Studies of subcutaneous oxygen tension in man (Joyner et al., 1960) showed a decrease coincident with postalimentary lipemia, with a slow rise to normal levels as the serum lipemia spontaneously cleared. Following a small dose of intravenous heparin the lowered skin oxygen tension rapidly rose. In the nonlipemia state heparin had no such effect. Lipemia prevented the usual enhancement of myocardial oxygen extraction after exercise (Regan et al., 1961). Oxygen availability in the brain tissue of hamsters was decreased after high-fat meals (Swank and Nakamura, 1960). Ear oximeter measurements showed an increase in the rate of oxygen transfer from the blood to the tissues after the injection of heparin in atherosclerotic patients (Engelberg, 1975). High-fat meals caused hypoxic changes in the electrocardiograms of patients with angina, and a decrease in the arteriovenous oxygen difference; both improved after heparin administration (Fukugaki et al., 1975). Heparin also corrected the decreased diffusion through pulmonary membranes found in some normal subjects after intravenous lipid emulsions (Greene et al., 1976). These various studies show that lipemia impairs oxygen diffusion and that this impediment is corrected by heparin.

There is considerable evidence which suggests how serum lipemia interferes with tissue oxygenation, but only a brief discussion of the mechanisms involved will be presented in this article. One such mechanism is by the formation of surface films. Such films of fat have been observed on red blood cells after fat ingestion (Fukugaki et al., 1975). Due to their property of combining with oxygen, fats affect the diffusion constant of oxygen (Davidson et al., 1952). The diffusion of oxygen through plasma was decreased by increasing concentrations of plasma protein and lipoproteins even over normal physiologic ranges (Chisolm et al., 1972). It is also probable that the increased adhesiveness, aggregation, and rouleaux formation of erythrocytes observed after fat meals (Cullen and Swank, 1954) limits the amount of cell surface available for oxygen diffusion. The increased cholesterol content in the red blood cells of hypercholesterolemic animals occurs primarily in the cell membrane where it may serve as a barrier to oxygen transfer (Steinbach et al., 1974).

Many years ago the early evidence supporting anoxemia as a major factor in the genesis of atherosclerosis was reviewed (Hueper, 1944). There is more recent evidence that hypoxia accelerates the atherosclerotic process (Astrup, 1969; Helin and Lorenzen, 1969). Hypoxia injures endothelium (Morrison et al., 1977), reduces enzymes in the arterial media (Adams et al., 1963), stimulates fibroblast production of mucopolysaccharides and collagen (Zemplenyi, 1974), and impairs lipid degradation by smooth muscle cells (Albers and Bierman, 1976), thereby impairing lipid removal from the arterial wall. It increases lipoprotein insudation into the vascular wall which in turn increases the metabolic requirements for oxygen, thus aggravating the hypoxia (Getz et al., 1969). Studies of human endothelial cells in tissue culture indicated that local hypoxia at the cellular level may accelerate atherogenesis by initiating a series of self-sustaining metabolic abnormalities (Robertson, 1968). Thus, decreasing the lipemic impediment to oxygen diffusion would improve the oxygen supply to the arterial wall and minimize the many harmful effects of hypoxia. It is probable that this action of heparin is one of its important effects relative to the prophylaxis of atherosclerosis (Engelberg, 1980).

7. Correction of Deficiency of Circulating Endogenous Heparin Activity

Many early studies using fat tolerance tests showed that there is a delay in the removal of ingested fats from the blood in atherosclerotic patients. Since it is highly probable that heparin plays a normal role in the function of lipoprotein lipase, one of the factors that could contribute to the delay would be an inadequate amount of endogenous heparin activity in the bloodstream or at the endothelial surface. Inhibitors of or interference with heparin could also impair the clearance of fat from the blood.

The normal presence of chemically defined heparin in the human bloodstream has not been proven yet. It has been shown in rat blood (Horner, 1975). However, extracts have been obtained from normal human plasma which showed the biologic activity of heparin. In 1947 such material was isolated from a large volume of human and horse plasma (Astrup, 1947). In 1954 an anticoagulant, heparin-like substance was extracted from small quantities of human plasma in two independent laboratories (Freeman et al., 1954; Nilsson and Wenckert, 1954). The next year increased yields were obtained by a new extraction method involving tryptic digestion of the plasma proteins, thus freeing protein-bound substances (Engelberg et al., 1955). The extracted material showed anticoagulant activity on recalcified sheep plasma, it was metachromatic and moved identically with commercial heparin on paper electrophoresis, and it was neutralized by protamine. A later study showed that the extract inhibited the generation of intrinsic plasma thromboplastin (Engelberg, 1962). It is possible that the plasma extract contained other mucopolysaccharides besides heparin. However, it was quantitated by anticoagulant activity and, of the possible contaminants, the one closest to heparin is heparan sulfate, which has only a small fraction of the anticoagulant activity of heparin. A small amount of heparan sulfate has been identified in human plasma (Calatroni et al., 1969). As noted earlier, X-ray diffraction studies have shown that 20% of the heparan molecule is identical to heparin (Atkins and Neiduszynski, 1977), and it is probable that the heparin-like activity of heparan resides in its heparin moiety. Also, from the standpoint of physiological function, it matters very little whether the material present in human plasma is heparin, heparan, or both. What is important is that normal human plasma contains biologic heparin activity.

However, some investigators do not believe that this has been proven. Oddly, even in more recent published studies in which circulating heparin activity was not demonstrated, the essential step of freeing heparin from its protein bonds was omitted (Jacobson and Lindahl, 1979). When one considers that heparin binds to many plasma proteins (McKay and Laurell, 1980), it is apparent that small quantities of heparin normally would not circulate in the blood as free or unbound heparin. When trypsin was used, endogenous heparin was found in normal dog blood (Eiber and Danishevsky, 1957). It was also demonstrated in ox blood after tryptic digestion (Charles and Scott, 1935). The presence of heparin in human plasma was confirmed in a study published in Turkey (Ozdamar, 1962). There is evidence that the anticoagulant activity of heparinoid drugs results from their displacement of endogenous circulating heparin from protein to which it is bound (Jacques, 1979; Pulver, 1965). After pronase digestion, a substance was demonstrated in human serum which competed with commercial heparin for attachment to a protamine–Sepharose column (Dawes and Pepper, 1982). Chondroitin sulfates were also present but were eliminated by digestion with chondroitinase ABC. The release of heparin from human platelets adherent to a solid surface, but not in suspension, was also shown (Vanucchi et al., 1982). The evidence that it was indeed heparin was the following: it prolonged thrombin times; the anticoagulant activity was not affected by boiling, proteolysis, testicular hyaluronidase, or chondroitinase ABC; it was mostly destroyed by nitrous acid; and it migrated like heparin on glycosaminoglycan electrophoresis. These findings add further evidence that endogenous heparin activity is present in the bloodstream of man. Macromolecular heparin has been found in the tissues of rats and monkeys (Horner, 1977). This higher molecular weight heparin must be depolymerized to a smaller molecule before it shows lipoprotein lipase activation and possibly anticoagulant activity. This investigator speculated that impairment of heparin depolymerization could reduce endogenous heparin activity. There are no published studies of the existence of macromolecular heparin in man.

The anticoagulant activity found in normal human plasma using the extraction method involving the tryptic digestion of the plasma proteins (Engelberg et al., 1955) ranged from 0.1 to 0.24 units of heparin per milliliter of plasma (Engelberg, 1961a), a value very close to that obtained earlier (Astrup, 1947). This low level of activity is physiologically significant, however, when it is considered that only 0.005–0.01 units of heparin per milliliter of blood facilitates thrombin inactivation (Machovich, 1975; Smith and Sundboom, 1981). More relevant to this discussion, however, was the finding of an inverse relationship in man between endogenous plasma heparin and the triglyceride-bearing, very low density (SP 12-400) lipoproteins (Engelberg, 1961b). The results were statistically significant ($p < 0.01$), but the relatively low correlation coefficient (-0.3) indicated that other factors besides plasma heparin activity affected the lipoprotein levels. It is unfortunate that these findings have not been investigated by others using reliable methods for the determination of circulating endogenous heparin activity, for they are both physiologically and clinically significant. The results indicate that hypertriglyceridemia, with its secondary hypercholesterolemia and accelerated atherogenesis, results in part from a deficiency of endogenous intravascular heparin activity, analogous to the insulin deficiency in some patients with diabetes mellitus. There was one confirmatory study in Turkey in which average plasma heparin was significantly lower in patients with cerebral atherosclerosis than in normal controls (Ozdamar, 1962). Studies of mast cells, which synthesize and store heparin, in experimental animals have also led to the suggestion that a high susceptibility to atherosclerosis might be related to a deficiency of endogenous heparin supply. (Constantinides, 1953; Sue and Jacques, 1976).

Apart from a low supply of endogenous heparin, there might be a relative deficiency in the face of an excessive functional demand or in the presence of inhibitors. Different substances which contribute to hypercoagulability or to hyperlipemia by blocking heparin activity were described earlier in this article. Thrombin (Pilgeram and Tu, 1957), extracts of white blood cells and platelets (Fekete et al., 1958), and tissue extracts (Klein et al., 1958), inhibit heparin-activated triglyceride lipolysis to some extent. Circulating clotting factors and lipoproteins compete for heparin (Hirsch et al., 1960). The addition of lipoproteins to a mixture of heparin and serum prevented the activation of antithrombin by heparin (Bleyl et al., 1975). The influence of serum lipoproteins and of fat meals on antithrombin activity (Winter et al., 1982) may be due to competition for binding to available heparin. Low-density lipoproteins, which specifically bind to high-molecular-weight heparin and form a soluble complex (Pan et al., 1978), inhibit the action of heparin in catalyzing the inactivation of activated factor X by antithrombin (Lane and MacGregor, 1981). Fibronectin-mediated macrophage phagocytosis, which requires heparin, is depressed during thrombin-induced intravascular coagulation, and it is corrected by low doses of heparin (Kaplan and Saba, 1981). The binding of low-density lipoproteins, of released platelet factors, of complement proteins, and of other normal plasma proteins (McKay and Laurell, 1980) to heparin clearly suggests that they may compete for available heparin and so impair its functions to some extent unless endogenous heparin supplies are abundant. Injection of exogenous heparin would correct any inadequacy of endogenous heparin production, or supplement its normal function (Engelberg, 1977) in the face of an excessive load. This reinforcement of normal function usually is the *modus operandi* when biologic substances such as insulin or thyroid are used therapeutically.

IV. Results of Heparin Therapy

It is not the intent of this article to present a thorough discussion of the results of heparin therapy in the prevention of atherosclerosis. This has been done elsewhere (Engelberg, 1978b; Engelberg, 1980). Briefly, however, heparin decreased the extent of atherosclerosis in experimental animals on an atherogenic diet. In patients who have recovered from an acute mycardial infarction, heparin has been used subcutaneously in long-term studies in doses of 30,000–40,000 units per week. Injection schedules varied from daily to twice a week. Except for the first postinfarction year, when the majority of deaths result from severe pump failure or unstable electrical mechanisms, the results have been excellent. There was a 75% reduction in mortality from coronary disease in the heparin-treated patients as compared to control groups, thus corroborating the findings of the initial study (Engelberg et al., 1956). Side effects have been minimal, and fractures or fatalities due to heparin have not occurred when heparin was given in small intermittent doses even over a 10–25 year period. In preinfarction or unstable angina the use of heparin has also markedly reduced mortality (Ruggiergo et al., 1982; Telford and Wilson, 1981). The benefits of heparin therapy were maintained for several months after heparin was stopped (Telford and Wilson, 1981). In view of these good results with heparin in controlled studies in experimental animals and in man, and the extensive rationale presented in this contribution, the use of heparin for the prophylaxis and therapy of atherosclerotic disease merits a wider application.

ACKNOWLEDGMENT

The work done in our laboratory was supported by the California Arteriosclerosis Research Foundation.

References

Adams, C. W. M., Bayliss, O. B., and Ibrahim, M. Z. M. (1963). *J. Pathol. Bacteriol.* **86,** 421–430.
Akbar, H., and Ardlie, N. G. (1978). *Br. J. Haematol.* **38,** 381–390.
Akman, N., and Ulutin, O. M. (1966). *Heomstase* **6,** 381–392.
Albers, J. S., and Bierman, E. L. (1976). *Biochim. Biophys. Acta* **424,** 422–439.
Alameda, S., Rosenberg, R. D., and Bing, D. H. (1983). *J. Biol. Chem.* **258,** 785–791.
Allen, C., Saba, T. M., and Molnar, J. (1973). *J. Reticuloendothel. Soc.* **13,** 410–423.
Anderson, N. G., and Fawcett, B. (1950). *Proc. Soc. Exp. Biol. Med.* **74,** 768–771.
Anderson, P., and Godal, H. C. (1977). *Haemostasis* **6,** 339–346.
Anderson, P., and Godal, H. C. (1979). *Thromb. Res.* **15,** 857–868.
Armstrong, M. L., Peterson, R. E., Hoak, J. C., Megan, M. B., Cheng, F. H., and Clarke, W. R. (1980). *Atherosclerosis* **36,** 89–100.
Asplund, J., Borell, U., and Holgren, H. (1939). *Z. Anat. Forsch.* **46,** 16–67.
Astrup, P. (1947). *Acta Pharmacol. Toxicol.* **3,** 165–178.
Astrup, P. (1969). *Ann. Int. Med.* **71,** 426–437.
Astrup, T., Albrechtsen, O. K., Classen, M., and Rasmussen, J. (1959). *Circ. Res.* **7,** 969–976.
Atkins, E. D., T., and Neiduszynski, I. A. (1977). *Fed. Proc. Fed. Am. Soc. Exp. Biol.* **36,** 78–83.
Aubrey, B. J., Hoak, J. C., and Owen, W. G. (1979). *J. Biol. Chem.* **254,** 4092–4095.
Baker, P. J., Lint, T. F., McLeod, B. C., Behrends, C. L., and Gewurz, H. (1975). *J. Immunol.* **114,** 554–558.
Barr, D. P., Russ, E. M., Eder, H. A., Rogmunt, J., and Aronson, R. (1951). *Am. J. Med.* **11,** 480–493.
Bengtsson, G., and Olivecrona, T. (1981). *FEBS Lett.* **128,** 9–12.
Bergheim, L. E., Ahlgren, L. T., Grundfeldt, M. B., Lahwborg, E., and Schildt, B. E. (1978). *J. Reticuloendothel. Soc.* **23,** 21–28.
Besterman, E. M. M., and Gillett, M. P. I. (1973). *Atherosclerosis* **17,** 503–513.
Bierman, E. L., and Albers, J. J. (1976). *Ann. N.Y. Acad. Sci.* **295,** 199–206.
Bierman, E. L., Eisenberg, S., Stein, O., and Stein, Y. (1973). *Biochim. Biophys. Acta* **329,** 163–169.
Biggs, R., Douglass, A. A., and MacFarlane, R. G. (1953). *J. Physiol. (London)* **122,** 554–569.
Blackwell, J., Schodt, K. P., and Gelman, R. A. (1977). *Fed. Proc. Fed. Am. Soc. Exp. Biol.* **36,** 98–100.
Bleyl, H., Addo, O., and Roka, L. (1975). *Thromb. Diath. Haemorrh.* **34,** 549.
Blumenstock, F. A., Saba, T. M., Weber, P., and Laffin, R. (1978). *J. Biol. Chem.* **253,** 4287–4291.
Blumenstock, F. A., Saba, T. M., Roccario, E., Cho, E., and DeLaughter, M. (1980). *Fed. Proc. Fed. Am. Soc. Exp. Biol.* **39,** 1034.
Boyle, E., Bragdon, J. H., and Brown, R. K. (1952). *Proc. Soc. Exp. Biol. Med.* **81,** 475–477.

Bragdon, J. H., and Havel, R. J. (1954). *Am. J. Physiol.* **177,** 128–133.
Braunwald, E. (1980). *New Engl. J. Med.* **302,** 290–292.
Brockner, J., Amris, C. J., and Larsen, V. (1965). *Acta Chir. Scand. Suppl.* **343,** 48–55.
Brown, W. D. (1952). *Q. J. Exp. Physiol.* **37,** 75–84.
Buchanan, M. R., De Jana, E., Cazenave, J. P., Mustard, J. F., and Hirsch, J. (1979). *Thromb. Res.* **16,** 551–555.
Buonassi, V. (1973). *Exp. Cell Res.* **76,** 363–368.
Buonassi, V. (1975). *Biochim. Biophys. Acta* **185,** 1–10.
Busch, C., Dawes, J., Pepper, D. S., and Wasteson, A. (1979a). *Thromb. Haemostasis* **42,** 43.
Busch, C., Ljungman, C., Heldin, C. M., Wastson, E., and O'Brink, B. (1979b). *Haemostasis* **8,** 142–148.
Calatroni, A., Donnelly, P. V., and DiFerrante, N. (1969). *J. Clin. Invest.* **48,** 332–343.
Carr, J. (1979). *Thromb. Res.* **16,** 507–516.
Castellot, J. J., Jr., Addonizio, M. L., Rosenberg, R. D., and Karnofsky, M. J. (1981). *J. Cell Biol.* **90,** 373–379.
Castellot, J. J., Jr., Favreaux, L. V., Karnofsky, M. J., and Rosenberg, R. D. (1982). *J. Biol. Chem.* **257,** 11256–11260.
Caughman, G. B., Boackle, R. J., and Vesely, J. (1982). *Mol. Immunol.* **19,** 287–295.
Cella, G., and Russo, R. (1977). *Thromb. Haemostasis* **38,** 696–700.
Chajek, T., Stein, O., and Stein, Y. (1978). *Biochim. Biophys. Acta* **528,** 456–462.
Charles, A. F., and Scott, D. A. (1935). *J. Biol. Chem.* **102,** 431–440.
Chen, L. B., and Buchanan, J. M. (1975). *Proc. Natl. Acad. Sci. U.S.A.* **72,** 131–135.
Chen, Y. C., and Wu, K. K. (1980). *Br. J. Haematol.* **46,** 263–268.
Chisolm, G. M., Gainer, J. L., Stoner, G. E., and Gainer, V. J., Jr. (1972). *Atherosclerosis* **15,** 327–343.
Choi, Y. C., Swack, N. S., and Hsiung, G. D. (1978). *Proc. Soc. Exp. Biol. Med.* **157,** 569–571.
Chong, B. H., Pitney, W. R., and Castaldi, P. A. (1982). *Lancet* **2,** 1246–1248.
Cines, D. B., Lyss, A. P., Bina, M., Corkey, R., Kefalides, N. A., and Friedman, H. M. (1982). *J. Clin. Invest.* **69,** 123–128.
Cochrane, C. G. (1971). *J. Exp. Med.* **134,** 75s–89s.
Cohn, Z. A., and Parks, E. (1967). *J. Exp. Med.* **125,** 213–230.
Collen, D., Semeraro, N., Telesforo, P., and Verstraete, M. (1978). *Br. J. Haematol.* **39,** 101–110.
Comp, P. C., Jacocks, R. M., and Taylor, F. B., Jr. (1979). *J. Lab. Clin. Med.* **93,** 120–127.
Constantinides, P. (1953). *Science* **117,** 505–507.
Cullen, C. F., and Swank, R. L. (1954). *Circulation* **9,** 335–346.
Curwen, K. D., Gimbrone, M. A., Jr., and Handin, K. A. (1980). *Lab. Invest.* **42,** 366–374.
Czervionki, R. L., Hoak, J. C., and Frey, G. L. (1978). *J. Clin. Invest.* **62,** 847–856.
Davidson, D., Eggleton, P., and Foggie, P. (1952). *Q. J. Exp. Physiol.* **37,** 91–105.
Davies, J. A., and Mengs, V. C. (1982). *Thromb. Res.* **26,** 31–41.
Davis, J. W. (1973). *Angiology* **24,** 391–397.
Dawes, J., and Pepper, D. S. (1979). *Thromb. Res.* **14,** 845–860.
Dawes, J., and Pepper, D. S. (1982). *Thromb. Res.* **27,** 388–396.
Day, C. E., Powell, J. R., and Levy, R. S. (1974–5). *Artery* **1,** 126–137.
Dolowitz, D. A., and Dougherty, T. F. (1965). *Ann. Allergy* **23,** 309–317.
Doran, J. E., Mansberger, A. R., Edmondson, H. T., and Reese, A. C. (1981). *J. Reticuloendothel. Soc.* **29,** 275–284.
Edson, J. R., Krivit, W., and White, J. G. (1967). *J. Lab. Clin. Med.* **70,** 463–470.

Egebert, O. (1963). *J. Lab. Clin. Med.* **15**, 833–841.
Egelrud, T. (1973). *Biochim. Biophys. Acta* **296**, 124–129.
Eiber, H. B., and Danishevsky, I. (1957). *Proc. Soc. Exp. Biol. Med.* **94**, 801–805.
Eika, C. (1973). *Thromb. Res.* **2**, 349–357.
Eika, L. (1971). *Scand. J. Haematol.* **8**, 216–222.
Ek, I., Thunell, S., and Blomback, M. (1982). *Scand. J. Haematol.* **29**, 185–191.
Eldor, A., Allan, G., and Weksler, B. B. (1980). *Thromb. Res.* **21**, 719–723.
Eldor, A., Falsone, D. J., Hajjar, D. P., Minick, C. R., and Weksler, B. B. (1982). *Am. J. Pathol.* **107**, 186–195.
Engelberg, H. (1952). *Am. J. Med. Sci.* **224**, 487–497.
Engelberg, H. (1955). *Am. J. Physiol.* **181**, 309–312.
Engelberg, H. (1956). *J. Biol. Chem.* **222**, 601–610.
Engelberg, H. (1957). *Proc. Soc. Exp. Biol Med.* **95**, 394–397.
Engelberg, H. (1958a). *Am. J. Med. Sci.* **236**, 175–182.
Engelberg, H. (1958b). *Circ. Res.* **6**, 266–270.
Engelberg, H. (1958c). *Metabolism* **7**, 172–178.
Engelberg, H. (1958d). *Proc. Soc. Exp. Biol. Med.* **99**, 489–493.
Engelberg, H. (1961a). *Circulation* **23**, 578–581.
Engelberg, H. (1961b). *Circulation* **23**, 573–578.
Engelberg, H. (1961c). *Metabolism* **10**, 439–455.
Engelberg, H. (1962). *Proc. Soc. Exp. Biol. Med.* **109**, 814–817.
Engelberg, H. (1963). "Heparin: Metabolism, Physiology and Clinical Application." Thomas, Springfield, Illinois.
Engelberg, H. (1964). *Proc. Soc. Exp. Biol. Med.* **116**, 422–425.
Engelberg, H. (1965). *J. Atheroscler. Res.* **6**, 240–246.
Engelberg, H. (1969). *J. Atheroscler. Res.* **10**, 353–358.
Engelberg, H. (1975). *Adv. Exp. Med. Biol.* **52**, 299–309.
Engelberg, H. (1977). *Fed. Proc. Fed. Am. Soc. Exp. Biol.* **36**, 70–72.
Engelberg, H. (1978a). *Pathobiol. Annu.* **8**, 85–104.
Engelberg, H. (1978b). *Monogr. Atheroscler.* **8**, 2–72.
Engelberg, H. (1980). *Am. Heart J.* **99**, 359–372.
Engelberg, H., Dudley, A., and Freeman, L. (1955). *J. Lab. Clin. Med.* **46**, 653–656.
Engelberg, H., Kuhn, R., and Steinman, M. (1956). *Circulation* **13**, 489–498.
Erdi, A., Thomas, D. P., Kakkar, V. V., and Lani, D. A. (1976). *Lancet* **2**, 342–343.
Essien, E. M., Cazenave, J. P., Moore, S., and Mustard, J. F. (1978). *Thromb. Res.* **13**, 69–78.
Evensen, S. A. (1979). *Haemostasis* **8**, 203–210.
Fabricant, C. G., Hajjar, D. P., Minick, C. R., and Fabricant, J. (1981). *Am. J. Pathol.* **105**, 176–185.
Fekete, L. L., Lever, W. F., and Klein, E. (1958). *J. Lab. Clin. Med.* **52**, 680–685.
Fielding, P. E., Shore, V. G., and Fielding, C. J. (1974). *Biochemistry* **13**, 4318–4323.
Florey, H. W., Poole, J. C. F., and Meek, G. A. (1959). *J. Pathol. Bacteriol.* **77**, 625–636.
Freeman, L., Engelberg, H., and Dudley, A. (1954). *Am. J. Clin. Pathol.* **24**, 599–606.
Friedman, H. M., Macarak, E. J., MacGregor, R. R., Wolfe, J., and Kefalides, N. A. (1981). *J. Infect. Dis.* **143**, 266–273.
Friedman, R. J., Moore, S., and Singal, D. P. (1975). *Lab. Invest.* **32**, 404–415.
Frigas, E., Loegering, D. A., and Gleich, G. L. (1980). *Lab. Invest.* **42**, 35–43.
Fry, G. L., Czervionka, R. L., Hoak, J. C., Smith, J. B., and Haycroft, D. L. (1980). *Blood* **55**, 271–275.
Fukugaki, H., Okamoto, R., and Matsuo, T. (1975). *Jpn. Circ. J.* **39**, 317–324.

Furst, G., Szondy, E., Szekely, J., Nanai, I., and Gero, S. (1978). *Atherosclerosis* **29**, 181–190.
Galdal, K. S., and Evenson, S. A. (1981). *Thromb. Res.* **21**, 273–284.
Gallagher, P. J., Jones, D. B., Casey, C. R., and Sharratt, E. P. (1982). *Atherosclerosis* **44**, 241–244.
Gamse, G., Fromme, H. G., and Kresse, H. (1978). *Biochim. Biophys. Acta* **535**, 544–550.
Gasic, G., and Baydack, T. (1962). *In* "Biologic Interactions in Normal and Neoplastic Growth" (W. Brennan and J. Simpson, eds.), pp. 709–717. Churchill, London.
Gautheron, P., and Renaud, S. (1972). *Thromb. Res.* **1**, 353–362.
Geertinger, P. (1980). *In* "Immunity and Atherosclerosis" (P. Constantinides ed.), pp. 151–157. Academic Press, New York.
Geertinger, P., and Sorensen, H. (1973). *Atherosclerosis* **18**, 65–71.
Geertinger, P., and Sorenson, H. (1974–1975). *Artery* **1**, 177–183.
Gero, S., Szondy, E., Furst, E., and Howath, M. (1980). *In* "Immunity and Atherosclerosis" (P. Constantinides, ed.), pp. 171–180. Academic Press, New York.
Getz, G. S., Vesselinovich, D., and Wissler, R. W. (1969). *Am. J. Med.* **46**, 657–673.
Gianturco, S. H., Eskin, S. G., Navarro, L. T., Lakart, C. J., Smith, L. C., and Gotto, A. M., Jr. (1980). *Biochim. Biophys. Acta* **618**, 143–149.
Gitel, S. N., Stephenson, R. C., and Wessler, S. (1978). *Haemostasis* **7**, 10–18.
Glenn, K. C., and Ross, R. (1981). *Cell* **25**, 603–615.
Glimelius, B., Busch, C., and Hook, M. (1978). *Thromb. Res.* **12**, 773–782.
Glueck, C. J., Garside, P. S., Steiner, P. M., Miller, M., Todhunter, T., Haaf, J., Terrana, M., Pollett, R. W., and Kashyap, M. L. (1977). *Atherosclerosis* **27**, 387–406.
Glueck, H. I., MacKenzie, M. R., and Glueck, C. J. (1972). *J. Lab. Clin. Med.* **79**, 731–744.
Glynn, M. F. Y. (1979). *Am. J. Clin. Pathol.* **71**, 397–400.
Godal, H. C. (1960). *Scand. J. Clin. Lab. Invest.* **12**, 56–65.
Goldberg, I., Stemerman, M. B., and Handin, R. I. (1980). *Science* **209**, 611–612.
Gordon, R. S., Jr., Boyle, E., Brown, R. K., Cherkes, A., and Anfinsen, C. B. (1953). *Proc. Soc. Exp. Biol. Med.* **84**, 168–170.
Graham, D. M., Lyon, T. P., Gofman, J. W., Jones, H. B., Yankley, A., Simonton, J., and White, S. (1951). *Circulation* **4**, 666–673.
Greene, H. L., Hazlett, D., and Demaree, R. (1976). *Am. J. Clin. Nutr.* **29**, 127–135.
Gregorius, F. K., and Rand, R. W. (1976). *Surgery* **79**, 584–589.
Griffith, M. J., Kingdon, H. S., and Lundblad, R. L. (1979). *Biochem. Biophys. Res. Commun.* **87**, 686–690.
Grossman, M. I. (1952). *J. Lab. Clin. Med.* **40**, 805–812.
Groves, H. M., Kinlough-Rathbone, R. L., Richardson, M., Jorgensen, L., Moore, S., and Mustard, J. F. (1982). *Lab. Invest.* **46**, 605–612.
Gruner, A., Hilden, K., and Hilden, T. (1953). *Scand. J. Clin. Lab. Invest.* **5**, 241–249.
Gudervich, P. N., Gebelman, L., and Molnar, J. (1980). *Fed. Proc. Fed. Am. Soc. Exp. Biol.* **39**, 547.
Guyton, J. R., Rosenberg, R. D., Clowes, A. W., and Karnofsky, M. J. (1980). *Circ. Res.* **46**, 625–634.
Hahn, P. F. (1943). *Science* **98**, 19–20.
Halkin, H., Goldberg, J., Modan, M., and Modan, B. (1982). *Ann. Intern. Med.* **96**, 561–564.
Ham, J. M. (1969). *Aust. J. Exp. Biol. Med. Sci.* **47**, 755–760.
Hammerschmidt, D. E., Greenberg, C. S., Yamada, O., Craddock, P. R., and Sclarof, H. (1981a). *J. Lab. Clin. Med.* **98**, 68–77.
Hammerschmidt, D. E., Harris, P. D., Wayland, J. H., Craddock, P. R., and Jacob, H. S. (1981b). *Am. J. Pathol.* **102**, 146–150.

Handin, R. I., Jordan, R., and Rosenberg, R. D. (1981). In "Chemistry and Biology of Heparin, Developments in Biochemistry" (R. L. Lundblad, W. V. Brown, K. E. Mann, and H. R. Roberts, eds.), Vol. 12, pp. 393–401. Elsevier, Amsterdam.
Hanno, P. M., Fritz, R. W., Mulholland, S. G., and Wein, A. J. (1981). *Urology* **28**, 273–276.
Hansen, M. S., and Ibsen, K. K. (1982). *Scand. J. Clin. Lab. Invest.* **42**, 383–386.
Harman, D. (1962). *Circ. Res.* **11**, 277–281.
Helin, P., and Lorenzen, I. (1969). *Angiology* **20**, 1–12.
Henriksen, T., Evensen, S. A., and Carlander, E. (1979a). *Scand. J. Clin. Lab. Invest.* **39**, 361–368.
Henriksen, T., Evensen, S. A., and Carlander, B. (1979b). *Scand. J. Clin. Lab. Invest.* **39**, 369–375.
Heughan, C., Ninikowski, J., and Hunt, T. K. (1973). *Atherosclerosis* **17**, 361–367.
Hewitt, J. E., Hayes, T. L., Gofwan, J. W., Jones, H. B., and Pierce, F. T. (1952). *Cardiologia* **21**, 353–361.
Hiebert, L. M. (1981). *Thromb. Res.* **21**, 383–390.
Hiebert, L. M., and Jaques, L. B. (1976a). *Thromb. Res.* **8**, 195–204.
Hiebert, L. M., and Jaques, L. B. (1976b). *Artery* **2**, 26–37.
Hirsch, R. L., Kellner, A., and Ireland, R. (1960). *J. Exp. Med.* **112**, 699–712.
Hladovec, J. (1979). *Thromb. Haematol.* **41**, 774–778.
Holger-Madsen, T. (1962). *Acta Haematol.* **27**, 157–165.
Hollander, W., Kramsch, D. M., Franzblau, C., Paddock, J., and Colombo, M. A. (1974). *Circ. Res.* **34**, 131–141.
Hollander, W., Colombo, M. A., Kirkpatrick, B., and Paddock, J. (1979). *Atherosclerosis* **34**, 391–405.
Hoover, R. L., Rosenberg, R. D., Haering, W., and Karnofsky, M. J. (1980). *Circ. Res.* **47**, 578–583.
Horlick, L. (1961). *Am. J. Cardiol.* **8**, 459–470.
Horner, A. (1975). *Adv. Exp. Med. Biol.* **52**, 85–91.
Horner, A. (1977). *Fed. Proc. Fed. Am. Soc. Exp. Biol.* **36**, 35–39.
Huang, D. H., and LeBlanc, P. (1981). *Prostaglandins Med.* **6**, 341–344.
Hueper, W. C. (1944). *Arch. Pathol.* **39**, 162–176, 245–258, 350–363.
Ishida, T., and Tanaka, K. (1982). *Atherosclerosis* **44**, 161–174.
Iverius, P. H. (1972). *J. Biol. Chem.* **247**, 2607–2613.
Iverius, P. H., and Lindahl, U. (1972). *J. Biol. Chem.* **247**, 6610–6616.
Jacob, H. S. (1980). *Arch. Pathol. Lab. Med.* **104**, 617–620.
Jacobson, K. J., and Lindahl, U. (1979). *Thromb. Haematol.* **42**, 84.
Jaques, L. B. (1967). *Prog. Med. Chem.* **5**, 139–181.
Jaques, L. B. (1979). *Science* **206**, 528–533.
Jaques, L. B. (1980). *Pharmacol. Rev.* **31**, 99–166.
Jokay, I., Kilemenics, K., Karczag, F., and Foldes, I. (1980). *Immunobiology* **157**, 390–400.
Jones, R. L., and Peterson, C. M. (1979). *J. Clin. Inv.* **63**, 485–493.
Joyner, C. R., Jr., Horwitz, O., and Williams, P. G. (1960). *Circulation* **22**, 901–907.
Kadish, J. L. (1979). *Atherosclerosis* **33**, 409–413.
Kaplan, J. E., and Saba, T. M. (1981). *J. Reticuloendothel. Soc.* **29**, 381–393.
Karnofsky, M. J. (1981). *Am. J. Pathol.* **105**, 200–206.
Kazatchkine, M. D., Fearon, D. T., Silbert, J. E., and Austen, K. F. (1979). *J. Exp. Med.* **150**, 1202–1215.
Kazatchkine, M. D., Fearon, D. T., Metcalfe, D. D., Rosenberg, R. D., and Austen, K. F. (1981). *J. Clin. Invest.* **67**, 223–228.
Kelley, V. E., and Cavallo, T. (1979). *Lab. Invest.* **39**, 547–553.

Klein, E., Lever, W. F., and Fekete, L. L. (1958). *J. Invest. Dermatol.* **30**, 41–47.
Klein, K., Sinzinger, H., Silberbauer, K., Stachelberger, H., and Leitchner, C. (1980). *Artery* **8**, 410–415.
Knicker, W. T., and Cochrane, C. G. (1968). *J. Exp. Med.* **127**, 119–130.
Korn, E. D. (1954). *Science* **120**, 399–400.
Korn, E. D. (1957). *J. Biol. Chem.* **226**, 827–832.
Korn, E. D., and Quigley, T. W., Jr. (1957). *J. Biol. Chem.* **226**, 833–839.
Kraemer, P. N. (1971). *Biochemistry* **10**, 1437–1445.
Kuo, P. T., and Joyner, C. R., Jr. (1955). *J. Am. Med. Assoc.* **158**, 1008–1013.
Kuo, P. T., and Joyner, C. R., Jr. (1957). *J. Am. Med. Assoc.* **163**, 68–75.
Kwaan, H. C. (1979). *Artery* **5**, 285–291.
Lagegren, H., Larson, R., Olsson, P., Radergran, K., and Swedenborg, J. (1975). *Thromb. Diath. Haematol.* **34**, 557.
Lahnborg, G., Bergham, L., Lagegren, H., and Schildt, B. (1976). *Ann. Chir. Gynaecol.* **65**, 376–382.
Lane, D. A., and MacGregor, I. R. (1981). *In* "Chemistry and Biology of Heparin, Developments in Biochemistry" (R. L. Lundblad, V. F. Brown, K. G. Mann, and H. R. Roberts, eds.), pp. 301–308. Elsevier, Amsterdam.
Larsson, R., Olsson, P., and Lindahl, U. (1980). *Thromb. Res.* **19**, 43–49.
Lazzarini-Robertson, A., Jr. (1961). *Angiology* **12**, 525–535.
Lazzarini-Robertson, A., Jr. (1963). *In* "Effects of Drugs on Synthesis and Mobilization of Lipids" (D. Kritchevsky, ed.), pp. 193–204. Macmillan, New York.
Lee, K. T., Scott, R. F., Kim, D. M., and Thomas, W. A. (1962). *Exp. Mol. Pathol.* **1**, 151–161.
Lin, C. I., Davis, J. W., Yue, K. T. N., and Phillips, P. E. (1977). *Am. Heart J.* **94**, 725–730.
Lindon, J., Rosenberg, R. D., Merrill, E., and Salzman, E. (1978). *J. Lab. Clin. Med.* **91**, 47–59.
Logue, G. L. (1978). *Blood* **50**, 239–247.
Lollar, P., and Owen, W. G. (1980). *J. Clin. Invest.* **66**, 1222–1230.
Loskutoff, D. J., and Levin, E. (1982). *N.Y. Acad. Sci. Symp. Endothel.*
Lough, J., and Moore, S. (1975). *Lab. Invest.* **38**, 130–135.
Lowe, D. G. O., Morrice, J. J., Forbes, C. D., Prentice, C. R. M., Fulton, A. J., and Barbend, J. C. (1979). *Angiology* **30**, 594–599.
Lowe, D. G. O., Drummond, M. M., Lorimer, A. R., Hutton, I., Forbes, C. D., Prentice, C. R. M., and Barbend, J. C. (1980). *Br. Med. J.* **280**, 673–676.
Lyford, C. I., Connor, W. E., Hoak, J. C., and Warner, E. D. (1967). *Circulation* **36**, 284–293.
McDonald, L., and Edgill, M. (1961). *Lancet* **2**, 844–847.
McGovern, V. J. (1955). *J. Pathol. Bacteriol.* **69**, 283–293.
McGovern, V. J. (1956). *J. Pathol. Bacteriol.* **71**, 1–6.
Machovich, R. (1975). *Thromb. Diath. Haemorrhag.* **34**, 867–868.
Machovich, R., Blasko, G., and Palos, A. (1975). *Biochim. Biophys. Acta* **379**, 193–201.
MacIntyre, D. E., Handin, R. I., Rosenberg, R. D., and Salzman, E. W. (1981). *Thromb. Res.* **22**, 167–175.
MacIomhair, M., and Lavelli, S. M. (1978). *Thromb. Haemostasis* **42**, 1018–1023.
McKay, E. J., and Laurell, C. B. (1980). *J. Lab. Clin. Med.* **95**, 69–80.
McKay, E. J., Johnson, U., Laurell, A. B., Mortensson, J., and Sjoholm, A. G. (1981). *Acta Pathol. Microbiol. Scand. Sect. C* **89**, 339–344.
McLean, M. R., and Hause, L. L. (1982). *Thromb. Haemostasis* **47**, 5–7.
Mahadoo, J., Wright, C. J., and Jaques, L. (1981). *Agents Actions* **11**, 335–338.

Majuro, G., and Palade, G. E. (1961). *J. Biophys. Biochem. Cytol.* **11**, 571–581.
Majuro, G., Shea, G. M., and Leventhal, M. (1969). *J. Cell Biol.* **42**, 647–655.
Malakkova, E. A., Bazasian, G. G., Levhuck, T. P., and Yakovlev, V. A. (1978). *Thromb. Res.* **12**, 209–218.
Marin, H. M., and White, H. S. (1961). *Fed. Proc. Fed. Am. Soc. Exp. Biol.* **22**, 619–624.
Markwardt, F., and Klocking, H. P. (1977). *Haemostasis* **6**, 370–379.
Maroko, P. R., Carpenter, C. B., Chiariello, C. M., Fishbein, M. C., Radway, P., Knortman, J. D., and Hale, S. L. (1978). *J. Clin. Invest.* **61**, 661–670.
Martin, B. M., Gimbrone, M. A., Jr., Unanne, E. R., and Cotran, R. S. (1981). *J. Immunol.* **126**, 1510–1515.
Martin, G. J., and Hueper, W. C. (1942). *Proc. Soc. Exp. Biol. Med.* **48**, 452–455.
Masana, L., Rubies-Prat, J., Nubiola, A. R., Masden, S., and de Sobugram, R. C. (1981). *Biomedicine* **35**, 42–51.
Mayne, E. E., Bridges, J. M., and Weaver, J. A. (1970). *Diabetologia* **6**, 436–442.
Meade, T. W., Chakrabarti, R., Haines, A. P., North, W. R. S., Stirling, Y., and Thompson, S. G. (1980). *Lancet* **1**, 1050–1053.
Meetinger, K. L. (1982). *Thromb. Res.* **26**, 183–190.
Michalski, R., Lane, D. A., and Kakkar, V. J. (1977). *Br. J. Haematol.* **37**, 247–256.
Mims, C. A. (1969). *Aust. J. Exp. Biol. Med.* **47**, 157–164.
Minick, C. R. (1980). In "Immunity and Atherosclerosis" (P. Constantinides, ed.), pp. 111–118. Academic Press, New York.
Minick, C. R., Fabricant, C. J., Fabricant, J., and Littrento, M. M. (1979). *Am. J. Pathol.* **96**, 673–700.
Mohammed, S. F., Anderson, W. H., Smith, J. B., Chuang, H. Y. K., and Mason, R. G. (1981). *Am. J. Pathol.* **104**, 132–141.
Molnar, J., McLain, S., Allen, C., Laga, H., Gara, A., and Gelder, F. (1977). *Biochim. Biophys. Acta* **493**, 37–45.
Moolten, S. E., Jennings. P. B., and Soden, A. (1963). *Am. J. Cardiol.* **11**, 290–300.
Mora, P. T., and Young, G. B. (1959). *Arch. Biochem. Biophys.* **82**, 6–20.
Morrison, A. D., Orci, L., Berwick, L., Perrelet, A., and Winegrad, A. I. (1977). *J. Clin. Invest.* **59**, 1027–1037.
Muir, J. R. (1968). *Clin. Sci.* **34**, 261–269.
Muller-Berghauf, G., Eckhardt, T., and Kramer, W. (1976). *Thromb. Res.* **8**, 725–730.
Nader, H. B., Strauss, A. H., Takahashi, H. K., and Dietrich, C. P. (1982). *Biochim. Biophys. Acta* **714**, 292–297.
Neri-Serneri, G. G., Gensini, G. F., Abbate, R., Mugnaini, C., Foolis, S., Brunelli, C., and Parodi, O. (1981). *Am. Heart J.* **101**, 185–194.
Nichols, A. V., Freeman, N. K., Shore, B., and Rubin, L. (1952). *Circulation* **6**, 456–457.
Nikkila, E. (1953). *Scand. J. Clin. Lab. Invest.* **5** (Suppl), 8.
Nikkila, E., Taskinen, M. R., and Kakki, M. (1978). *Atherosclerosis* **29**, 497–501.
Nilsson, I. M., and Wenckert, A. (1954). *Acta Med. Scand.* **150** (Suppl.), 207.
Ninikowski, J., Heughan, C., and Hunt, T. K. (1973). *Atherosclerosis* **17**, 353–359.
Nordoy, A., Hamlin, J. T., Chandler, A. B., and Newland, H. (1968). *Scand. J. Haematol.* **5**, 458–473.
Nordoy, A., Swenson, B., Wiebe, D., and Hook, J. C. (1978). *Circ. Res.* **43**, 527–534.
O'Brien, J. R. (1974). *Thromb. Diath. Haemorrhag.* **32**, 116–123.
O'Brien, J. R. (1975). *Curr. Ther. Res.* **18**, 79–90.
O'Brien, J. R., Etherington, M. D., and Adams, C. M. (1980). *Thromb. Res.* **19**, 55–61.
Ofosu, F., Blajckman, M. A., and Hirsh, J. (1980). *Thromb. Res.* **20**, 391–403.
Ohta, G., Susaki, H., Fugitsugu, M., Tanishima, K., and Watanabe, S. (1962). *Proc. Soc. Exp. Biol. Med.* **109**, 298–303.

Oldberg, A., Wasterson, A., Busch, C., and Hook, M. (1980). *Biochemistry* **19**, 5755–5762.
Oosta, G. M., Favreux, L. V., Becker, D. L., and Rosenberg, R. D. (1982). *J. Biol. Chem.* **257**, 11249–11255.
Owen, W. G., and Esmon, C. T. (1981). *J. Biol. Chem.* **256**, 5532–5535.
Owens, G. K., and Hollis, T. M. (1979). *Atherosclerosis* **34**, 365–373.
Ozdamar, E. (1962). Docentship thesis, Istanbul University.
Packham, M. A., and Mustard, J. F. (1980). *Circulation* **62** (Suppl. V), 26–40.
Pan, Y. T., Krueski, A. W., and Elbeim, A. D. (1978). *Biochem. Biophys.* **189**, 231–237.
Parsons, C. L., Stuffer, C., and Schmidt, J. D. (1981). *J. Infect. Dis.* **144**, 180.
Patten, R. L., and Hollenberg, C. H. (1969). *J. Lipid Res.* **10**, 374–380.
Paul, D., Niewiarowski, S., Varma, K. G., and Rucker, S. (1980). *Thromb. Res.* **18**, 883–888.
Peck, S. D. (1979). *Thromb. Haemostasis* **42**, 764–777.
Pelikan, D., Gimbrone, M. A., Jr., and Cotran, R. S. (1979). *Atherosclerosis* **32**, 69–80.
Petrassi, G. M., Vettor, R., Padovan, D., and Girolami, R. (1982). *Eur. J. Clin. Invest.* **12**, 307–312.
Philp, R. B., Francey, I., and Warren, B. A. (1978). *Haemostasis* **7**, 282–293.
Pilgeram, L. O., and Tu, A. T. (1957). *J. Appl. Physiol.* **11**, 450–458.
Pomerantz, M. W., and Owen, W. G. (1978). *Biochim. Biophys. Acta* **535**, 66–73.
Pulver, R. (1965). *Arzneim. Forsch.* **15**, 1320–1327.
Rao, S. N., Magill, P. J., Miller, N. E., and Lewis, B. (1980). *Clin. Sci.* **59**, 309–315.
Redgrave, T. G. (1970). *J. Clin. Invest.* **49**, 465–475.
Regan, T. J., Timmis, G., Gray, M., Binak, K., and Hellems, H. K. (1961). *J. Clin. Invest.* **40**, 624–630.
Reidy, M. A., and Bowyer, D. P. (1978). *Atherosclerosis* **29**, 459–466.
Renaud, S., Kinlough, R. L., and Mustard, J. R. (1970). *Lab. Invest.* **22**, 339–343.
Rent, R., Nyhrman, R., Fiedel, B. A., and Gewurz, H. (1976). *Clin. Exp. Immunol.* **23**, 264–271.
Rigdon, R. H., and Wilson, H. (1941). *Arch. Surg.* **43**, 64–73.
Robertson, A. L., Jr. (1968). *Prog. Biochem. Pharmacol.* **4**, 305–316.
Robinson, D. S., and French, J. E. (1953). *Q. J. Exp. Physiol.* **38**, 233–239.
Ronsaglione, M. C., Minno, G. D., Gaetano, G. D., and Donatti, M. O. (1980). *Thromb. Res.* **18**, 895–903.
Rosenberg, R. D. (1977). *Fed. Proc. Fed. Am. Soc. Exp. Biol.* **36**, 10–18.
Ross, R., and Glomsett, J. A. (1976). *New Engl. J. Med.* **295**, 369–375.
Ruddy, S., Gigli, I., and Austen, K. F. (1972). *New Engl. J. Med.* **287**, 489–495, 545–549, 592–596, 642–646.
Ruggiergo, H. A., Castellanos, H., Caprini, L. F., and Capressi, E. S. (1982). *Clin. Cardiol.* **5**, 215–222.
Ruoslahti, E., and Engvall, E. (1980). *Biochim. Biophys. Acta* **631**, 350–359.
Saba, T. M. (1970). *Arch. Intern. Med.* **126**, 1031–1052.
Saba, T. M., Fillius, J. P., and DiLuzio, N. R. (1966). *J. Reticuloendothel. Soc.* **3**, 398–410.
Saba, H. I., and Saba, S. R. (1981). "Chemistry and Biology of Heparin, Developments in Biochemistry" (R. L. Lundblad, W. V. Brown, K. E. Mann, and H. R. Roberts, eds.), Vol. 12, pp. 417–425. Elsevier, Amsterdam.
Saba, H. I., Roberts, H. R., and Herion, J. C. (1968). *Blood* **31**, 369–380.
Salzman, E. W., Rosenberg, R. D., Smith, M. H., Lindon, J. N., and Favreaux, L. (1980). *J. Clin. Invest.* **65**, 64–73.
Samuels, P. B., and Webster, D. R. (1952). *Am. Surg.* **136**, 422–438.

Sauar, J., Skrede, S., Erickssen, J., and Blowhoff, J. P. (1980). *Acta Med. Scand.* **208**, 199–203.
Sawyer, P. N., Stanczewski, B., Pomerance, A., Lucas, T., Stover, G., and Srinivasan, S. (1973). *Surgery* **74**, 263–275.
Shanberge, J. N., Kambayashi, J., and Nakagawa, M. (1976). *Thromb. Res.* **9**, 595–609.
Simionescu, M., Simionescu, N., Silbert, J. E., and Palade, G. E. (1981). *J. Cell Biol.* **90**, 614–627.
Skutelsky, E., and Danon, D. (1976). *J. Cell Biol.* **71**, 232–241.
Smith, G. F. (1977). *Biochem. Biophys. Res. Commun.* **77**, 111–117.
Smith, G. F., and Sundboom, J. L. (1981). *Thromb. Res.* **22**, 115–133.
Sortie, P. D., Garcia-Polmiceri, M. R., Costas, R., Jr., and Havlik, R. J. (1981). *Am. Heart J.* **101**, 456–461.
Spitzer, J. J. (1953). *Am. J. Physiol.* **174**, 551–572.
Srinivasan, S., Aaron, R., Chopra, P. S., Lucas, T., and Sawyer, P. N. (1968). *Surgery* **64**, 827–837.
Steele, P., and Rainwater, J. (1978). *Circulation* **58**, 365–367.
Steinbach, J. H., Blackshear, P. L., Varco, R. L., and Buchwald, H. (1974). *J. Surg. Res.* **16**, 134–139.
Stone, A. L. (1977). *Fed. Proc. Fed. Am. Soc. Exp. Biol.* **36**, 101–105.
Sue, T. K., and Jaques, L. B. (1976). *Atherosclerosis* **25**, 137–139.
Swank, R. L., and Nakamura, H. (1960). *Am. J. Physiol.* **198**, 217–220.
Taskinen, M. R., Glueck, C. J., Kashyap, M. L., Srivastava, L. S., Hyud, B. A., Perisutti, E., Robinson, K., Kinnunen, P. J., and Kuusi, T. (1980). *Atherosclerosis* **37**, 247–256.
Teien, A. N., Abildgaard, U., and Hook, M. (1976). *Thromb. Res.* **8**, 859–867.
Telford, A. M., and Wilson, C. (1981). *Lancet* **1**, 1225–1228.
Thomson, C., Forbes, C. D., and Prentice, C. R. M. (1973). *Clin. Sci. Mol. Med.* **45**, 485–492.
Thonnard-Neumann, E. (1981). *Ann. Allergy* **47**, 328–332.
Tiffany, M. L., and Penner, J. A. (1979). *Thromb. Haemostasis* **42**, 322–328.
Todd, A. S. (1972). *Atherosclerosis* **15**, 137–140.
Vaheri, A. (1964). *Acta Pathol. Microbiol. Scand.* Suppl. 171, 7–98.
Vairel, E. G., Brouty-Borge, H., and Toulernonde, F. (1982). *Haemostasis* **11** (Suppl. 1), 79.
Van de Water, L., Schroeder, S., Creashaw, E. B., and Hynes, R. O. (1981). *J. Cell Biol.* **90**, 32–39.
Van Eck, W. F., Peters, J. P., and Man, E. B. (1952). *Metabolism* **1**, 383–393.
Vanucchi, S., Fibbi, G., Pasquali, F., Del Rosso, M., Cappelletti, R., and Chiarugi, V. (1982). *Nature (London)* **296**, 352–353.
Velican, C. (1967). *J. Atheroscler. Res.* **7**, 517–527.
Velican, C., and Velican, D. (1980). *Atherosclerosis* **37**, 33–46.
Vejayagopal, P., and Ardlie, N. E. (1978). *Thromb. Res.* **12**, 721–733.
Vinazzer, H., Stemberger, A., Haas, S., and Blumel, G. (1982). *Thromb. Res.* **27**, 341–351.
Wallis, J., Moses, J. W., Borer, J. S., Weksler, B., Goldberg, H. L., Fisher, J., Kase, M., Taek-Goldman, K., Carter, J., and Calle, S. (1982). *Circulation* **66** (Suppl. II), 263.
Walsh, P. N. (1976). In "Heparin: Chemistry and Clinical Usage" (V. J. Kakkar and D. P. Thomas, eds.), pp. 125–131. Academic Press, New York.
Wasteson, A., Glimelius, B., Busch, C., Westermark, B., Heldin, C. H., and Norbing, B. (1977). *Thromb. Res.* **11**, 309–317.
Wautier, J. L., and Caen, J. P. (1979). *Semin. Thromb. Haematol.* **5**, 293–315.
Weiber, J. M., Yurt, R. W., Fearon, D. T., and Austen, K. F. (1978). *J. Exp. Med.* **147**, 409–421.

Weld, C. B. (1944). *Can. Med. Assoc. J.* **51,** 578-582.
Wessler, S., Gitel, S. U., Wan, L. S., and Pasternack, B. S. (1976). *J. Am. Med. Assoc.* **236,** 2179-2183.
Williams, A. V., Higginbotham, A. C., and Knisely, M. H. (1957). *Angiology* **8,** 29-40.
Winter, J. H., Bennett, B., McTaggart, F., and Douglas, A. S. (1982). *Thromb. Haemostasis* **47,** 236-438.
Witte, L. D., Kaplan, K. L., Nossel, H. L., Lages, B. A., Weiss, H. J., and Goodman, D. S. (1978). *Circ. Res.* **42,** 402-409.
Wolf, P., and Williams, S. (1976). *Thromb. Res.* **9,** 209-215.
Workman, E. F., Jr., White, G. C., and Lundblad, R. L. (1977). *Biochem. Biophys. Res. Commun.* **75,** 925-931.
Workman, E. F., Jr., White, G. C., and Lundblad, R. L. (1978). *Biochim. Biophys. Acta* **544,** 514-520.
Yamada, K. M., Kennedy, D. W., Komata, K., and Pratt, R. M. (1980). *J. Biol. Chem.* **255,** 6055-6063.
Zemplenyi, T. (1974). *Med. Clin. North Am.* **58,** 293-321.
Zilversmit, D. B. (1973). *Circ. Res.* **33,** 633-638.
Zugibe, F. T. (1962). *J. Histochem. Cytochem.* **10,** 448-461.

Lipids of Actinomycetes

J. N. VERMA AND G. K. KHULLER

Department of Biochemistry
Postgraduate Institute of Medical Education and Research
Chandigarh, India

I.	Introduction	257
II.	Lipid Composition	258
	A. Phospholipids	258
	B. Glycolipids	265
	C. Lipoamino Acids	270
	D. Glycerides	271
	E. Fatty Acids	272
	F. Mycolic Acids	274
	G. Sterols	275
III.	Subcellular Distribution	275
	A. Membrane Lipids	276
	B. Cell Wall Lipids	276
	C. Cytosolic Globular Lipids	277
IV.	Taxonomic Significance	277
	A. Fatty Acids	277
	B. Mannophosphoinositides	279
V.	Influence of Environmental Factors on Lipid Composition	280
	A. Effect of Age	280
	B. Variation of Carbon and Nitrogen Sources	282
	C. Growth Temperature	284
	D. Effect of Various Growth-Medium Additives	285
VI.	Metabolism of Lipids	288
	A. Biosynthesis	288
	B. Catabolism—Lipolytic Enzymes	292
	C. Lipid Turnover: Physiological Significance	298
VII.	Immunological Properties of Lipids	300
	A. Phospholipids	301
	B. Glycolipids	303
VIII.	Lipids and Production of Antibiotics	305
IX.	Prospects	307
	References	307

I. Introduction

The functions and metabolism of phospholipids have been extensively studied in plant and animal systems; nevertheless, their role in cellular physiology is not clearly defined. In recent years, the increase in our understanding of lipids in bacteria has been rapid and phenomenal, but their exact functions in these simple organisms are still not completely clear. *Mycobacteria, Nocardia,* and *Streptomyces* (order Actinomycetales, as classified by Asselineau, 1966) possess a close immunological relationship (Kwapinsky, 1972), and their physiologies have been the subject of exhaustive investigations in various laboratories, probably due to their clinical and industrial importance. Microbial lipidologists have studied the structures, compositions, metabolisms, and immunological properties of lipids in the actinomycetes. These bacteria are distinguished from others by the presence of a unique class of phospholipids, the phosphatidylinositol mannosides (PIMs), and Shaw (1974) suggested that the lipid composition of these organisms is of taxonomic significance. In spite of the visible phylogenetic relationship between these bacterial classes, a definite correlation of lipid composition to their evolutionary emergence has not been possible. This is only because systematic studies on the lipids of these bacteria are not available, since some facets of lipid metabolism have been exhaustively studied in one genera and not in others. In this article a close phylogenetic (and immunological) relationship between the three genera of the order Actinomycetales will form the basis for comparison of various aspects of lipid metabolism in these microorganisms.

II. Lipid Composition

A. PHOSPHOLIPIDS

A considerable portion of the microbial cell consists of lipids, in addition to the other cellular components. Although high lipid content is an auxiliary character of gram-negative bacteria (Kates, 1964), actinomycetes, which are gram-positive bacteria, possess an exceptionally high lipid content in comparison to other bacteria of this group. Lipids form up to 40% of the dry weight of the mycobacteria cell (Anderson, 1941). In general, phospholipids may form 2–4% of the dry weight of *Mycobacteria* and taxonomically related microorganisms (Kataoka and Nojima, 1967). Nevertheless, these values may differ in shaking and surface cultures.

Based on their chemical composition, the various phospholipids of actinomycetes can be classified as (1) the simplest phospholipid, i.e., phosphatidic acid; (2) phosphatidylglycerol and related phospholipids; (3) amino-group-containing phospholipids and their derivatives; or (4) phosphatidylinositol and its mannose-containing derivatives. The simplest phospholipid, phosphatidic acid, occurs in small amounts and is sometimes completely absent in bacteria, probably due to its rapid turnover. The presence of small amounts of phosphatidic acid is indicated in mycobacteria (Faure and Marechal, 1962; Subrahmanyam, 1965; Akamatsu *et al.*, 1967). The presence of this lipid component has not been reported in any other actinomycetes.

Phosphatidylglycerol is present in almost all bacteria, and usually cardiolipin coexists with it, possibly because of their precursor–product relationship (Short and White, 1972). Although the presence of phosphatidylglycerol in *Nocardia leishmanii* was demonstrated only recently (Yano *et al.*, 1970; Yribarren *et al.*, 1974), diacylphosphatidylglycerol was found in all strains of mycolic acid-containing nocardia and mycobacteria (Minnikin *et al.*, 1977). Some strains of the genus *Streptomyces* have also been found to contain phosphatidylglycerol (Kovalchuk *et al.*, 1973). The absence of phosphatidylglycerol from most other bacteria belonging to the order Actinomycetales, and the abundance of cardiolipin in these organisms, suggests that phosphatidylglycerol may be rapidly utilized for the biosynthesis of cardiolipin. Usually cardiolipin constitutes less than 10% of the total phosphatides in bacteria, and the proportion of this lipid is a unique feature of actinomycetes. However, Batrakov and Bergelson (1978) reported that it constitutes more than 65% of the total cellular lipids in members of the genus *Streptomyces*. A very unusual and previously unknown phospholipid was characterized as the butane-2,3-diol analog of phosphatidylglycerol in *Streptomyces olivaceus*, and it accounted for 20% of the total cellular lipids (Batrakov *et al.*, 1973, 1974a).

Phosphatidylserine (PS) is an obligatory intermediate for phosphatidylethanolamine (PE) biosynthesis in all microbial systems studied and is normally a minor or a trace component of bacterial phospholipids (Finnerty, 1978). The absence of phosphatidylserine decarboxylase activity in *Mycobacterium smegmatis* (Nandedkar, 1975) and *Nocardia polychromogenes* (A. K. Trana and G. K. Khuller, unpublished observations) is suggestive of the formation of PE through some alternative pathway in these organisms. The presence of PE was demonstrated in *Mycobacterium phlei, M. avium, M. tuberculosis* $H_{37}Rv$, and *M. bovis* (Akamatsu and Nojima, 1965; Subrahmanyam *et al.*, 1962). However, Barbier and Lederer (1952) identified hydroxylysine as the nitrogenous constituent of *M. phlei* phosphatides. Phosphatidylethanolamine is now known to be a common phosphatide constituent of all actinomycetes (Shaw, 1974; Batrakov and Bergelson, 1978). In addition to PE, the presence of lysophosphatidylethanolamine (LPE) in *Mycobacterium* 607 and *M. tuberculosis* $H_{37}Ra$ and $H_{37}Rv$ strains was reported by Subrahamanyam and Singhvi (1965a).

One or more methyl derivatives of phosphatidylethanolamine, that is, phosphatidyl-N-methylethanolamine, phosphatidyl-N,N'-dimethylethanolamine, and phosphatidylcholine, have been detected in several microorganisms (Goldfine, 1972). On the contrary, phosphatidylcholine and its biosynthetic intermediates are believed to be absent in actinomycetes. However, *Nocardia coeliaca* and some strains of the genus *Streptomyces* (Mitra and Maitra, 1966; Yano et al., 1969) were reported to contain phosphatidylcholine, whereas no reports are available indicating the presence of this lipid in *Mycobacteria* (Pangborn, 1968). Isolation of 2-aminoethylphosphoric acid in *Mycobacteria* (Sarma et al., 1970) suggested the presence of phosphoric acid analogs of PE in this microorganism.

The most striking feature of the phospholipid composition of actinomycetes is the presence of phosphatidylinositol and mannophosphoinositides (Ballou and Lee, 1964; Goren, 1972; Khuller, 1976, 1977; Talwar and Khuller, 1977a; Batrakov and Bergelson, 1978). However, phosphatidylinositol remained undetected in members of the genus *Streptomyces* (Kwanami, 1971) until its identification in *Streptomyces griseus* (Talwar and Khuller, 1977a). The occurrence of phosphatidylinositol in these organisms is noteworthy, since it has been shown to be the precursor for mannophosphoinositide biosynthesis (Hill and Ballou, 1966). The only exception to this generalization was an aureolic acid-producing strain of *Streptomyces* which contained only phosphatidylinositol, and not its mannoside derivatives (Batrakov and Bergelson, 1978).

A clear identification of phosphatidylinositol mannosides in actinomycetes was not possible because of the exclusive availability of inadequate and cumbersome techniques which could not clearly separate the different species of these lipids. Detailed studies on the characterization of these glycophospholipids became possible after Banerjee et al. (1974) and Khuller and Banerjee (1978) reported an improved thin-layer chromatography method for the distinct separation of these lipid components.

Several reports concerning the characterization of mannophosphoinositide from *Mycobacteria* revealed the complex structure of these lipids (Pangborn and McKinney, 1966; Brennan and Ballou, 1967; Khuller and Subrahmanyam, 1968), and most of the mannosides were found to have a high fatty acid-to-phosphorus ratio. Of these, two fatty acids were considered to be attached to glycerol, as in phosphatidylinositol, and the remaining fatty acids could be located on any hydroxyl group of inositol–mannose. The results of Brennan and Ballou (1967) suggested the possible attachment of at least one additional fatty acid to a mannose residue (see Fig. 1).

The mannophosphoinositides will be presented in this contribution in order of increasing molecular size, beginning with the monomannosides.

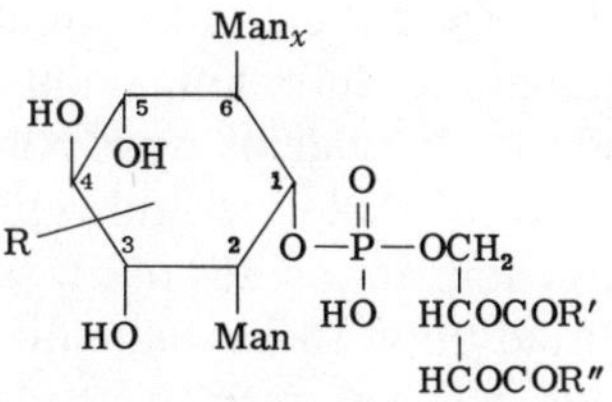

Lipids of Actinomycetes

$$\text{Structure diagram: } \text{Man}_x\text{-pyranose ring with HO groups at 4,5,6; R at 3; linked via 1-O to P(=O)(OH)-OCH}_2\text{-CH(OCOR')-CH(OCOR'')}$$

R, R', R″ = fatty acids

Man = mannose, $x = 0 - 5$ Man residues

$x = 0$ in monomannoside (PIM_1)
$x = 1$ in dimannoside (PIM_2)
$x = 2$ in trimannoside (PIM_3)
$x = 3$ in tetramannoside (PIM_4)
$x = 4$ in pentamannoside (PIM_5)
$x = 5$ in hexamannoside (PIM_6)

FIG. 1. Structure of phosphatidylinositol mannosides of mycobacteria.

The monomannosides have been reported in some *Streptomyces* species; however, this component was not detected in *S. griseus* (Talwar and Khuller, 1977a). Monomannosides were believed to be present in mycobacteria (Vilkas, 1960; Vilkas and Lederer, 1960; Ballou *et al.*, 1963; Pigretti *et al.*, 1965; Khuller and Subrahmanyam, 1968), until Banerjee and Subrahmanyam (1978) critically analyzed mycobacterial lipids and confirmed their absence. Besides their presence in the unclassified *Mycobacterium* P_6 (Motomiya *et al.*, 1969), monomannosides were shown to be present in corynebacteria (Khuller and Brennan, 1972a), and are the principal mannosides of propionibacteria (Brennan and Ballou, 1968b; Prottey and Ballou, 1968; Shaw and Dinglinger, 1969).

Two types of dimannophosphoinositides, differing in their fatty acid content, were identified in *S. griseus*. Triacyldimannosides are the major mannosides of this organism, whereas the tetraacylated analogs are present in minor amounts (Talwar and Khuller, 1977a). Mycolic acid-containing strains of nocardia and mycobacteria (named earlier), along with corynebacteria, were found to contain both monoacyl- and diacyl-phosphatidylinositol dimannosides (Minnikin *et al.*, 1977). Investigations of Brennan and Ballou (1967) and Banerjee and Subrahmanyam (1978) revealed the presence of three distinct types of dimannosides in *M. phlei* and *Mycobacterium* 607, whereas in *M. tuberculosis* $H_{37}Rv$ only diacyl- and tetraacyldimannosides were reported (Pangborn and McKinney, 1966; Sasaki, 1975). Similar to *Mycobacteria*, di-, tri-, and tetraacyldimannosides have been characterized in *Nocardia* also (Khuller, 1976; Trana *et al.*, 1980a).

Mycobacterial lipids were indicated to contain tri-, tetra-, penta-, and hexamannosides (Khuller and Subrahmanyam, 1968). Further, hexamannosides with three and four fatty acids were confirmed in *Mycobacteria* by Sasaki (1975), as well as by Banerjee and Subrahmanyam (1978). However, the presence of pentamannosides in mycobacteria was ruled out by these investigators. In addition to hexamannosides, *S. griseus* contained three other mannosides which remain to be characterized for their mannose and fatty acid content (Talwar and Khuller, 1977a).

Taking into account the observations on mannoside composition, along with the proposal of Brennan and Lehane (1971), an interesting picture of the evolutionary pattern of the order Actinomycetales emerges in which the more advanced organisms contain higher mannosides. The proposed scheme, illustrated in Fig. 2, is quite in agreement with the evolutionary pattern given by Lechevalier and Lechevalier (1967). Based on this scheme, *Mycobacteria* and *Nocardia* may be considered to be the most advanced groups of bacteria among the Actinomycetales.

Fatty Acids and Their Positional Distribution

The fatty acid composition of *M. phlei* was based on the observations of Lennarz *et al.* (1962) until palmitic, hexadecanoic, octadecanoic, and

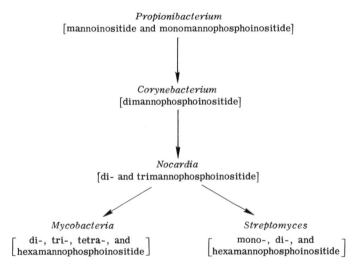

FIG. 2. Interrelationship of mannoside composition and evolutionary pattern of actinomycetes.

tuberculosteric acids were shown to be the major fatty acid constituents of mycobacterial phospholipids (Okuyama et al., 1967; Khuller and Subrahmanyam, 1974). However, Campbell and Nawaral (1969a,b) showed that the organism produced a more complex mixture of fatty acids than previously reported. Apart from the range of normal (saturated) and monoenoic fatty acids, the most striking observation was the occurrence of large amounts of monomethyl-substituted fatty acids, providing evidence for the occurrence of iso- and anteiso-fatty acids in mycobacteria. These branched-chain fatty acids consist of seven methyl-substituted fatty acids, some of which are known to be constituents of lipids of some mycobacterial species (Cason and Miller, 1963; Campbell and Nawaral, 1969a,b). Long-chain fatty acids, such as phthienoic, mycoceroic, or mycolic acid, are either absent or found in traces in purified phospholipids of *Mycobacteria* (Okuyama et al., 1967).

The composition and positional distribution of fatty acids were investigated in cardiolipin and phosphatidylinositol and its mannoside derivatives isolated from *Mycobacterium butyricum, M. tuberculosis*, and *M. phlei* (Okuyama et al., 1967). Substantial differences in fatty acid composition were observed among individual phospholipids from different strains of mycobacteria, although the fatty acid composition of individual phospholipids showed, in general, similar patterns in the same bacterial species. Fatty acid distribution in phospholipids was different from usual in that palmitic acid was located mainly at the α position, whereas octadecanoic and tuberculostearic acids were present at the β position.

The positional distribution of fatty acids in phospholipids and triglycerides of *M. smegmatis* and *M. bovis* BCG (*Bacillus Calmette-Guérin*), elucidated by Walker and his co-workers (1970), is summarized in Table I.

Table I
POSITIONAL DISTRIBUTION OF FATTY ACIDS IN LIPIDS OF *Mycobacteria*[a]

Position (carbon number)	M. bovis BCG		M. smegmatis	
	Triglycerides	Phospholipids	Triglycerides	Phospholipids
1	16:0, 18:0, 18:1	18:0, 18:1, 19 Br	18:0, 18:1, 19 Br	18:0, 18:1, 19 Br
2	16:0, 18:0	16:0, 16:1	16:0, 16:1	16:0, 16:1
3	20:0–33:0	—	20:0–26:0	—

[a] From Walker et al. (1970).

The fatty acid components given in Table I are similar to the findings of various laboratories (Tuboly, 1968; Dhariwal et al., 1976; O'Neil and Gershbein, 1976; Nesterenko et al., 1980), and their positional distribution is in accordance with other mycobacterial strains (Subrahmanyam, 1965; Okuyama et al., 1967). Based on the similarity between acyl chain composition of triglycerides and phospholipids, Walker et al. (1970) suggested the possibility that the biosynthesis of these lipids could be through a common precursor.

Phospholipids of nocardia have a similar fatty acid composition as those of mycobacteria. Studies in our laboratory revealed the presence of myristic, palmitic, palmitoleic, stearic, and oleic acids in total phosphatides of *N. polychromogenes* (saprophytic) and *Nocardia asteroides* (pathogenic) (Trana et al., 1982). Palmitic and oleic acids were found to be the most abundant fatty acids during all stages of growth in both strains of nocardia, whereas myristic and stearic acids are present in minor amounts, similar to the earlier observations of Kataoka and Nojima (1967). Further, fatty acid analysis of individual phosphatide fractions from *N. coeliaca* revealed that phosphatidylethanolamine, phosphatidylinositol, and phosphatidylcholine have similar fatty acid compositions, whereas cardiolipin had a characteristic higher palmitic acid content (Yano et al., 1969). Minnikin et al. (1977) reported the presence of C_{16} and C_{18} saturated and C_{18} unsaturated fatty acids in mannosides of *N. asteroides*.

Fatty acids of *S. griseus* phospholipids have been extensively investigated in our laboratory (Verma and Khuller, 1981a). Our observations revealed the occurrence of normal, odd-numbered, monoenoic, iso-, and anteiso-fatty acids ranging in chain lengths from C_{12} to C_{18} in total and individual major phospholipids in *S. griseus*; however, variations were noted in their acyl-chain compositions. Carbon$_{16}$ iso-acids and C_{15} anteiso-acids were the major components of all the phospholipids. A distinctive feature of the phospholipid fatty acid composition is the high content and wide range of branched-chain fatty acids not detected earlier by Kataoka and Nojima (1967). The presence of branched-chain fatty acids was also demonstrated in *Streptomyces erythrus* and *Streptomyces halstedii* (Hofheinz and Grisebach, 1965). These studies, along with those of Ballio et al. (1965), suggested that all of the streptomycetes may have a common fatty acid pattern in which saturated C_{14}, C_{15}, and C_{16} iso-acids, and saturated C_{15} anteiso-acids, are most abundant.

The occurrence of branched-chain fatty acids in nocardial lipids has yet to be established. Nevertheless, their presence in *Mycobacteria* and *Streptomyces* suggests a close taxonomical relationship between the Actinomycetales and Eubacteriales (Kaneda, 1977).

B. GLYCOLIPIDS

Several classes of glycolipids, varying in chemical structure and biological function, have been identified in actinomycetes. These include acylated sugars, glucosylglycerides, mycosides, and trehalose glycolipids consisting of acyltrehalose, cord factor and sulfatides, and wax D. Some of these aspects have been extensively reviewed with special reference to *Mycobacteria* (Lederer, 1967; Shaw, 1970; Goren and Brennan, 1979).

1. Acylated Sugars

Acylated sugars are lipids that do not contain glycerol as the backbone: fatty acids are esterified to carbohydrate residues. These are structurally the simplest glycolipids and have been identified in several microorganisms grown on glucose-containing media.

Acylglucoses were identified as the major components of *M. smegmatis* and *M. bovis* BCG grown in presence of glucose (Brennan *et al.*, 1970). Mass spectrometry of the methylated lipids and periodiate oxidation indicated that the fatty acids are esterified to position 6 of the α-D-glucopyranose ring. The major fatty acid of acylglucoses from *M. smegmatis* is corynomycolic acid (Brennan *et al.*, 1970). Khuller and Brennan (1972b) identified di- and triacylated glucoses from three strains of *Nocardia*. Tri- and tetraacylated glucoses were identified in *S. griseus* (Talwar and Khuller, 1977a).

Itoh and Suzuki (1974) and Suzuki *et al.* (1974) examined the effects of sucrose- and fructose-rich media on the qualitative changes in acylated sugars of some actinomycetes. They selected n-alkane-utilizing bacteria, which included *Nocardia rubra, Nocardia butanica, Nocardia convoluta, M. avium, Mycobacterium koda, M. tuberculosis,* and *Corynebacterium alkanum,* for their studies. The occurrence of sucrose and fructose lipids was observed in hydrocarbon-utilizing *Nocardia* and *Corynebacteria* when these were grown in their respective media. It was noted that fructose lipids were formed by *Mycobacteria* also, but that this organism was incapable of forming sucrose lipids, possibly due to growth inhibition in sucrose medium. Sucrose lipid SL-1 was identified as 6-O-mono-fatty acylglucosyl-β-fructoside, while SL-2 was partially characterized as a sucrose ester of at least two fatty acids. Two of the fructose lipids, FL-1 and FL-2, were characterized as 6-O-acylfructose and 1,6-O-diacylfructose, respectively (Itoh and Suzuki, 1974). The basic fatty acid structure of all these acylated sugars is the α-branched, β-hydroxy fatty acid; nevertheless, differences in the chain lengths are related to group characteristics of the bacteria; for example, nocardic, nocardomycolic, and mycolic acids could be referred to the individual genera.

These studies provided interesting evidence that the sugar moiety of glycolipids could be interchanged, depending on the carbon source available. Further, fatty acyl constituents of acylated sugars can provide a significant amount of information on bacterial taxonomy.

2. Glycosylglycerides

Glycosylglycerides are widely distributed in gram-positive bacteria (Shaw, 1970), but little is known about their presence in actinomycetes. During early 1970s their presence was reported in *Streptomyces* (Bergelson *et al.*, 1970), nocardia (Khuller and Brennan, 1972b), and mycobacteria (Schulz and Elbein, 1974). Batrakov and Bergelson (1978) detected two unusual glycosyldiglycerides in the cellular lipids of *Streptomyces* LA 7017. The sugar moieties of these glycolipids proved to be isoaldobionic acids (Fig. 3a,b). One of them (Fig. 3a) contained an additional fatty acid attached to the carbohydrate moiety. According to Stern and Tietz (1973a,b) the immediate precursors of isoaldobionosyldiglycerides are the corresponding uronosyldiglycerides. Batrakov and Bergelson (1978) succeeded in detecting such a glycolipid in an aureolic acid-producing *Streptomyces* strain.

Khuller and Brennan (1972b) identified a diglucosyldiglyceride from two strains of *N. polychromogenes,* while Wilkinson and Bell (1971) identified a phosphatidyl derivative of monoglucosyldiglyceride in *Pseudomonas diminuta*. Findings of Khuller and Brennan (1972b) indicate that in some nocardial membranes, the neutral lipid requirements are met by diglucosylglycerides. The presence of diglucosylglycerides in some nocardial species, compared to their complete absence in *Mycobacteria,*

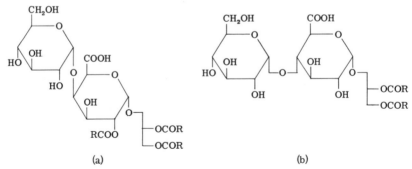

FIG. 3. Structure of unusual glycosyldiglycerides from *Streptomyces* LA 7017. [From Batrakov and Bergelson (1978).]

was considered to be a feature of some phylogenetic significance. In contrast to this speculation, Schulz and Elbein (1974) reported a particulate enzyme preparation from *M. smegmatis* which catalyzed the transfer of [^{14}C]galactose from UDP–[^{14}C]galactose and of [^{14}C]glucose from UDP–[^{14}C]glucose into chloroform-soluble products. In this study the major products were identified to be monoglucosyldiglycerides and diglucosyldiglycerides.

3. Acyltrehalose and Cord Factor

Trehalose esters with different combinations of acyl substituents were detected in mycobacteria as well as in nocardia (Vilkas and Rojas, 1964; Vilkas *et al.*, 1968; Lederer, 1976); trehalose was unsymmetrically substituted, but both the acyl chains (mostly palmitic and tuberculostearic acids) were present on the same glucose moiety. Kato and Maeda (1974) isolated a 6-monómycolate of trehalose from *M. tuberculosis* $H_{37}Rv$ which was shown to be a precursor of 6,6'-dimycolate (Lederer, 1976). In the same year, Takayama and Armstrong (1976) described a symmetric diester of trehalose isolated from the avirulent $H_{37}Ra$ strain. It was 6-monomycolate, that is, 6'-monoacetate trehalose, which was involved in the transfer of newly synthesized mycolic acids into the cell wall. Some unusual fully acylated trehalose esters were recognized in *M. phlei* and *M. smegmatis* (Goren and Brennan, 1979). Asselineau and co-workers (1969) characterized them to be substituted with unusual polyunsaturated "phleic" acids, the principal homolog being a hexatriaconta-4,8,12,16,20-pentanoic acid (for further details on this subject and mycobacterial cord factor, readers may refer to Goren and Brennan, 1979).

Ioneda *et al.* (1970) isolated and characterized the cord factors of *N. asteroides* and *Nocardia rhodochrous*. At about the same time, Yano *et al.* (1971) reported the incorporation of low- to intermediate-molecular-weight α-branched, β-hydroxy acids into trehalose derivatives that had biological properties similar to the mycobacterial cord factor. Lederer (1976) isolated a trehalose diester containing nocardomycolic acids with chain lengths of C_{40} to C_{46} from *N. rhodochrous*. However, in *N. asteroides* fatty acids of the isolated cord factor were a mixture of corynomycolic, corynomycolenic, and corynomycoladienoic acids ranging from C_{28} to C_{36} (Ioneda *et al.*, 1970). In the free lipids of this strain only C_{32} corynomycolic acids were found, whereas lipids of the cell wall contained nocardomycolic acids. It is clear from these observations that nocardomycolic acids are specifically used for glycolipids of the cell wall, as observed earlier by Bordet and Michel (1969).

4. Sulfolipids

Middlebrook et al. (1959) described an anionic sulfur-containing lipid in the pathogenic $H_{37}Rv$ (human) and Vallee (bovine) strains of M. tuberculosis. Subsequently, Ito et al. (1961) partially purified the sulfolipid from $H_{37}Rv$, and later Gangadharam et al. (1963) proposed that the levels of sulfolipids in different strains of M. tuberculosis could be used as determinants for the degree of virulence in these organisms. Further, Goren (1970a,b) showed the sulfolipids of M. tuberculosis $H_{37}Rv$ to consist of several distinct and closely related sulfatides. Another report of mycobacterial sulfatide appeared when these lipids were observed to be involved in the cell division of M. avium (McCarthy, 1976). In 1976, Goren et al. characterized the sulfolipid as 2,3,6,6'-tetraacyl-α-α'-trehalose 2'-sulfate from Mycobacteria (Fig. 4).

These discoveries led to the belief that sulfolipids were constituents of only pathogenic bacterial strains. However, in contrast to this generalization their presence was observed in several nonpathogenic mycobacterial strains, for example, M. tuberculosis $H_{37}Ra$, M. phlei, and M. smegmatis in our laboratory (Khuller et al., 1982d). Similar to pathogenic strains, sulfolipid I (SL-I) was the principal sulfolipid component of these Mycobacteria. Sulfolipid II (SL-II) and sulfolipid III (SL-III) were not detected in any of the saprophytic or avirulent strains of mycobacteria, while in virulent strains, all three sulfolipids were identified (Prabhudesai et al., 1981b). These groups of sulfolipids vary in number and position of acyl substituents on the trehalose core. Sulfolipids have been identified in several rapidly growing nonphotochromogenic strains of mycobacteria and have been thought to be taxonomically significant (Michio and Shoji, 1981).

Sulfolipids were detected in Nocardia species in our laboratory

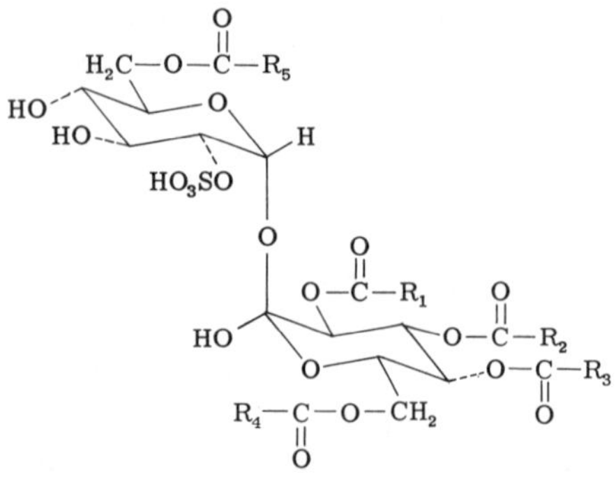

FIG. 4. General structure of sulfolipids of M. tuberculosis $H_{37}Rv$, R_1, R_2, R_3, R_4, and R_5 are different acyl substituents.

(Prabhudesai *et al.*, 1981a). A recent review on the distribution and biological significance of sulfolipids in actinomycetes may be referred to for further details on this subject (Khuller *et al.*, 1982d).

5. *Mycosides*

Mycosides are glycolipids or peptidoglycolipids of mycobacteria that have a common terminal saccharide moiety, that is, rhamnose, which is O-methylated at different positions. The distribution of these lipids was reported to be limited to a single species of mycobacteria (Smith *et al.*, 1960). However, Nacash and Vilkas (1967) suggested that this species specificity was no longer rigidly applicable. Extensive reviews dealing with the mycosides have appeared (Smith *et al.*, 1957, 1960; Lederer, 1967, 1968; Goren, 1972; Goren and Brennan, 1979).

The mycosides are divided into two categories. Category I includes those lipidic aglycons which consist of p-phenolglycosides, having a branched glycolic chain whose hydroxyl groups are esterified with straight- and branched-chain fatty acids. Gastambide-Odier and Sarda (1970) established the structures of the phenolglycerol ethers of mycoside A from *Mycobacterium kansasii* and of mycoside B from *M. bovis*. Mycosides G and G' from *Mycobacterium marinum* also fall into this category (Ville and Gastambide-Odier, 1973; Gastambide-Odier, 1973). Category II is made up of those peptidoglycolipids which contain a sugar moiety, a short peptide, and a fatty acid. Mycoside C_1, isolated from *M. avium*, was also of a similar type (Jolles *et al.*, 1961; Voiland *et al.*, 1971), and structurally distinct mycosides of this category have been characterized in *Mycobacterium* 1217 (Laneelle and Asselineau, 1968), *M. farcinogenes* (Laneelle *et al.*, 1971), and other mycobacterial species (Asselineau, 1966; Goren, 1972). It is assumed that the peptidoglycolipids occupy a surface position on the mycobacterial cell (Barksdale and Kim, 1977) and probably prevent loss of water from the cell during periods of drying.

6. *Waxes*

Chloroform-extractable, ether-soluble, but acetone-insoluble peptidoglycolipid components, derived from old cultures of *M. tuberculosis*, are designated as "waxes" (Anderson, 1943; Jolles *et al.*, 1962; Asselineau, 1966; Lederer, 1971). These are essentially composed of mycolic acids. A polysaccharide linked through a galactosamine residue to peptides and muramic acid was detected in *Mycobacteria* (Asselineau, 1966). Wax D may serve as a building block of cell wall polymers and is responsible for

Freund's adjuvant activity in the enhancement of antibody production, the stimulation of delayed hypersensitivity (Raffel, 1968; Smith, 1977), the production of allergic disseminate encephalomyelitis, and the elicitation of allergic arthritis in rats (Goren, 1972).

C. Lipoamino Acids

A number of ornithine- or lysine-containig lipids have been isolated from *M. paratuberculosis* (Laneelle and Asselineau, 1962), *Mycobacterium* 607 (Khuller and Subrahmanyam, 1970), *Streptomyces sioyaensis* (Kawanami *et al.*, 1968; Kimura and Otsuka, 1969), and other *Streptomyces* species (Batrakov *et al.*, 1971; Batrakov and Bergelson, 1978). These ornithino- and lysinolipids are widely distributed among bacterial species and are characterized by a high degree of structural heterogeneity (Finnerty, 1978). They all contain a common structural unit, a 3-hydroxy fatty acid residue linked to the α-amino group of ornithine or lysine through an amide bond. The second fatty acyl constituent may be esterified in one of three possible ways, thus forming the basis of their classification into three groups (Batrakov and Bergelson, 1978; see Fig. 5): (1) type a, esterifica-

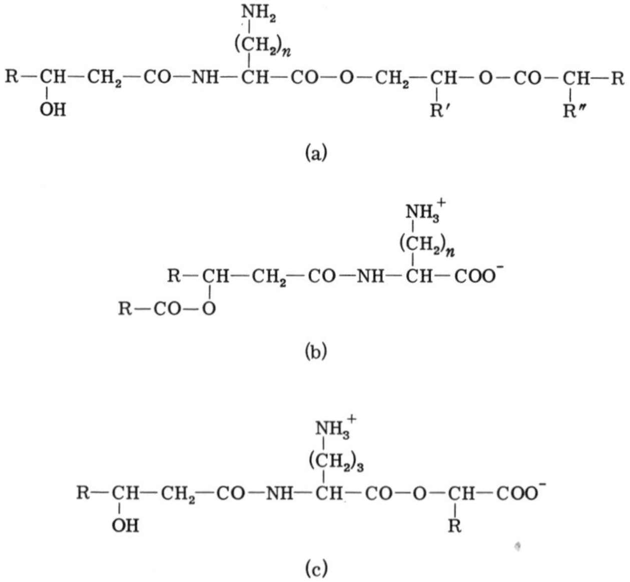

$$R-\underset{OH}{CH}-CH_2-CO-NH-\underset{\underset{NH_2}{(CH_2)_n}}{CH}-CO-O-CH_2-\underset{R'}{CH}-O-CO-\underset{R''}{CH}-R$$

(a)

$$\underset{R-CO-O}{R-CH}-CH_2-CO-NH-\underset{\underset{NH_3^+}{(CH_2)_n}}{CH}-COO^-$$

(b)

$$R-\underset{OH}{CH}-CH_2-CO-NH-\underset{\underset{NH_3^+}{(CH_2)_3}}{CH}-CO-O-\underset{R}{CH}-COO^-$$

(c)

R = alkyl; R' = H or methyl; R" = H or OH;   = 3 or 4

Fig. 5. Structure of different classes of ornithinolipids. R, Alkyl; R', H or methyl; R", H or OH. [From Batrakov and Bergelson (1978).]

tion through a connector diol to the carboxyl group of ornithine; (2) type b, esterification of a fatty alcohol to the carboxyl group of ornithine; and (3) type c, esterification via the 3-hydroxyl group of the 3-hydroxy fatty acid linked to the α-amino group. The fatty acid composition of the ornithinolipids varies considerably, depending on the organism. The acyl constituents are normal, unsaturated, iso-, anteiso-, cyclopropane, 2-hydroxy, or 3-hydroxy fatty acids. However, iso- and anteiso-fatty acids are the major constituents.

Various classes of lipoamino acids have been found in different strains of *Streptomyces* species (Kimura *et al.*, 1967a,b; Kawanami, 1971; Kawanami *et al.*, 1968; Batrakov *et al.*, 1972). Batrakov and co-workers (1974b, 1977) isolated zwitterionic ornithinolipids of type b from cells of *Streptomyces globisporus* and *Streptomyces aureoverticillus*. Ornithinolipids of both the streptomycetes were identical and form a major portion of the polar lipids. Nonetheless, in *S. aureoverticillus* ornithinolipids are the only zwitterionic lipids and appear to substitute for phosphatidylethanolamine, which is absent in this organism. The same group of investigators (Batrakov *et al.*, 1971, 1972) had isolated a type c ornithinolipid from *Streptomyces* 660-75, which produces the antibiotic albofungine. Type c lipoamino acids have not been identified in any other organism.

Laneelle *et al.* (1963) noticed the presence of L-ornithine in lipids of atypical mycobacteria (strain 1217). Though the solubility and adsorbing properties of this lipid made its separation from phospholipids difficult, it was evident from the work that this ornithinolipid contained phosphorus, suggesting it to be different from the lipoamino acids of *Streptomyces*. When grown at acidic pH or on exposure to ultraviolet light, *Mycobacterium* 607 was found to accumulate ornithinyl esters of phosphatidylglycerol (Khuller and Subrahmanyam, 1970).

D. GLYCERIDES

Attempts to study the neutral lipid composition of bacteria have been limited; however, the few reports available have been reviewed by Kates (1964), Asselineau (1966), and O'Leary (1967). Unlike the case in higher organisms, glycerides do not seem to be major lipid components in bacteria.

The acetone-soluble lipids constitute about 2–10% of the tubercle bacilli dry weight (Long, 1958). The presence of glycerides in *Mycobacteria* has been reported by various workers (Bloch *et al.*, 1957; Michel, 1957; Asselineau, 1966; Antonie and Tepper, 1969a; Dhariwal *et al.*, 1977), and mono-, di-, and triacylglycerols are the major neutral lipids of this organism (Asselineau, 1966; Barksdale and Kim, 1977). Triacylglycerols were earlier reported in other *Streptomyces* species (Kimura *et al.*, 1967b; Ballio *et al.*, 1965).

Gas–liquid chromatography of triglyceride fatty acids of *M. avium* indicated palmitic, palmitoleic, and oleic acids to be the major constituents. Fatty acids with chain lengths above C_{18} and below C_{16} were also detected. Walker *et al.* (1970) reported that triglycerides of *M. bovis* BCG and *M. smegmatis* were esterified with C_{18} fatty acids at position 1, C_{16} fatty acids at position 2, and C_{20} or higher chain length fatty acids at position 3 (Table I). The fatty acids from acetone-soluble lipids of *N. asteroides, Nocardia brasiliensis, Nocardia lurida,* and *Nocardia rugosa* consisted of straight-chain saturated and unsaturated and branched-chain iso- or anteiso-fatty acids (Bordet and Michel, 1963). Straight-chain fatty acids in these bacteria varied from C_{14} to C_{19}; however, palmitic acid was the major component. The unsaturated fatty acids consisted of 9-hexadecenoic, 10-hexadecenoic, and oleic acids. The iso- and anteisoacid series consisted of saturated branched-chain C_{15}, C_{16}, and C_{17} fatty acids, along with C_{18} isoacid and 10-methyloctadecanoic acid.

Triacylglycerols have been observed to be of physiological significance. Weir *et al.* (1972) and McCarthy (1971, 1974) suggested that triglycerides, stored in the mycobacterial cell, are a ready source of energy in the cell. Similarly, triglycerides in *Streptomyces* have been suggested to be another source of energy in the cell (Verma and Khuller, 1980). The importance of the physiological state of the bacterial cell with regard to triglyceride storage and turnover was studied by Brennan and co-workers (1970). Thus more studies on bacterial glycerides are required in order to be able to understand the exact function of these lipid components.

E. Fatty Acids

Fatty acids are the simplest constituents which act as precursors for the synthesis of most lipids. The complexity of fatty acids produced by mycobacteria has been documented by Cason and Miller (1963), Campbell and Naworal (1969a,b), Hung and Walker (1970), and Walker *et al.* (1970). Studies of Campbell and Naworal (1969a,b) revealed the presence of normal straight-chain, unsaturated, methyl-branched, iso-, and anteiso-fatty acids in *M. phlei*. Cyclopropane fatty acids were reported to be absent in all the lipids of mycobacteria (O'Neil and Gershbein, 1976), and a multiplicity of methyl-branched homologs were identified. Mycobacteria and some other members of the order Actinomycetales contain appreciable amounts of 10-methyl octadecanoic acid, i.e., tuberculostearic acid (Lennarz *et al.*, 1962; Larsson *et al.*, 1979). Other major fatty acids in mycobacterial phospholipids are myristic, palmitic, palmitoleic, stearic, and oleic acids (Nesterenko et al., 1980).

Lennarz et al. (1962) have shown that the synthesis of tuberculostearic acid proceeds from stearate through the methylation of oleate at C_{10}. Akamatsu and Law (1970) showed the formation of tuberculostearic acid on a phospholipid oleyl residue and recovered a soluble enzyme from *M. phlei* that transferred methyl groups from *S*-adenosylmethionine to the carboxyl group of fatty acids. The role of this carboxyl-group alkylation in *M. phlei* is not known.

The presence of phthioic acids in tubercle bacilli was shown by Canetti (1955). Other types of fatty acids present in mycobacteria are phthienoic and mycoceroic, which are multiple (3 or 4) methyl-branched acids, of which 2,4,6,8-tetramethyl octacosanoic acid is a typical example. Propionyl-CoA was reported to be a precursor for these fatty acids in *M. smegmatis* (Stjernholm et al., 1962), *M. tuberculosis* $H_{37}Ra$ (Gastambide-Odier et al., 1963, 1966), and *M. bovis* (Yano and Kusunose, 1966).

Polyenoic acids were reported to be synthesized by *Mycobacteria* (Asselineau et al., 1969). Asselineau et al. isolated *M. phlei* polyenoic acids which had the general formula $CH_3-(CH_2)_n-(CH=CH-CH_2-CH_2)_m-COOH$ (where $n = 12$ or 14 and $m = 5$ or 6). These fatty acids were, however, absent in the lipids of *M. tuberculosis* $H_{37}Rv$.

The presence of C_{16} to C_{32} fatty acids in *Nocardia* was shown for the first time by Guinand et al. (1958). Further analysis of nocardial fatty acids from several species revealed that apart from the high-molecular-weight (C_{50}) nocardic acids (Michel et al., 1960), normal saturated acids (C_{14} to C_{19}, palmitic acid predominating), branched-chain saturated acids (C_{15}, C_{16}, and C_{17} iso-acids and anteiso-acids, C_{18} iso-acid, and 10-methylstearic acid), and unsaturated acids (9-hexadecenoic, 10-hexadecenoic, 9-heptadecenoic, and oleic acids) were also present in these bacteria (Bordet and Michel, 1963; Kataoka and Nojima, 1967). Yano et al. (1969) found a characteristic fatty acid composition in *N. coeliaca*. The most commonly occurring iso- and anteiso-fatty acids in gram-positive bacteria were observed to be 12- and 13-methyl tetradecenoic acids (Kaneda, 1967; Moss et al., 1967; Tornabene et al., 1967), whereas in *N. coeliaca*, 14-methyl pentadecenoic acid was predominant.

Fatty acids of *Streptomyces* have been extensively studied in several laboratories (Hofeheinz and Grisebach, 1965; Ballio et al., 1965, 1968; Kataoka and Nojima, 1967; Ballio and Barcellona, 1968; Efimova and Tsyganov, 1969; Verma and Khuller, 1981a), and the findings have been very consistent. A common observation of all these studies was the predominance of branched-chain iso- and anteiso-fatty acids in all species of *Streptomyces*, with some variation in their chain lengths. Hofheinz and

Grisebach (1965) demonstrated the occurrence of branched-chain fatty acids in *S. erythrus* and *S. halstedii*. At the same time, Ballio and coworkers (1965) provided structural details of fatty acids from these bacteria. In another report, Ballio and Barcellona (1968) demonstrated the predominance of branched-chain fatty acids in the family Streptomycetaceae and suggested that the iso-/anteiso-fatty acid ratio and the nature of the predominating fatty acids were characteristic of the genus.

Phospholipid fatty acids of *S. griseus* were analyzed (Verma and Khuller, 1981a). Our findings revealed the occurrence of normal straight-chain, odd-numbered, monoenoic, iso-, and anteiso-fatty acids ranging from C_{12} to C_{18} in this bacteria. C_{16} iso-acid and C_{15} anteiso-acid were the major fatty acid components, similar to the fatty acids of triglycerides (Ballio *et al.*, 1965), suggesting a common pathway for glyceride and phospholipid biosynthesis in streptomyces. A rather unique observation was made by Ballio *et al.* (1968) in a survey of fatty acids from several strains of actinomycetes. Up to 6% of the total fatty acids in *Streptomyces* was constituted by an unidentified component which was characterized as 9,10-methylenehexadecanoic acid, a cyclopropane fatty acid.

Taking into account the fatty acid compositions of mycobacteria, nocardia, and streptomyces, it can be generalized that branched-chain fatty acids are common constituents of the genera of Actinomycetales, and hence this order is taxonomically closely related to the order Eubacteriales, whose members typically contain higher amounts of branched-chain fatty acids. The appreciable content of branched-chain fatty acids may also account for the absence of polyenoic fatty acids in a majority of the organisms from the order Actinomycetales.

F. Mycolic Acids

Cell walls of *Nocardia, Mycobacteria,* and *Corynebacteria* have been shown to contain α-branched, β-hydroxylated fatty acids called mycolic acids (Minnikin and Goodfellow, 1976; Yano *et al.*, 1978) since these fatty acids were actually first identified in *Mycobacteria* by Anderson (1929). Mycobacterial mycolic acids contain 60–90 carbon atoms, whereas the nocardial mycolic acids, called nocardomycolic acids, contain 36–66 carbon atoms with 0–4 double bonds, in addition to the 3-hydroxy ester system (Bordet and Michel, 1969; Minnikin and Goodfellow, 1976). Corynomycolic acid, from *Corynebacteria,* contains smaller (26–30 carbon atoms) carbon chains which are usually saturated or monoenoic (Pudles and Lederer, 1954; Etemadi *et al.*, 1965; Minnikin *et al.*, 1977).

Steck *et al.* (1978) isolated almost 24 different mycolic acids from each of the several mycobacterial species examined. The entire structural profile of the mycolic acids was resolved by Etemadi (1967a,b) and Asselineau *et al.* (1969, 1970). Mycolic acids were characterized and their biosynthesis was studied in *M. kanasasii, M. tuberculosis,* and *M. smegmatis* by several workers (Etemadi *et al.*, 1964; Etemadi and Lederer, 1965; Etemadi and Pinte, 1966; Wong *et al.*, 1979). Kanetsuna and Bartoli (1972) observed the specificity of mycolic acid structure in the genus *Mycobacterium.* Three types of mycolic acids, that is, cyclopropane α, methoxy β, and keto γ, were reported in different strains of mycobacteria by various workers (Etemadi, 1967a; Minnikin and Polgar, 1967; Asselineau *et al.*, 1969), and the β type was shown to be absent in avirulent strains. Mycolic acids were suggested to be related to the virulence of pathogenic *Mycobacteria* (Strain *et al.*, 1977; Toubiana *et al.*, 1979).

Long-chain nonmycolate fatty acids ($C_{26}-C_{56}$) have been purified from *Mycobacteria,* and though they are thought to be precursors of mycolic acids (Qureshi *et al.*, 1980), the exact functions of these fatty acids are not yet known. Such long-chain fatty acids are absent in *Streptomyces* (Shaw, 1974).

G. Sterols

Sterols are usually absent in bacteria (Goldfine, 1982). However, some bacteria contain sterols only when grown in the presence of sterols, confirming the lack of sterol-synthesizing ability in these organisms (Lettre *et al.*, 1954; Asselineau, 1966). Schubert *et al.* (1968) examined some glucose-grown bacterial strains, including members of the order Actinomycetales, for the presence of sterols, and the only actinomycete containing traces of sterols was *S. olivaceus.*

III. Subcellular Distribution

Knowledge about the subcellular distribution of lipids may be of help in understanding the lipid-associated functions of cells. The main subcellular bacterial organelles which have been of interest are the cytoplasmic or protoplasmic membrane and the cell wall. Analysis of cellular components is complex in *Mycobacteria* and related organisms as compared to other actinomycetes, since the multilayered cell wall and the cytoplasmic membrane are not distinctly differentiated. In general, bacterial membranes (except in members of the order Actinomycetales) account for most of the cellular lipids, whereas the cell wall contributes only minor amounts and the cytosol is usually devoid of these components. Actinomycetes have a relatively high lipid content in their cell wall and thus differ from other gram-positive bacteria in this respect. Yet another striking feature of this group of bacteria is the presence of some lipids in the cytosol of these organisms, a feature not encountered in other bacteria.

A. MEMBRANE LIPIDS

The cell membrane contributes a major amount of lipid to the total cellular lipid content in *Mycobacteria* (Akamatsu *et al.*, 1966), *Streptomyces* (Verma and Khuller, 1981b), and *Nocardia* (unpublished observations). In *M. phlei* (Akamatsu *et al.*, 1966) and *S. griseus* (Verma and Khuller, 1981b), about 50% of the membrane lipids are phospholipids, whereas in *N. polychromogenes* (unpublished observations) phospholipids represent only about one-fourth of the membrane lipids. The phospholipid composition of the protoplasmic membrane of *N. polychromogenes* is similar to that of whole cells of *M. phlei* (Akamatsu *et al.*, 1966) and *S. griseus* (Verma and Khuller, 1981b). Cell membranes of *S. griseus* and *N. polychromogenes* also contain minor amounts of acylglycerols and acylated sugars.

B. CELL WALL LIPIDS

A high cell wall lipid content is the most interesting feature of actinomycetes. Cell wall phospholipids are represented qualitatively by the phospholipids present in the whole cell. Phosphatidylinositol mannosides are the major cell wall phospholipid constituents of *M. phlei* (Akamatsu *et al.*, 1966), *S. griseus* (Verma and Khuller, 1981b), and *N. polychromogenes* (unpublished observations). A detailed analysis of the *M. tuberculosis* cell wall revealed that these cellular structures contain mostly dimannosides and only a limited amount of monomannosides (Goldman, 1970).

The subcellular distribution of phospholipids during different stages of growth of *Mycobacterium* 607 was examined (Penumarti and Khuller, 1982c). Mannosides were mainly concentrated in the cell wall as compared to the cell membrane during the early logarithmic growth phase of this organism. However, by the midlogarithmic phase, the proportion of PIMs in the cell wall and cell membrane became equal.

More complex, but characteristic, mycobacterial cell wall lipids are wax D, cord factor, mycosides, and sulfolipids (Lederer *et al.*, 1975). Wax D of the human pathogenic strain of tubercle bacilli contained an arabinogalactan mycolate linked to a mucopeptide containing N-acetylglucosamine, N-glycosylmuramic acid, L- and D-alanine, meso-α-diaminopimelic acid, and D-glutamic acid (Amar and Vilkas, 1973; Kanetsuna and San Blas, 1970). A tetrasaccharide heptapeptide constituent of wax D was reported to be present in human pathogenic strains by Migliore and Jolles (1968, 1969).

Cord factor (see Goren, 1972) and sulfolipids (see Khuller *et al.*, 1982d) are major trehalose-containing lipids of mycobacterial and nocardial cell walls. A trehalose 6-monomycolate was isolated from *M. tuberculosis* (Kato and Maeda, 1974), whereas *M. phlei* and *M. smegmatis* were shown to contain trehalose esterified completely with polyunsaturated "phleic" acids, which were predominantly hexatriaconta-4,8,12,16,20-pentaenoic acid (Asselineau *et al.*, 1972).

C. Cytosolic Globular Lipids

The presence of a variety of lipids in the cytosolic fractions of all actinomycetes (Akamatsu *et al.*, 1966; Goldman, 1970; Verma and Khuller, 1981b) is consistent with the earlier observations of fat globules in the cytoplasm of mycobacteria and streptomyces (Kates, 1964). The cytosolic fraction of these organisms has phospholipid and neutral lipid compositions qualitatively similar to the whole cell. In *M. phlei*, phospholipids were present only in traces in the cytoplasm (Akamatsu *et al.*, 1966). Phosphatidylinositol monomannoside was found in substantial quantities in the cytoplasm of *M. tuberculosis* (Goldman, 1970). Our findings on *S. griseus* and *N. polychromogenes* revealed that the cytoplasm in these organisms contributed to more than half of the total cellular lipids. Triacylglycerols were the most abundant in the cytosol of *S. griseus* and *N. polychromogenes*.

IV. Taxonomic Significance

A. Fatty Acids

1. Straight-Chain Fatty Acids

Taxonomic classification of microorganisms with the aid of their fatty acid compositions was first proposed by Able *et al.* (1963). Subsequently, Kates (1964) reviewed the information on bacterial lipids and extended Able's correlations to various other bacteria. In an extensive review by Shaw (1974), lipid composition was used as a guide in the classification of several bacterial classes. However, Okami (1975) reviewed the taxonomy of actinomycetes and indicated the problem of placing these organisms in an evolutionary or phylogenetic relationship with other microorganisms. An enormous amount of literature has been published on the taxonomic significance of lipids of actinomycetes (Lechevalier et al., 1971; Minnikin et al., 1977; Steck et al., 1978).

Actinomycetes are not only distinguished by their higher cellular lipid content, but also by their atypical lipid constituents, that is, mannophosphoinositides, mycolic acids, and iso- and anteiso-fatty acids. While the presence of mannophosphoinositides may be used to identify the actinomycetes, it is mycolic acids that provide a basis for the taxonomic classification of various bacteria into different genera. Nonetheless, the occurrence of branched-chain fatty acids in genera of the order Actinomycetales places these organisms close to those of the order Eubacteriales on the phylogenic map.

2. Branched-Chain Fatty Acids

Until lately, branched-chain fatty acids were considered to be limited to the genus *Bacillus* of the order Eubacteriales (Kaneda, 1977). However, with the improvement of techniques for isolation and characterization, a number of other bacteria, including several actinomycetes, were found to contain these fatty acids (Kataoka and Nojima, 1967; Bordet and Michel, 1963; Campbell and Naworel, 1969a,b; Verma and Khuller, 1981a). Although the iso- and anteiso-fatty acids have not been detected in all species of *Mycobacteria, Nocardia,* and *Streptomyces,* their consistent presence in some species of these organisms warrants a reinvestigation of the strains in which they have not been detected.

3. Mycolic Acids and Analogous Fatty Acids

Once an organism has been established to be an actinomycete, its genus can be determined by its content of high-molecular-weight fatty acids collectively known as "mycolic acids." Those present in *Mycobacteria* are the most complex, with carbon atoms numbering around 80 (in a strict sense mycolic acids). Mycolic acids of *Nocardia*, the nocardic acids, are smaller, with carbon numbers around 50. Nocardiac acids are thus intermediate between the mycolic acids from *Mycobacteria* and *Corynebacteria*. *Streptomyces* are characterized by the absence of these long-chain fatty acid analogs, but they possess α-hydroxy fatty acids of normal chain length. Several pathogenic and nonpathogenic *Mycobacteria* have been examined so far, including *M. bovis* BCG, *M. bovis* bovinus I, *M. smegmatis,* and *M. tuberculosis* $H_{37}Rv$, and approximately two dozen different mycolic acids were detected in each (Steck et al., 1978). Ever since their discovery, mycolic acids have been considered to be of value for the classification of actinomycetes (Alshamaoni *et al.*, 1976a,b; Lechevalier *et al.*, 1971; Maurice *et al.*, 1971; Minnikin and Goodfellow, 1976).

Lechevalier and his associates (1971) conducted a detailed survey of 96 aerobic actinomycetes for their mycolic and nocardomycolic acid contents and provided some interesting information. In general, the strains which were claimed to be *Mycobacteria* contained mycolic acid, confirming their generic assignment. However, since *Mycobacterium brevicale, M. rhodochrous,* and *Mycobacterium thamnophoeos* contained nocardomycolic acids, it was suggested that these organisms be placed in the genus *Nocardia*. Due to the dubious mycolic acid content in *Nocardia farcinia*, it could not be placed in either genera and was considered a *nomen dubium*. Detailed studies have shown that the mycolic acid structure is species specific within the genus *Mycobacterium*. Mycolic acids of *M. tuberculosis, M. avium, M. smegmatis,* and *M. phlei* differ in structure, which further shows the taxonomic significance of these lipids.

B. Mannophosphoinositides

From the data available on the phospholipid composition of microorganisms, it appears that phosphatidylinositol is not a common constituent of bacterial lipids. However, phosphatidylinositol derivatives containing mannose, that is, phosphatidylinositol mannosides, are present in mycobacteria and related organisms (Shaw, 1974). The discovery of this interesting group of glycophospholipids in human mycobacteria was first made by Anderson in 1930. Subsequently, these lipids were characterized and structurally elucidated by Vilkas and Lederer (1960), as well as by Ballou and his associates (1963). Studies of Pangborn and McKinney (1966), Brennan and Ballou (1967), and Khuller and Subrahmanyam (1968) elucidated the complex structure of these glycophospholipids. It is now clear that all species of *Mycobacteria, Nocardia,* and *Streptomyces* contain phosphatidylinositol mannosides, though they vary significantly in qualitative and quantitative composition. The complete absence of these mannosides in all other microorganisms is equally well established. Thus, this unique lipid class may be used for the definite identification of actinomycetes.

In addition to the other distinctive lipid components, the absence of phosphatidylethanolamine methyl derivatives in members of the order Actinomycetales (with the exception of *N. coeliaca* and some *Streptomyces* species mentioned earlier) is a noteworthy feature. In 1967, Ikawa proposed a close natural relationship between the orders Eubacteriales and Actinomycetales. Later, the detection of branched-chain fatty acids in bacteria belonging to both the orders further strengthened the belief that the two are related.

V. Influence of Environmental Factors on Lipid Composition

It is well established that variations in culture conditions of microorganisms are often reflected in their qualitative or quantitative lipid compositions. Phospholipid and fatty acid composition change with growth temperature, media composition, pH, and even age of culture. Oxygen tension in the culture medium and supplements added to the medium are other factors affecting the lipid composition of bacteria (Farrell and Rose, 1967; Sinensky, 1971; McElhaney, 1976; Cronan, 1978; Silvius and McElhaney, 1978; Ingram et al., 1980). The mechanism by which environmental factors influence bacterial physiology may vary to different extents, but these factors do alter the functions of bacterial membrane lipids.

A. Effect of Age

The lipid content of microorganisms is age dependent: generally, the total lipid content increases with culture age. The effect of age on the lipid composition of *Mycobacteria* was studied (Frouin and Guillaumie, 1928; Asselineau, 1951; Tepper, 1965; Donetz et al., 1970; Dhariwal et al., 1976), and it was observed that lipids were characteristic for different growth phases. Due to increased neutral lipids, the total lipid content of several mycobacterial strains increased during growth; however, the phosphatide content of these microorganisms remained virtually unaltered during this period (Asselineau, 1951). In *Mycobacterium album* and *Mycobacterium rubrum*, an increase in phospholipids was observed as log-phase cultures entered into the stationary phase, whereas triglycerides decreased with aging (Donetz et al., 1970). Chandramauli and Venkitasubramanian (1974) examined the lipid content of aging pathogenic and saprophytic mycobacterial strains. They observed that the total lipids and phospholipids remained unchanged during the different growth phases of the organisms. Thus, it is difficult to generalize about the pattern of changes in total lipid and phosphatide content of mycobacteria during the various phases of its growth cycle. Similar studies have been carried out in only a few species of *Nocardia* and *Streptomyces*. In *N. polychromogenes* (Khuller and Trana, 1977) and *S. griseus* (Talwar and Khuller, 1977b), total lipids and phospholipids decreased significantly with increased culture age. This inverse relationship between culture age and lipid content suggests a higher rate of lipid synthesis in actively multiplying cells. Variations in lipids during aging also indicate that besides their role in cell structure, these components are required for other cellular processes.

Changes in relative proportions of the major phospholipids, that is, phosphatidylethanolamine, cardiolipin, and mannophosphoinositides, appeared to be similar in aging cultures of the three genera. Phosphatidylethanolamine content decreased, whereas the cardiolipin content increased, in *Mycobacteria* (Khuller et al., 1972; Chandramouli and Venkitasubramanian, 1974), *Nocardia* (Khuller and Trana, 1977), and *Streptomyces* (Talwar and Khuller, 1977b).

Mannophosphoinositides, in general, increase in parallel with growth. Penumarti and Khuller (1982c) studied the changes in lipid composition of subcellular fractions in addition to intact cells, of *M. smegmatis* ATCC 607. The increase in the phospholipid content of the mycobacterial cell wall was approximately fivefold as compared to the membrane fraction in the bacteria during the stationary phase. Mannophosphoinositides, which were concentrated in the cell wall during the early logarithmic phase of growth, gradually increased in the membrane fraction as the culture aged, and during the stationary phase their mannoside content became higher than that of the cell wall. Thus, it can be assumed that mannophosphoinositides control the physiological activity of the bacilli as growth of the culture progresses. These observations support the dynamic membrane flow hypothesis of Jelsema and Moore (1978), who suggested an interconversion of membranes, including the phospholipid bilayer, between different cellular membrane systems and organelles.

Fatty acids of bacterial lipids, in general, exhibit an increased saturation with the advancement of culture age (Kates, 1964). Lennarz et al. (1962) observed analogous changes in proportions of C_{18} and C_{19} fatty acids from *M. phlei* during the growth cycle; they reported a high proportion of 10-methylstearic acid as the culture passed from active growth into the stationary phase, and a low proportion of this acid as the culture passed from the stationary phase to active growth. Palmitic and oleic acids were the most abundant fatty acids during all stages of growth in the pathogenic and saprophytic strains of nocardia (Trana et al., 1982). Interestingly, short-chain fatty acids in *N. polychromogenes* and *N. asteroides* disappeared when the cells entered into the stationary phase. This phenomenon may take place because of utilization of short-chain fatty acids for energy purposes (Trana et al., 1982).

Similar changes in fatty acid composition were obtained during our studies on aging cultures of *S. griseus* (Verma and Khuller, 1981a), and presumably this was due to the cessation of chain elongation in older cultures as compared to cultures in the logarithmic phase of growth. Another interesting feature was the decrease in anteiso-fatty acid levels, with a simultaneous increase in unsaturated fatty acids. This phenomenon may be a regulatory mechanism for membrane functions since branched-chain fatty acids have been suggested to play an important role in the maintenance of membrane fluidity (Kaneda, 1977). Similarly, the ratio of saturated and unsaturated fatty acids remained almost constant in growing *Nocardia* cells (Trana *et al.*, 1982). Maintenance of membrane fluidity is important for the growth of microorganisms and is generally achieved by changing the fatty acid composition, which probably depends on cellular requirements. Melting points of the anteiso series are 25°–35°C lower than those of their normal analogs (Kaneda, 1977). Thus changes observed in the anteiso-fatty acids and compensation by unsaturated fatty acid composition may be responsible for regulating the fluidity of membranes in *S. griseus*.

B. VARIATION OF CARBON AND NITROGEN SOURCES

Induced alterations in cellular lipids, as a function of medium composition, are valuable in delineating metabolic pathways. Some studies have been conducted to investigate the effects of carbon and nitrogen sources on the lipid compositions and cellular metabolisms of actinomycetes. Hexoses, glycerol, amino acids, and alkanes were used as growth-medium supplements most frequently. In a study on several strains of *N. asteroides* (Farshtchi and McClung, 1970), differences were noticed in the fatty acid compositions of cells grown on either glucose, glucose plus amino acids, glycerol, or Dubos oleic albumin complex. A greater amount of 10-methyl C_{18} acid was produced in cells grown in the medium containing methionine (Lennarz *et al.*, 1962).

The cellular lipid and phospholipid contents of mycobacteria were quantitatively affected when growth was on different carbon and nitrogen substrates. In *M. phlei,* the total lipids and phospholipids increased in nitrogen-limited medium. Amino acids alanine, glutamine, and asparagine decreased the amount of cellular lipids mainly by their effect on glyceride levels (Dhariwal *et al.*, 1977). Similarly, in *M. tuberculosis* R_1Rv and *M. phlei*, glycogen and lipid reserves accumulated in nitrogen (asparagine)-limited cultures (Antonie and Tepper, 1969a,b). Nitrogen limitation favored the accumulation of reserve compounds in several bacteria (Dawes and Ribbons, 1964); however, opposite observations were made with *M. smegmatis* grown on glycine-supplemented medium (Subramoniam and Subrahmanyam, 1981). Changes in the mannoside contents of these organisms were parallel to those of total phospholipids, and these changes may be related to the levels of their precursors. Surprisingly, the level of nitrogen-containing lipid, phosphatidylethanolamine, in *Mycobacteria* remained unaffected by the carbon or nitrogen supply of the medium. Mycobacteria grown on nitrogen-limited medium contained higher proportions of saturated fatty acids, whereas growth on carbon-limited medium resulted in higher proportions of unsaturated fatty acids. The influence of carbon limitation on other lipid components in *Mycobacteria* was opposite when glycerol was replaced by glucose, and a marked decrease in the major phospholipid, PIM_2, was observed under these conditions. Preferential production of certain lipids in a specified medium may be due either to direct regulation of biosynthesis or to enhancement of lipid catabolism by the substituent in the medium (Dhariwal et al., 1977).

Limited work has been done concerning the effect of alkanes on the cellular lipids of alkane-utilizing actinomycetes, though a lot of information is available for yeasts (Thorpe and Ratledge, 1972) and other bacteria (Foster, 1962; Dunlap and Perry, 1967). The only actinomycetes which have been studied in this regard are *Mycobacterium* species, and the effect of alkanes has been investigated on *M. rhodochrous* (Dunlap and Perry, 1967), *Mycobacterium vaccae* (Vestal and Perry, 1971; King and Perry, 1975), and *Mycobacterium convolutum* (Hallas and Vestal, 1978; Ascenzi and Vestal, 1979). Growth rate and cell yields of *M. convolutum* grown on a wide range of odd and even carbon-numbered alkanes decreased as the alkane chain length increased. All the mycobacterial strains examined by different workers consistently showed a 2- to 2.5-fold higher lipid content following growth on hydrocarbon substrates. The major difference between the cells grown on hydrocarbon and nonhydrocarbon substrates was in neutral lipid content, and mainly in triglyceride levels (Vestal and Perry, 1971). The increase in triglycerides was suggested to be due to their involvement in the assimilation of hydrophobic hydrocarbon substrates in the cell (Vestal and Perry, 1971). In addition, triglycerides can serve as storage materials and may form a part of lipoidal micelles in the cell (Kavanau, 1965). Phospholipids, in general, decreased in *Mycobacteria* grown on alkanes; however, their composition was largely unaffected, with the exception of a minor increase in phosphatidylserine and a corresponding decrease in phosphatidylglycerol in *M. convolutum* (Hallas and Vestal, 1978). Short-chain alkanes, gaseous alkanes (ethane, propane, and *n*-butane), were observed to have a more pronounced effect on the lipid levels than alkanes of higher chain lengths (Vestal and Perry, 1971). Johnson (1964) postulated that hydrocarbon growth supplement ultimately becomes a part of the phospholipid micelle in the cellular membrane.

Extending their studies, King and Perry (1975) analyzed the fatty acid patterns in *M. vaccae* grown on various types of alkanes. The *de novo* synthesis of fatty acids was inhibited in bacilli grown with hydrocabon substrates, and the cellular fatty acids were derived from the products of substrate degradation resulting from subterminal oxidative cleavage and β-oxidation. The parent fatty acids, produced by oxidative cleavage and products of β-oxidation, were later incorporated into lipids. Long-chain fatty acids underwent desaturation; however, elongation did not occur. Growth on 1-alkenes resulted in the incorporation of fatty acids that were products of two modes of oxidation. Cells grown on 2- and 3-methyl octadecane contained the corresponding iso- and anteiso-fatty acids in significant quantities. Similarly, in *M. convolutum* grown on C_{13} through C_{16} alkanes, the predominant fatty acids were of the same chain length as the growth substrate (Ascenzi and Vestal, 1979). Of particular interest was the observation that hexadecane, hexadecanoic acid, and hexadecanoyl–CoA inhibit *de novo* fatty acid synthesis, with hexadecanoyl–CoA being the most potent inhibitor. These findings support the hypothesis that the hydrocarbon substrate is oxidized to its corresponding fatty acid and is finally converted to the CoA thioester. The resulting fatty acyl–CoA is then incorporated into various lipid components. *De novo* synthesis of fatty acids may be decreased because of inhibition or repression of the fatty acid synthetase enzyme complex, particularly acetyl–CoA carboxylase, as has been shown in several eukaryotic microorganisms (Weeks and Wakil, 1970; Makula and Finnerty, 1972).

C. Growth Temperature

Temperature is one of the most important factors affecting the growth and metabolism of microorganisms. Extensive studies have been carried out in this area with microbial lipids (Marr and Ingram, 1962; Farrell and Rose, 1967; Cronan and Vagelos, 1972; McElhaney, 1976; Okuyama *et al.*, 1977; Veerkamp, 1977; Dhariwal *et al.*, 1977; Crowe and Urban, 1978; Rottem *et al.*, 1978; Watanabe *et al.*, 1979). These studies indicated increased fatty acid unsaturation in microorganisms as their growth temperature was reduced. Sinensky (1974) suggested that the phospholipid fatty acid composition changed over a wide range of temperatures and was regulated to help the organism survive at a particular temperature.

The total lipid content of mycobacteria (Taneja *et al.*, 1979) remained unchanged when they were grown either at 37° or 27°C, whereas under the same growth conditions in nocardia (Trana *et al.*, 1980a), an increase in total lipids was observed. Lowering the growth temperature resulted in an increased phospholipid content in *M. smegmatis* (Taneja *et al.*, 1979), whereas in *N. polychromogenes* (Trana *et al.*, 1980a) the phospholipid content decreased significantly. Although PIMs decreased in *M. phlei* (Dhariwal *et al.*, 1977) and *M. smegmatis* (Taneja *et al.*, 1979) at the reduced growth temperature, their levels increased in *N. polychromogenes* (Trana *et al.*, 1980). The PE level in *M. smegmatis* increased, while in *N. polychromogenes* it decreased on lowering the temperature. Cardiolipin levels of *M. smegmatis* remained unchanged at both temperatures, but an increase was observed in *N. polychromogenes* as the growth temperature was reduced. Phospholipid biosynthesis at low temperature in *M. smegmatis* ATCC 607 seemed to be regulated partly at the phosphatidic acid level (Khuller *et al.*, 1982a).

Depending on the temperature, microorganisms are known to change their contents of saturated, branched-chain, or unsaturated fatty acids (Farrell and Rose, 1967). Similar to other microorganisms, the relative proportion of unsaturated fatty acids increased markedly at the lower growth temperature in both of the actinomycetes mentioned previously (Taneja *et al.*, 1979; Trana *et al.*, 1980a), though the individual fatty acids were affected differently. Increased unsaturation at lower growth temperatures has been suggested to be necessary to maintain membrane functions (Cronan and Gelman, 1975).

D. EFFECT OF VARIOUS GROWTH-MEDIUM ADDITIVES

1. Alcohols

Alcohols are amphipathic molecules that alter the polar and nonpolar environments of biomembranes, thereby affecting numerous processes associated with the latter cellular organelle (Lee, 1976; Sullivan *et al.*, 1979). A membrane with an altered lipid composition in response to ethanol is considered to be adapted to the presence of this additive (Cossins, 1977; Ingram, 1976, 1977a,b), much as with the changes observed in response to growth temperature (Shaw and Ingram, 1965). Ethanol was investigated for its effects on membrane lipids of microorganisms (Littleton *et al.*, 1979; Berger *et al.*, 1980; Buttke and Ingram, 1980; Rigmoier *et al.*, 1980), and among the actinomycetes, mycobacteria is the only organism which has been studied for its response to the presence of alcohol.

Ethanol supplementation affects the total lipid content of *M. smegmatis* (Teneja and Khuller, 1980) more at the lower growth temperature (27°C) in the presence of alcohol than at the higher (37°C). At 37°C in the presence of ethanol there was a significant decrease in total phospholipid content, whereas under the same conditions at 27°C there was no decrease in the level of these components. Among individual phospholipids, variations were greater in alcohol-grown cells at 37°C than in cells grown at the lower temperature. Ethanol addition decreased the unsaturated-to-saturated fatty acid ratio in the cells grown at 37°C; however, this ratio increased for cells grown at 27°C. The observed increase in the unsaturated/saturated fatty acid ratio may result from ethanol-induced inhibition of saturated fatty acid synthesis, as reported earlier for *Escherichia coli* (Buttke and Ingram, 1978, 1980).

2. Fatty Acids

Bacterial lipid composition is altered when these organisms are grown in the presence of various fatty acids (Silbert *et al.*, 1973; McElhaney, 1974; Melchoir and Steim, 1977), and this is a useful approach to modify the physical state of biological membranes (King and Spector, 1978; Sanderman, 1978). Exogenous fatty acids affect membranes in a more or less similar manner as growth at low temperature or in the presence of ethanol, thereby influencing membrane-mediated functions. Fatty acids added to the growth medium during the same stage of culture growth may produce different effects on the organisms, depending on the fatty acid chain length and the concentration used (Fay and Farias, 1975; Silvius and McElhaney, 1978).

The effects of fatty acid supplementation on the lipids of actinomycetes were investigated in our laboratory (Trana and Khuller, 1981; Taneja *et al.*, 1982; Khuller *et al.*, 1983). Supplementation of unsaturated fatty acids into the growth medium of *N. polychromogenes* increased the total lipids and phospholipids of this organism (Trana and Khuller, 1981). Among the individual phospholipids of this bacteria, only PE levels were elevated markedly, whereas other phosphatide levels were influenced less significantly. The fatty acid composition of *N. polychromogenes* phospholipids did not change under these growth conditions.

In mycobacteria, the simultaneous effects of growth-temperature variation and fatty acid supplementation were investigated (Khuller *et al.*, 1982e; Taneja *et al.*, 1982). In contrast to the observations on *M. smegmatis* (Khuller *et al.*, 1983), the total lipid content of *M. phlei* (Taneja *et al.*, 1982) increased at 27°C when grown in the presence of unsaturated fatty acids. The effect on total phosphatides of *M. phlei* varied depending on the fatty acid supplemented, whereas in *M. smegmatis* the total phosphatide content decreased at the normal and lower growth temperatures with all fatty acid supplements. However, the individual phosphatide compositions of both mycobacterial species changed differently with the respective fatty acid supplements at the normal as well as at the lower growth temperatures. Modification of mycobacterial lipids is a reflection of the organisms' ability to adapt to changed growth conditions.

Fatty acid supplementation studies on *Streptomyces* species have been found to be of great commercial value (Arima *et al.*, 1973; Okazaki *et al.*, 1973). Neomycin is produced by the parent strain of *Streptomyces fradiae*

(3123), but a mutant strain requires oleic acid for neomycin formation. Palmitic acid supplementation can satisfy the oleic acid requirement and enhance antibiotic formation in *S. fradiae*. Among other fatty acids, linoleic and linolenic acids had the same effect as oleic acid, but saturated fatty acids such as acetic, butyric, caproic, caprylic, capric, lauric, myristic, stearic, and arachidic acids could not be substituted for palmitic acid. In addition to inducing neomycin formation in the mutant strain, palmitic and oleic acids also changed its cellular fatty acid profile, making it similar to the parent strain. The major fatty acid component of the parent strain was iso-acid 16:0, whereas that of the mutant strain was anteiso-acid 15:0. The fatty acid composition of the mutant strain changed from anteisoacid 15:0 to the parental isoacid 16:0 or normal 16:0 when oleic or palmitic acid was supplemented into the growth medium. Whether these changes in fatty acid composition were the primary effect on neomycin formation is not known.

3. Antibiotics

The discussion in this section, though meager due to the paucity of information available, points to an interesting query, i.e., how a secondary metabolite (streptomycin) of one actinomycete (*Streptomyces*) affects the lipid composition and metabolism of another actinomycete (*Mycobacterium*). The emergence of antibiotic resistance in tubercle bacilli has been a problem in the therapy of tuberculosis. Phthioic acid, a mixture of α,β-unsaturated fatty acids, was reported to be associated with the pathogenicity of mycobacteria, and it was shown to be absent in the avirulent tubercle bacilli $H_{37}Ra$ (Cason *et al.*, 1956). Considering the importance of α,β-unsaturated fatty acids of tubercle bacilli, Chandrashekar *et al.* (1958) attempted a comparative study of the bacilli sensitive and resistant to streptomycin. An increase in lipid, fatty acid, and acetate content was noticed in streptomycin-resistant cells. Subrahmanyam (1965) examined the effect of dihydrostreptomycin on the phospholipids of *Mycobacterium* 607 and observed that the antibiotic did not affect phospholipid biosynthesis at low concentrations (9 μg/ml growth medium). These studies, however, do not precisely reveal whether streptomycin therapy for tuberculosis is associated with mycobacterial lipids; further work in this direction is warranted.

4. Detergents

Mycobacteria and related actinomycetes are usually grown as aerobic surface cultures of interwoven filaments. Certain experiments require uniformly submerged and dispersed cultures, which is achieved by the addition of detergents into the mycobacterial growth medium. One detergent which has been tried is Tween-80 (a polyoxyethylene derivative of sorbitol monooleate). Tween-80 enhanced the growth of mycobacteria by acting as a nontoxic reservoir for oleic acid (Power and Hanks, 1965), which is a potent carbon and energy source for this organism (Dubos, 1950; Youmans and Youmans, 1954). Tween-80 undergoes enzymatic hydrolysis to provide free oleic acid and polyoxyethylated sorbitol in the growing mycobacterial cells (Andrejzew et al., 1960). These released fatty acids are incorporated into the cell structure (Stinson and Solotrovsky, 1971) and bring about morphological changes in the organisms (Schaefer and Lewis, 1965). When grown in the presence of Tween-80, *M. avium* contained twice the amount of lipid usually present in glycerol-grown cells (Stinson and Solotrovsky, 1971). Similarly, Dhariwal and Venkitasubramanian (1978) observed that Tween-80 supplementation increased the phospholipid levels, particularly that of cardiolipin, in *M. phlei*. The increase in cardiolipin, remarkably, changed the overall phospholipid composition, affecting the proportion of mannosides. The metabolic fate of Tween-80, or any other detergent, and its secondary effects on the cellular lipids of microorganisms are largely unknown as yet.

VI. Metabolism of Lipids

Most of the work which has been done on the metabolism of lipids in actinomycetes has been confined to the genus *Mycobacterium*, and has been reviewed extensively (Ramakrishnan et al., 1972; Barksdale and Kim, 1977; Goren and Brennan, 1979). In the other two genera, *Nocardia* and *Streptomyces*, metabolic studies have been limited mainly to phospholipids. Studies *in vitro* have delineated the enzymatic pathways of the biosynthesis and catabolism of phospholipids in these organisms, and this topic forms the first part of this section, followed by a discussion on the significance of observations from studies *in vivo*.

A. Biosynthesis

During the past 15 years, greater emphasis has been placed on the functional aspects of lipids with respect to cellular metabolism. Interestingly, work on pathways of phospholipid biosynthesis in microorganisms developed rapidly during this period. Accumulated data from studies *in vivo* as well as *in vitro* illustrated the biosynthetic pathways of common phospholipids in gram-positive and gram-negative bacteria. However, relatively few of the enzymes involved in phospholipid biosynthesis have been purified to a significant degree of homogeneity as yet. Several reviews have appeared on various aspects of lipid metabolism, in particular on the phospholipid metabolism of bacteria (Ambron and Pieringer, 1973; Gatt and Barenholz, 1973; van den Bosch, 1974; Finnerty and Makula, 1975; Raetz, 1978; Finnerty, 1978).

The initial reaction in phospholipid biosynthesis is the sequential acylation of *sn*-glycerol 3-phosphate; this molecule therefore occupies a key position as the *de novo* precursor for most of the subsequent phospholipid synthesized in microbial systems.

The formation of phosphatidylethanolamine in bacteria occurs by a pathway that is uniquely different from the pathway in animal systems. In bacteria, Kanfer and Kennedy (1964) demonstrated phosphatidylserine to be an obligatory intermediate for the formation of phosphatidylethanolamine through a decarboxylation reaction. This reaction has been thoroughly studied in some representative gram-negative (Kanfer and Kennedy, 1964) and gram-positive bacteria (Patterson and Lennarz, 1971). However, in *Mycobacterium* 607, phosphatidylserine decarboxylase is reported to be absent and phosphatidylethanolamine is biosynthesized through the CDP–ethanolamine pathway (Nandedkar, 1975). The reactions involved in phosphatidylethanolamine biosynthesis in bacteria can be summarized as follows:

1. The pathway in most bacterial systems:

$$\text{CDP-diglyceride} + \text{L-serine} \rightarrow \text{phosphatidylserine} + \text{CMP}$$
$$\text{Phosphatidylserine} \rightarrow \text{phosphatidylethanolamine} + \text{CO}_2$$

2. The pathway in *Mycobacterium* 607:

$$\text{Ethanolamine} + \text{ATP} \rightarrow \text{ethanolamine phosphate} + \text{ADP}$$
$$\text{Ethanolamine phosphate} + \text{CTP} \rightarrow \text{CDP-ethanolamine} + \text{PP}_i$$
$$\text{CDP-ethanolamine} + \text{diacylglycerol} \rightarrow \text{phosphatidylethanolamine} + \text{CMP}$$

The second major and metabolically most active phospholipid of mycobacteria is cardiolipin (Akamatsu *et al.*, 1967). The findings of Mathur *et al.* (1976) revealed the existence of two pathways for cardiolipin biosynthesis in *M. smegmatis*. The first pathway was designated as the CDP–glycerides pathway and was first detected in *E. coli* (Stanacev *et al.*, 1967), while the second pathway is the phosphatidylglycerol pathway and was initially observed to operate in *Micrococcus lysodeikticus* (De Siervo and Salton, 1971). These two pathways may be written as follows:

Phosphatidylglycerol + CDP–diacylglycerol → cardiolipin + CMP
2-Phosphatidylglycerol → cardiolipin + glycerol

The biosynthesis of phosphatidylinositol mannosides has drawn the attention of several workers, yet there are many controversies regarding the biosynthesis of these components. The first report concerning the biosynthesis of mannosides was made by Hill and Ballou (1966) using intact and cell-free extracts of *M. phlei*. They demonstrated that these glycophospholipids are derived from phosphatidylinositol, and also identified an enzyme system capable of forming phosphatidylinositol monomannoside by a glycosyl transfer from GDP–mannose to PI. Synthesis of higher mannoside homologs would involve a stepwise addition of mannose to the lipid acceptors. This pathway was later confirmed by Brennan and Ballou (1967).

Takeyama and Goldman (1969) observed that a cell-free particulate preparation from *M. tuberculosis* $H_{37}Rv$ was capable of synthesizing mono- as well as dimannosides from GDP–[^{14}C]mannose. Their report suggested that phosphatidylinositol monomannoside required some modification, perhaps by acylation, before the second mannose moiety was added. Studying the problem of acylation, Brennan and Ballou (1968a) showed that a particulate fraction from *M. phlei* was able to incorporate a variety of fatty acyl–CoA substrates into dimannophosphoinositides, resulting in the formation of tri- and tetraacylated dimannophosphoinositides.

The biosynthesis of monomannosides was clearly demonstrated in *Propionibacterium shermanii* (Brennan and Ballou, 1968b) and was analogous to that suggested by Hill and Ballou (1966) for *M. phlei*. In view of the contradictory findings of Takeyama and Goldman (1969) regarding the biosynthesis of mannosides, Shaw and Dinglinger (1969) proposed an alternate pathway involving diacylinositol mannosides, which were identified in propionic acid bacteria (Prottey and Ballou, 1968; Shaw and Dinglinger, 1968) for the synthesis of mannosides.

Since *Corynebacterium aquaticum* has PIM_1 as its major mannophosphoinositide, it became the organism of choice to demonstrate the first step in mannoside biosynthesis (Khuller and Brennan, 1972a); this biosynthetic step could not be demonstrated in *M. phlei* because of low mannoside content (Lee and Ballou, 1965). However, the biosynthesis of PIM_1 could not be demonstrated in cell-free extracts of *C. aquaticum*

(Khuller and Brennan, 1972b) as proposed by Hill and Ballou (1966; later reviewed by Ballou, 1972). A cell-free system of *M. smegmatis* was found to yield PIM_1, PIM_2, PIM_4, and PIM_5, with PIM_4 being the most abundant mannoside synthesized (Takeyama and Armstrong, 1971). By pulse–chase experiments with $^{32}P_i$ as a precursor, Hackett and Brennan (1975) demonstrated a precursor–product relationship between PI and monomannosides of *C. aquaticum*. Pathways for the synthesis of mono- and dimannophosphoinositides are depicted in Fig. 6.

Penumarti and Khuller (1982a) investigated the biosynthesis of hexa-mannophosphoinositides in *M. smegmatis* ATCC 607. They observed that on supplementation of ^{32}P-labeled or [^3H]myoinositol-labeled PIM_2-4F, significant amounts of radioactivity were recovered in the PIM_2-3F fraction of mannosides which could have been due to spontaneous deacylation of the tetraacylated analogs. In addition, high amounts of radioactivity were incorporated into hexamannosides and PIM_2-2F + $PIM_3$3F. It is pertinent that the incorporation of radioactivity in hexamannosides was negligible when the bacilli were grown in the presence of labeled PIM_2-3F; furthermore, the possibility of labeled hexamannosides being formed and then hydrolyzed is ruled out since α-mannosidase could not be detected in *Mycobacteria*. These studies suggested the existence of two pools of PIM_2-3F: Pool I, synthesized by mannosylation and acylation of PI, appeared to be metabolically stable, while pool II, formed by the deacylation of PIM_2-4F, was metabolically active and possibly served as a precursor for the synthesis of hexamannosides. The proposed scheme for the interconversion of mannosides in *Mycobacteria* is given in Fig. 7.

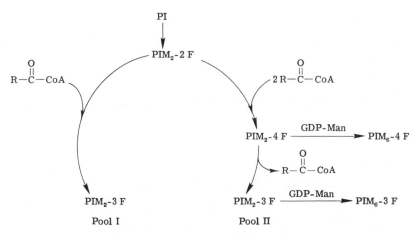

FIG. 6. Pathways for the biosynthesis of phosphatidylinositol mannosides.

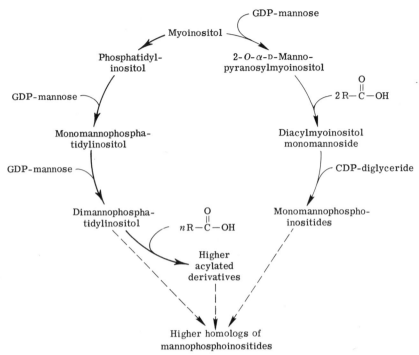

FIG. 7. Proposed scheme for hexamannoside biosynthesis. Thick line, major pathway; thin line, alternate pathway; dashed line, not detected.

B. CATABOLISM—LIPOLYTIC ENZYMES

Since most phospholipid components of bacteria are in a dynamic state, catabolism must be considered to be an important aspect of their metabolism. While the enzymology of phospholipid biosynthesis has been studied in some detail, enzymes involved in the catabolism of phospholipids have received attention only during the last few years (Finnerty and Makula, 1975; Avigad, 1976; Finnerty, 1978; Möllby, 1978). Within the order Actinomycetales, the lipolytic enzymes of *Streptomyces* and *Mycobacteria* have been identified and characterized, whereas no such study is available on *Nocardia*. The role of phosphatidic acid phosphatase, glycerylphosphorylethanolamine diesterase, and lipases, in addition to the phospholipases group, is described here.

The phospholipases are designated with the letters A, B, C, or D according to their specific sites of actions on the phospholipid molecule (Table II). As is also evident from Table II, there is a great need for an improved

Table II
DESIGNATIONS OF PHOSPHOLIPASES ACCORDING TO THE INTERNATIONAL UNION OF BIOCHEMISTRY (*Enzyme Nomenclature*, 1973)

Recommended name	Systematic name	Classification number
Phospholipase A	Phosphatidate acylhydrolase	
Phospholipase A_1	Phosphatidate-1-acylhydrolase	EC 3.1.1.32
Phospholipase A_2	Phosphatidate-2-acylhydrolase	EC 3.1.1.4
Lysophospholipase	Lysophosphatidate acylhydrolase	EC 3.1.1.5
Phospholipase B	Phosphatidate and lysophosphatidate acylhydrolase	
Phospholipase C	Phosphatidylcholine cholinephosphohydrolase	EC 3.1.4.3
Phospholipase D	Phosphatidylcholine phosphatide hydrolase	EC 3.1.4.4

nomenclature of phospholipid deacylases. For example, phospholipase B has been under debate for a long time, since this enzyme has been found to consist of two separate entities, i.e., phospholipase A and lysophospholipase (Kawasaki and Saito, 1973; Nishijima *et al.*, 1974). Möllby (1978) has discussed in detail the discrepancies existing in the classification of phospholipases.

1. *Phospholipid Deacylases—Phospholipase A and Lysophospholipase*

Among the phospholipases, phospholipase A has been extensively studied in *Mycobacteria* and *Streptomyces*. Deacylating enzymes were reported to be located in the membrane fraction of *M. phlei*, and were activated by heat treatment (Ono and Nojima, 1969a). The membrane preparation hydrolyzed a variety of substrates like cardiolipin, phosphatidylethanolamine, and phosphatidylcholine to give the products which indicated the presence of phospholipase A and lysophospholipase activity. The partially purified phospholipase A, free from lysophospholipase activity, was an allosteric protein with Fe^{3+} as an effector (Ono and Nojima, 1969b). Phospholipase A of *M. phlei* was in the inactive form when iron was in the ferric state, and was activated only when iron was reduced to the ferrous state. It is not known, however, whether ferric ions induce polymerization or conformational changes of the enzyme. Subsequently, Nishijima et al. (1974) purified phospholipase A_1 of M. phlei to homogeneity, and concluded that it was a nonspecific lysophospholipase with phospholipase A_1 activity, resembling phospholipase B of Penicillium notatum (Nishijima and Nojima, 1977). The enzyme consistently showed two molecular weights, 45,000 and 27,000, on repeated fractionation with DEAE–cellulose columns. The enzyme hydrolyzed acidic phospholipids, that is, phosphatidylglycerol and cardiolipin, more effectively as compared to the neutral, that is, choline- and ethanolamine-containing, phospholipids. At an alkaline pH it exhibited slight lipase activity.

Kashiwabara et al. (1980) identified phospholipid-deacylating activity in the Hawaiian-Ogawa strain of Mycobacterium lepraemurium. In addition to the phospholipids hydrolyzed by the M. phlei enzymes (Nishijima et al., 1974), the particulate fraction from M. lepraemurium was shown to deacylate phosphatidylinositol oligomannosides. This enzyme hydrolyzed 1-acyl and 2-acyl lyso compounds more rapidly than phosphatidylcholine. These activities, from the two respective mycobacterial species, exhibited different properties. Calcium and Mg^{2+} stimulated the M. phlei (Ono and Nojima, 1969a) enzymes, whereas Ca^{2+} had no effect on either the diacyl or the monoacyl hydrolase activities of M. lepraemurium (Kashiwabara et al., 1980). Appropriate concentrations of Triton X-100 had a pronounced stimulatory effect on the M. phlei enzymes, whereas Triton X-100 had little effect on the M. lepraemurium enzymes. Iron, which is a modulator of M. phlei phospholipase A, inhibited the enzyme of M. lepraemurium. These observations suggest distinctive properties of phospholipase A and lysophospholipase in different species of the same genus.

Phospholipase A from S. griseus was identified and characterized in our laboratory (Verma et al., 1980a). The enzyme activity was present mainly in the membrane fraction, though it was also present in the cell wall and cytosol. In M. phlei, phospholipase A activity was, however, absent in the cytosol (Ono and Nojima, 1969a). The S. griseus phospholipase A was a sulfhydryl group-dependent enzyme and required bivalent metal ions for optimal activity. Hydrophobic solvents like ether and chloroform inhibited the enzyme, possibly because of a lesser availability of substrate to the enzyme in the presence of these solvents. In general, water-miscible organic solvents and/or detergents enhanced phospholipase A hydrolysis of phosphatidylethanolamine. Lysophosphatidylethanolamine, formed on hydrolysis of phosphatidylethanolamine by phospholipase A, was further hydrolyzed by a lysophospholipase in S. griseus (Verma and Khuller, 1982).

2. Phospholipase C

Phospholipase C enzymes are the most extensively studied bacterial enzymes and are usually extracellular (Finnerty and Makula, 1975). Both extracellular as well as intracellular phospholipase C have been characterized from *Streptomyces hachijoensis* (Okawa and Yamaguchi, 1975a) and *S. griseus* (Verma and Khuller, 1982), respectively. Phospholipase C has not been reported in any other organism belonging to the order Actinomycetales. Phospholipase C from *S. hachijoensis* consists of two active fractions, each having a molecular weight of 18,000. They were tentatively designated as phospholipase C-I and phospholipase C-II. The intracellular phospholipase C of *S. griseus* is primarily localized in the membrane fraction. The intracellular enzyme showed an optimal activity between pH 7 and pH 8, similar to the extracellular C-I of *S. hachijoensis*. Both the enzymes were inhibited by sodium deoxycholate and Triton X-100. However, sodium taurocholate and sodium dodecyl sulfate stimulated the *S. griseus* enzyme, while they inhibited the enzyme from *S. hachijoensis*. The extracellular enzyme hydrolyzed a wide range of phospholipids.

The responses of phospholipase C enzymes from *S. hachijoensis* (Okawa and Yamaguchi, 1975a) and from *S. griseus* (Verma and Khuller, 1982) were different for various divalent cations. The *S. griseus* enzyme behaved in a manner similar to other bacteria phospholipase C enzymes (Möllby, 1978). Surprisingly, the extracellular phospholipase C of *S. hachijoensis* was markedly inhibited by Ca^{2+} ions, whereas Mg^{2+} stimulated the enzyme activity. Inhibitory effects of Zn^{2+}, Ba^{2+}, Cu^{2+}, and Fe^{2+} were observed in the extracellular as well as the intracellular enzymes of streptomyces; however, Fe^{2+} was about 10 times more inhibitory for the intracellular enzymes.

3. Phospholipase D

Phospholipase D enzymes were considered to be of plant origin. Knowledge about bacterial phospholipase D is meager even today, but a few available reports of mammalian (Saito and Kanfer, 1975), fungal (Uehara *et al.*, 1979), and bacterial systems (Cole *et al.*, 1974) show its ubiquitous occurrence in nature. Similar to phospholipase C, phospholipase D has been reported only from streptomyces among the actinomycetes (Verma and Khuller, 1982). Phospholipase D from *S. griseus* was found to be a cytosolic enzyme (Verma and Khuller, 1982). Extracellular phospholipases D enzymes were purified from the respective culture filterates of *Streptomyces chromofuscus* (Imamura and Horiuti, 1979) and

S. hachijoensis (Okawa and Yamaguchi, 1975b), and have been characterized. Phospholipase D of *S. hachijoensis* and *S. chromofuscus* had molecular weights of 16,000 and 50,000, and isoelectric points at pH 8.6 and 5.1, respectively. The *S. chromofuscus* enzyme hydrolyzed lysolecithin, phosphatidylcholine, phosphatidylethanolamine, and sphingomyelin, whereas the *S. hachijoensis* enzyme hydrolyzed cardiolipin and phosphatidylserine in addition to the above substrates. Imamura and Horiuti (1979) observed that the enzyme from *S. chromofuscus* has a hydrophobic site distinct from its catalytic site. The hydrophobic site of this extracellular enzyme is presumed to be occupied by some lipophilic substances in the growth culture, and under *in vitro* conditions it has the ability to bind Triton X-100 and albumin, which influence the enzyme's activity and stability.

4. Phosphatidic Acid Phosphatase

Phosphatidic acid phosphatase, which is well documented in plants (Kates, 1955) and animals (Hill and Lands, 1970), was first reported in bacteria by van den Bosch and Vagelos (1970). The presence of phosphatidic acid phosphatase has been demonstrated in *Mycobacterium* 607 (Nandedkar, 1975), and the enzyme has been suggested to play an important role in the biosynthesis of phosphatidylethanolamine. However, the properties, significance, and distribution of phosphatidic acid phosphatase in actinomycetes or in any other bacteria are still obscure.

5. Glycerylphosphorylethanolamine Diesterase

While investigating the enzymatic degradation of phosphatidylethanolamine in *S. griseus*, we identified an enzyme, glycerylphosphorylethanolamine diesterase (GPE diesterase) which hydrolyzed the deacylated product of phosphatidylethanolamine to give ethanolamine and glycerol phosphate (Verma and Khuller, 1982). This enzyme has been reported in *E. coli* (Albright *et al.*, 1973).

6. Lipases

Lipase activity in bacteria was demonstrated with certainty for the first time by Nantel and Proulx (1973). A triacylglycerol lipase (EC 3.1.1.3) was purified from *M. phlei* by Paznokas and Kaplan (1977). This enzyme was associated with the wall–membrane complex of the organism. The lipase was suggested to be responsible not only for the turnover of triacylglycerols during early growth, but also for the breakdown of triacylglycerols in the aged cultures. Cell-free extract of *S. griseus* hydrolyzed triacylglycerols and monoacylglycerols equally; however, the activity on diacylglycerols was comparatively lower (Verma and Khuller, 1982). These lipases play an important role in the intermediary metabolism of lipids in *S. griseus*.

Our results (Verma and Khuller, 1982) of intracellular phosphatidylethanolamine degradation in *S. griseus* suggested that the phospholipid can be catabolized by three different pathways (Fig. 8). First, it may be deacylated successively by phospholipase A and lysophospholipase, followed by removal of ethanolamine by GPE diesterase. In the second pathway it may be hydrolyzed by phospholipase C, and the resulting diacylglycerol could be further deacylated by the lipases. Phospholipase D action on phosphatidylethanolamine would yield phosphatidic acid in the third route of catabolism. The disappearance of phosphatidic acid, formed by phospholipase D action, suggested that it is reutilized instantaneously for the biosynthesis of other lipids. It is clear that the catabolism of phospholipids is catalyzed not only by phospholipases, but by additional enzymes required for their complete degra-

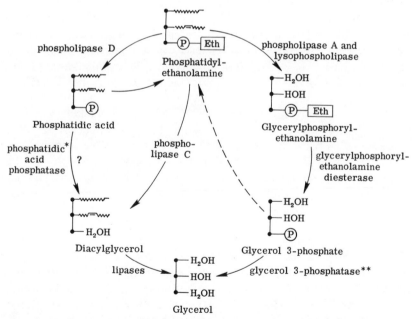

FIG. 8. Pathways for phosphatidylethanolamine catabolism in *S. griseus*. *, Absent; **, not detected; dashed arrow, recycling in biosynthesis.

dation as well. The first ester bond cleaved in the phospholipid determines the pathway of its further degradation.

C. LIPID TURNOVER: PHYSIOLOGICAL SIGNIFICANCE

Lipids are considered to be essential for biomembrane formation and function (Getz, 1970), and their presence in membranes is universal, although their species and quantity may change from one living system to another. Phospholipids respond to external stimuli and, depending on cellular requirements, their structure is modified to help cell survival under changed conditions. Catabolism and turnover of phospholipids has been suggested to be essential for membrane adaptation to environmental changes (Raetz, 1978). The turnover of phospholipids, to a great extent, would reflect the turnover of membranes; however, this may not hold true for actinomycetes since these bacteria contain a high proportion of lipids even in the cell wall. Nevertheless, phospholipids are essentially an integral part of actinomycete membranes too (Akamatsu et al., 1966; Verma and Khuller, 1981b). Studies on lipid turnover in mycobacteria (Akamatsu et al., 1967; Subrahmanyam, 1965; Dhariwal et al., 1978), *N. polychromogenes* (Trana et al., 1980a; Trana and Khuller, 1980), and *S. griseus* (Verma et al., 1980b; Verma and Khuller, 1980; Khuller et al., 1981b) have been conducted in logarithmically growing cells. These studies indicate the presence of highly active membrane components in actinomycetes.

Phospholipid turnover in mycobacteria was studied with labeled phosphate (Akamatsu et al., 1967; Subrahmanyam, 1965) and palmitate (Dhariwal et al., 1978). In *Mycobacterium* 607 (Subrahmanyam, 1965), LPE was observed to have the maximum turnover rate and PI turned over only slightly, while other phospholipids (CL, PIMs, PE, and PG) of this organism exhibited no turnover. In an earlier study (Subrahmanyam, 1964), the LPE of this bacteria was shown to have considerable turnover rate. The study by Akamatsu et al. (1967) on *M. phlei* demonstrated that CL had the maximum turnover, whereas the inositol-containing phospholipids turned over at a relatively slow rate. Little loss of radioactivity from PE in this organism was observed. In this study, phosphatidic acid was suggested to be the metabolic precursor of phospholipids in *M. phlei*.

In a comparative study on *M. phlei* and *M. smegmatis* (Dhariwal et al., 1978) with labeled palmitate as the precursor, the pattern of loss of radioactivity from phospholipids of the two strains was almost similar; however, the extent of loss varied. In both strains, CL turned over rapidly, whereas PE and PIMs exhibited relatively low turnovers. Glyceride fractions from both organisms turned over rapidly.

Phospholipid metabolism in *M. smegmatis* was shown to be influenced by its growth temperature (Taneja and Khuller, 1981). With labeled phosphate as the precursor, the rate of synthesis and degradation of phospholipids in this organism was lower in cells grown at 27°C as compared to those grown at 37°C. Cardiolipin showed a higher turnover rate than PE at 27° versus 37°C. Phosphatidylinositol mannosides were metabolically more active in cells grown at 37°C than in cells grown at 27°C.

The effect of ethanol on phospholipid metabolism of *M. smegmatis* was studied in our laboratory (Taneja and Khuller, 1982). Our investigations indicated a higher cardiolipin turnover in ethanol-grown cells, whereas in controls PE had the maximum degradation. The incorporation of the radioactive precursor into phospholipids of cells grown with ethanol was much lower than in the control cultures. Cells grown in the presence of ethanol were therefore assumed to have low metabolic activity.

We investigated the effect of antibodies against mycobacterial phospholipids on phospholipid synthesis (Penumarti and Khuller, 1982b). Antiserum to mannophosphoinositides was found to have a significant inhibitory effect on phospholipid synthesis in *M. smegmatis* ATCC 607 during different phases of growth; however, the inhibition was more pronounced in exponential-phase cells. The rate of synthesis of total and individual phospholipid, that is, of cardiolipin, phosphatidylethanolamine, and mannophosphoinositides, was observed to decrease in the presence of antiserum as compared to normal serum. The maximum inhibitory effect of antiserum was exerted on the synthesis of mannophosphoinositides, particularly tetraacylated dimannophosphoinositide.

In *N. polychromogenes* (Trana *et al.*, 1980b), PIMs had a higher turnover rate, as compared to either CL or PE, when $^{32}P_i$ was used as the phospholipid precursor. Of the latter two phospholipids, PE was observed to be an almost stable component in this organism. Phospholipids of *N. polychromogenes* were found to exist in two pools, one turning over faster than the other. In another study on *N. polychromogenes*, with [^{14}C]acetate as a precursor (Trana and Khuller, 1980), although PIMs had the maximum turnover, among the other two phospholipids PE turned over faster than CL. A comparison of the studies with labeled phosphate and labeled acetate suggests that the phosphoryl moiety and the acyl chains of PIMs turned over simultaneously; however, in PE and CL the turnover of these moieties was at different rates.

Streptomyces griseus is the most extensively studied actinomycete as regards its lipid metabolism. Three different labeled precursors, phosphate (Verma *et al.*, 1980b), acetate, and glucose (Khuller *et al.*, 1981b; Verma and Khuller, 1980), were used to investigate the metabolism of lipids in this organism. With labeled phosphate, PIMs were observed to have the maximum rate of synthesis and turnover, followed by CL and PE. As in *N. polychromogenes* (Trana *et al.*, 1980b), PE was a stable phospholipid component in *S. griseus* (Verma *et al.*, 1980b). In *S. griseus* phospholipids were observed to exist in two different metabolically active pools, as in *N. polychromogenes* (Trana *et al.,* 1980b). With labeled acetate and glucose as precursors (Khuller *et al.,* 1981b), different phospholipids and their individual moieties were found to have different turnover rates. Inositol-containing phospholipids had the maximum turnover rate with either precursor. However, PE exhibited turnover only with acetate, and CL was almost stable with both the precursors.

The metabolism of neutral lipids and acylated sugars in *S. griseus* was studied with labeled acetate as the precursor (Verma and Khuller, 1980). Triacylglycerols turned over the fastest, whereas fatty acids and acylated sugars showed comparatively less turnover. This study indicated a recycling of the label between the polar and nonpolar lipids of *S. griseus,* suggesting the existence of a common pathway for the synthesis of the two lipid classes.

Although phospholipid catabolism and turnover are not essential for growth, these phenomena may affect some membrane functions which are not necessarily related to cell division (Raetz, 1978). Several explanations have been offered for different turnover rates of submolecular phospholipid moieties. Glaser *et al.* (1970) suggested that different turnover rates were observed due to different orientations of the phospholipid moieties in the cellular environment, that is, polar head groups are in a hydrophilic environment, whereas the fatty acids are oriented toward a hydrophobic surrounding. Differences in the pool size of certain biosynthetic intermediates would also result into different turnover rates. Variable activities of the enzymes involved in the assembly and breakdown of phospholipids would also result in different turnover rates of the moieties. Differences in the rates of biosynthesis and degradation may help regulate cellular phospholipid levels, which in turn could influence lipid-associated functions of the cell. Moreover, phospholipase activity has been suggested to be responsible for the turnover of phospholipids in an organism (Avigad, 1976; Möllby, 1978; van den Bosch, 1981). Except in *Nocardia,* phospholipase activities have been reported in *Mycobacteria* (Avigad, 1976; Möllby, 1978), as well as in *Streptomyces* (Verma and Khuller, 1982). Several other explanations for the phospholipid turnover phenomenon have been given by Dawson (1966).

VII. Immunological Properties of Lipids

The order Actinomycetales consists of a large group of bacteria which have very complex serological relationships among families, genera, species, and types. Various representatives of the order Actinomycetales seem to be linked together biologically due to an overlapping in morphological (Lechevalier and Lechevalier, 1967) and immunological (Prevet, 1961) properties. The immunological properties of lipids were discovered during the early part of this century when Forssman (1911) observed lysis of sheep erythrocytes by antibodies to guinea pig organs in rabbits. The antigenic component was later found to be soluble in alcohol (Landsteiner and Simms, 1923). Our present knowledge of lipid immunochemistry is largely the result of the elegant work of Rapport and co-workers (1964). Fatty acids (including complex mycolic acids), phospholipids, glycolipids, and waxes have been recognized to be the immunologically active lipids of actinomycetes (see Goren, 1972).

Proteins (Boyden, 1958) and polysaccharides (Stacey and Kent, 1948) of tubercle bacilli have for a long time been implicated in immediate and delayed hypersensitivity reactions. However, the biological activity of lipids from mycobacteria came into prominence only when the antigenic methanol extract of *M. tuberculosis*, "antigenic methylique" (Bouquet and Negre, 1922), was found to be lipoidal in nature (Wadsworth *et al.*, 1925). Subsequent reports by Witebsky *et al.* (1931) and Albertweil (1931) revealed various lipid extracts of the tubercle bacilli to be antigenic in nature. Lipids of mycobacteria were later shown to be immunologically active and were recognized to confer some protection against tuberculosis (Crowle, 1958).

As for other members of the order Actinomycetales, purified lipid fractions from *Nocardia brasiliensis* were serologically inactive, as were those of *Actinomyces israelii* or firmly bound lipids of *M. tuberculosis* (Kwapinsky, 1963). Similarly, lipids from *N. asteroides* (Mordarska, 1966) as well as those from *S. griseus* (Krzywy, 1963) were serologically inactive.

A. Phospholipids

Phospholipids are the most active antigenic substances among lipids of tubercle bacilli. Since the work of Sabin and co-workers (1930), phospholipids have gained importance in relation to tuberculosis. Macheboenf *et al.* (1935) observed that phosphatides alone possessed the ability to function as haptens. Later, while studying the morphology of mycobacteria, Youmans *et al.* (1955) implicated the phospholipid-rich "red fraction" of tubercle bacilli in the stimulation of resistance against tuberculosis. Stoss and Herrmann (1965) also reported phosphatides to be immunogenic in nature. Antibodies against phosphatides from *M. tuberculosis* $H_{37}Rv$ and *M. bovis* BCG were detected in sera of tuberculous animals and humans (Takahashi *et al.*, 1961; Takahashi, 1962). Common phosphatide antigens were present in all virulent, avirulent, and saprophytic strains of mycobacteria (Subrahmanyam et al., 1964), and the intact lipid moiety appeared to be essential for serological activity. Phosphatides of N. polychromogenes were shown to be antigenic in nature (Trana and Khuller, 1977). Furthermore, live N. asteroides, when injected intraperitoneally into guinea pigs, resulted in the formation of antibodies to phospholipids of nocardia, whereas antibodies were not detected in animals injected with heat-killed N. asteroides (Trana and Khuller, 1978). Phospholipids of nocardia were implicated in the serodiagnosis of pulmonary nocardiosis (Sehgal et al., 1979). There are no reports available on the antigenicity of phospholipids from other members of the order Actinomycetales.

1. Cardiolipin

The immunological properties of cardiolipin have been studied extensively because of its role in the serodiagnosis of syphilis, although antibodies reactive with cardiolipin were also present in the sera of patients suffering from spirochete leprosy, systemic lupus erythematosus, and acute viral infections (Sparling, 1971). Antibodies against cardiolipin were raised in rabbits by using methylated bovine serum albumin as the carrier protein or with lecithin–cholesterol as auxiliary lipid (Inoue and Nojima, 1967; Guarnieri, 1974). The antigenic site of cardiolipin has been found to reside mainly in the polar head group of the molecule (Faure and Conlon-Morelec, 1963; De Bruijn, 1966; Inoue and Nojima, 1967, 1969). Motomiya and co-workers (1969) reported that cardiolipin recovered from various strains of Mycobacteria behaved like beef-heart cardiolipin by reacting with the Wasserman antibody, "reagin," of syphilitic serum. Studies on the unclassified, constitutively chromogenic Mycobacterium P_6 (Motomiya et al., 1968a) showed that cardiolipin from this organism behaved as a satisfactory antigen in flocculation tests for the diagnosis of syphilis and lepromatous leprosy. No protection was afforded by cardiolipin when it was used for immunizing guinea pigs against live M. tuberculosis $H_{37}Rv$ (Motomiya et al., 1968b).

2. Phosphatidylethanolamine

Reports on the antigenic properties of phosphatidylethanolamine are scanty. However, antibodies to phosphatidylethanolamine have been demonstrated in rabbits (Uemura and Kinsky, 1972). Phosphatidylethanolamine was used for immunizing guinea pigs prior to challenge with *M. tuberculosis* $H_{37}Rv$ infection (quoted from Barksdale and Kim, 1977).

3. Phosphatidylinositol Mannosides

Phosphatidylinositol mannosides have been reported to be the principal lipid antigens in tuberculous patients (Subrahmanyam and Singhvi, 1965b; Pangborn and McKinney, 1966). Khuller and Subrahmanyam (1971) observed that PIM_x stimulated humoral antibodies in rabbits when administered in Freund's incomplete adjuvant either alone or complexed with methylated bovine serum albumin (mBSA). Precipitating, agglutinating, and complement-fixing antibodies resembling those recognized in human tuberculous serum were characterized. The nature of the antigenic determinant in mycobacterial mannosides was elucidated by Banerjee and Subrahmanyam (1978), and antigenicity in these molecules was observed to reside on the mannose units. Although dimannosides with two, three, and four fatty acid esters showed complete antigenic cross-reactivity, these components did not cross-react with hexamannosides. Two dimannoside antigens, isolated from *M. tuberculosis,* were found to have different serological specificities in the passive hemagglutination reaction (Sasaki, 1974). Pangborn and McKinney (1966) showed that different combinations of di- and pentamannosides exhibited complement-fixation properties with human sera from tuberculosis patients. The higher phosphatidylinositol mannosides (including hexamannosides) are probably the most active sensitizing components of the Takahashi antigen (Sasaki and Takahashi, 1974). Mannosides of *Mycobacteria* were shown to enhance the resistance of guinea pigs against tuberculous infection (Khuller *et al.,* 1982e). Among the phosphatides of *Nocardia*, also, mannosides were observed to be antigenic (Trana and Khuller, 1978), and the intact lipid moiety was involved in serological reactions.

B. GLYCOLIPIDS

1. Cord Factor

Cord factor (6,6'-dimycoloyl-α-α'-trehalose) is not antigenic, but its haptenic properties help stimulate antibody formation when injected into animals with swine serum as a carrier protein (Ohara *et al.,* 1957). Complexes of cord factor with methylated bovine serum albumin stimulated the production of 19 S IgM antibodies in mice and rabbits. These antibodies cross-reacted with trehalose-6,6'-dicorynomycolate of *C. diphtheriae,* suggesting the presence of a common cord factor antigen in these organisms (Kato, 1972, 1973a). Active or passive immunization with cord factor–mBSA complex protected mice against the toxic action of cord factor and also against infection with *M. tuberculosis* $H_{37}Rv$ (Kato, 1972, 1973b). This protective effect was specific for the *M. tuberculosis* infection, since mice vaccinated with this complex were not protected against infection with listeria, brucella, or salmonella organisms.

Cord factor solutions in mineral oil, when emulsified in Tween–saline and injected intravenously into mice, stimulated the formation of pulmonary granulomas (Bekierkunst *et al.*, 1969). It was observed that the pulmonary granulomatous response to cord factor was much greater in mice infected intraperitoneally with *M. bovis* BCG. However, guinea pigs as well as rabbits were relatively insensitive to intravenous injections of cord factor (Moore *et al.*, 1972). In mice, cord factor was observed to confer protection against an intravenous challenge with virulent tubercle bacilli (McLaughlin *et al.*, 1976).

Oil solutions of cord factor emulsified in Tween–saline also exhibited adjuvant properties. Bekierkunst *et al.* (1971b) showed that cord factor, but not wax D, elicited an enhanced humoral response in mice to sheep erythrocytes. This was also confirmed by Saito *et al.* (1976), who showed that in mice, cord factor in Freund's complete adjuvant was more active than killed mycobacterial cells. Cord factor elicited delayed hypersensitivity to protein antigens in rats, but not in guinea pigs (Granger *et al.*, 1976).

Cord factor was shown to be responsible for the antitumor activity of a variety of preparations. Yarkoni *et al.* (1973) reported that trehalose-6,6′-dimycolate, trehalose-6,6′-dipalmitate, and sucrose-6,6′-dimycolate, when administered to mice challenged with Ehrlich ascites tumor cells, brought about a significant inhibition of tumor development; similar to these observations, urethane-induced lung adenomas were also inhibited (Bekierkunst *et al.*, 1971a). The induction of antitumor activity by cord factor was similar to the effects attributed to whole *M. bovis* BCG cells (Zbar and Tanaka, 1971; Zbar *et al.*, 1972; Meyer *et al.*, 1974). However, cord factor used alone was ineffective in suppressing or regressing various experimental tumors, although it was shown to possess marked antitumor activity against syngeneic murine leukemia (Lecrec *et al.*, 1976) and the 1023 syngeneic murine fibrosarcoma (Yarkoni *et al.*, 1977).

2. Sulfolipids

These sulfur-containing lipids (2,3,6,6′-tetraacyl-α-α′-trehalose 2′-sulfate) were observed to elicit antibodies in rabbits immunized with sulfolipid–mBSA complex (Khuller *et al.*, 1982c). The antiserum gave positive precipitin reactions against SL-I and SL-III. Antiserum to sulfolipids of *M. tuberculosis* $H_{37}Rv$ gave a positive serological test to sulfolipids from nonpathogenic mycobacterial strains, that is, *M. smegmatis, M. phlei*, and *M. bovis* BCG, indicating that these strains have common sulfolipid antigens. Antisulfolipid antibodies did not cross-react with other mycobacterial lipid antigens such as mannophosphoinositides, cardiolipin, and phosphatidylethanolamine. However, cross-reactivity between sulfolipids and cord factor was suggested to be due to the common trehalose core in these lipids. This was confirmed by partial inhibition of the serological response of the antiserum with trehalose, whereas no inhibition was observed with glucose and maltose tetrapalmitate. Further, increased inhibition observed with *p*-nitrocatechol sulfate emphasizes the role of the sulfate group at the 2' position of the trehalose moiety. This suggested that the antigenicity of sulfolipids was associated with the α-D-trehalose moiety and the sulfate group. Both IgG and IgM antibodies against sulfolipids were detected.

Antibodies to sulfolipids were shown to protect guinea pigs against experimental tuberculosis (Khuller *et al.*, 1982c). Similar to the observations of Yarkoni *et al.* (1979), sulfolipids were shown to decrease the toxicity of cord factor in mice (Malik and Khuller, 1982a). Moreover, antibodies to sulfolipids were demonstrated in guinea pigs injected with live *M. tuberculosis* $H_{37}Rv$ (Malik and Khuller, 1982b); however, the heat-killed cells failed to elicit these antibodies. The antibody titer increased with degree of infection, but it decreased on streptomycin treatment. Antibodies to sulfolipids were also observed in patients suffering from tuberculosis and leprosy (Khuller *et al.*, 1981a, 1982b).

VIII. Lipids and the Production of Antibiotics

Within the order Actinomycetales, Streptomycetaceae are the best producers of antibiotics, and 66% of the 5000 antibiotics known to date are produced by these organisms (Malik, 1979). Earlier investigations and efforts to enhance antibiotic production revealed that the yield of certain antibiotics, particularly polyene antibiotics, was increased in submerged cultures of *Streptomyces* by the addition of lipid materials. These findings have been summarized in Table III.

The yield of polyene antibiotics (1 to 8 in Table III) is lipid dependent, probably because of the utilization of lipid metabolites as precursors for antibiotic synthesis. Biosynthesis of polyene antibiotics generally involves the condensation of acetate and propionate units (polyketide pathway) (Hamilton-Miller, 1973), which can be easily obtained by the degra-

Table III

LIPID SUPPLEMENTS WHICH ENHANCE ANTIBIOTIC PRODUCTION

Antibiotic	Producing organism	Lipid supplement enhancing production	Reference
Fungichromin	*Streptomyces cinnamoneus*	Oleic acid	McCarthy et al. (1955)
	Streptomyces roseoluteus		
	Streptomyces cellulosae		
Sistomycosin	*Streptomyces viridosporus*	Soybean meal	Ehrlich et al. (1955)
Lagosin	*Streptomyces cinnamomeus*	Palm oil	Bessel et al. (1961)
	Streptomyces roseoluteus	Maize oil	
	Streptomyces cellulosae	Lard oil, palmitic acid	
Aureofungin	*Streptomyces cinnamomeus* var. *terricola*	Soybean meal	Thirumalacher et al. (1964)
Filipin	*Streptomyces filipensis*	Palmitic acid	Brock (1956)
Perimycin	*Streptomyces coelicolor* var. *aminophilus*	Mevalonate	Mohan et al. (1964)
Antimycosin A	*Staphylococcus aureus*	Mevalonate	Mohan et al. (1964)
DJ 400 series, DJ 400 B$_1$, DJ 400 B$_2$	*Streptomyces surinam*	Olive oil, coconut oil	Siewert and Kieslich (1971)
Neomycin	*Streptomyces fradiae*	Oleic acid,[a] palmitic acid	Okazaki et al. (1974)

[a] Necessary for antibiotic formation.

dation of lipid supplements. Besides acting as precursors, lipid supplements influence antibiotic formation also. A β-lysine-type streptothricin antibiotic was produced by *Streptomyces* 362 only when oleic acid, palmitic acid, or a high concentration of L-glutamic acid (or L-glutamine) was supplemented into the growth medium of this organism (Okazaki et al., 1974). Supplemented palmitic or oleic acid altered the cellular fatty acid profile, and these changes in membrane structure brought about changes in the influx and efflux of various amino acids. Enlargement of the amino acid pool, especially the L-glutamic acid and hexosamine pools, ultimately induced the formation of neomycin (Okazaki et al., 1973).

IX. Prospects

The absence of phosphatidylinositol mannosides (mannosides) in bacteria, other than those belonging to the order Actinomycetales, is intriguing, and even their location in the cell is ambiguous and needs further exploration. Although mannosides are known to be immunologically active, their exact biological roles are unknown. Another unsolved problem is the role of lipids in the virulence and drug resistance of mycobacteria. Can the modification of lipid composition render drug-resistant organisms sensitive? These are only a few of the several unanswered questions about the lipids of actinomycetes, and much work will have to be done to answer any one of them. The use of mutants to study bacterial physiology is common these days, but studies on lipid mutants of actinomycetes have yet to be reported. Research with lipid mutants of actinomycetes would greatly facilitate progress in this field and could unravel many of the as yet unknown functions of bacterial lipids.

ACKNOWLEDGMENTS

Work carried out in our laboratory was funded by grants to G.K.K. by the Indian Council of Medical Research and the Council of Scientific and Industrial Research, New Delhi, India.

References

Able, K., de Schmertzing, H., and Peterson, J. I. (1963). *J. Bacteriol.* **85,** 1039.
Akamatsu, Y., and Law, J. H. (1970). *J. Biol. Chem.* **245,** 709.
Akamatsu, Y., and Nojima, S. (1965). *J. Biochem.* **57,** 430.
Akamatsu, Y., Ono, Y., and Nojima, S. (1966). *J. Biochem.* **59,** 176.
Akamatsu, Y., Ono, Y., and Nojima, S. (1967). *J. Biochem.* **61,** 96.
Albertweil, J. (1931). Balliere et Fils, Paris.
Albright, F. R., White, D. A., and Lennarz, W. J. (1973). *J. Biol. Chem.* **248,** 3968.
Alshamaoni, L., Goodfellow, M., and Minnikin, D. E. (1976a). *J. Gen. Microbiol.* **92,** 188.
Alshamaoni, L., Goodfellow, M., Minnikin, D. E., and Mordraska, H. (1976b). *J. Gen. Microbiol.* **92,** 183.
Amar, C., and Vilkas, E. (1973). *C.R. Acad. Sci. Ser. D* **277,** 1949.
Ambron, R. T., and Pieringer, R. A. (1973). *In* "Form and Functions of Phospholipids" (G. B. Ansell, J. N. Hawthorne, and R. M. C. Dawson, eds.), 2nd ed., p. 289. Elsevier, Amsterdam.
Anderson, R. J. (1929). *J. Biol. Chem.* **85,** 351.
Anderson, R. J. (1930). *J. Am. Chem. Soc.* **52,** 1607.
Anderson, R. J. (1941). *Chem. Rev.* **29,** 225.
Anderson, R. J. (1943). *Yale J. Biol. Med.* **15,** 311.

Andrejzew, A., Gernex-Rieux, Ch., and Tacquet, A. (1960). *Ann. Inst. Pasteur (Paris)* **99**, 56.
Antonie, A. D., and Tepper, B. S. (1969a). *J. Gen. Microbiol.* **55**, 247.
Antonie, A. D., and Tepper, B. S. (1969b). *J. Bacteriol.* **100**, 538.
Arima, K., Okazaki, H., Ono, H., Yamada, K., and Beppu, T. (1973). *Agric. Biol. Chem.* **37**, 2313.
Ascenzi, J. M., and Vestal, J. R. (1979). *J. Bacteriol.* **137**, 384.
Asselineau, J. (1951). *Ann. Inst. Pasteur (Paris)* **81**, 306.
Asselineau, J. (1966). "The Bacterial Lipids." Holden-Day, San Francisco, California.
Asselineau, C., Montrozier, H., and Prome, J. C. (1969). *Eur. J. Biochem.* **10**, 580.
Asselineau, C., Tocanne, G., and Tocanne, J. F. (1970). *Bull. Soc. Chim. Fr.* p. 1455.
Asselineau, C. P., Montrozier, H. L., Prome, J. C., Savagnac, A. M., and Welby, M. (1972). *Eur. J. Biochem.* **28**, 102.
Avigad, G. (1976). In "Mechanisms in Bacterial Toxinology" (A. W. Bernheimer, ed.), p. 99. Wiley, New York.
Ballio, A., and Bracellona, S. (1968). *Ann. Inst. Pasteur (Paris)* **114**, 121.
Ballio, A., Barcellona, S., and Boniforti, L. (1965). *Biochem. J.* **94**, 11C.
Ballio, A., Barcellona, S., and Salvatori, T. (1968). *J. Chromatogr.* **35**, 211.
Ballou, C. E. (1972). *Methods Enzymol.* **28**, 493.
Ballou, C. E., and Lee, Y. C. (1964). *Biochemistry* **3**, 682.
Ballou, C. E., Vilkas, E., and Lederer, E. (1963). *J. Biol. Chem.* **238**, 69.
Banerjee, B., and Subrahmanyam, D. (1978). *Immunochemistry* **15**, 359.
Banerjee, B., Jain, S. K., and Subrahmanyam, D. (1974). *J. Chromatogr.* **74**, 342.
Barbier, M., and Lederer, E. (1952). *Biochim. Biophys. Acta* **8**, 590.
Barksdale, L., and Kim, K. S. (1977). *Bacteriol. Rev.* **41**, 217.
Batrakov, S. G., and Bergelson, L. D. (1978). *Chem. Phys. Lipids* **21**, 1.
Batrakov, S. G., Pilipenko, T. V., and Bergelson, L. D. (1971). *Dokl. Akad. Nauk SSSR* **200**, 226.
Batrakov, S. G., Pilipenko, T. V., and Bergelson, L. D. (1972). *Khim. Prir. Soedin.* 145.
Batrakov, S. G., Panosyan, A. G., Konova, I. V., and Bergelson, L. D. (1973). *Dokl. Akad. Nauk SSSR* **211**, 722.
Batrakov, S. G., Panosyan, A. G., Konova, I. V., and Bergelson, L. D. (1974a). *Biochim. Biophys. Acta* **337**, 29.
Batrakov, S. G., Shub, M. M., Rosynov, B. V., and Bergelson, L. D. (1974b). *Khim. Prir. Soedin.* **3**.
Batrakov, S. G., Pridachina, N. N., Kruglyak, E. D., Martyakov, A. T., and Bergelson, L. D. (1977). *Bioorg. Khim.* **3**.
Bekierkunst, A., Levij, I. S., Yarkoni, E., and Lederer, E. (1969). *J. Bacteriol.* **100**, 95.
Bekierkunst, A., Levij, I. S., Yarkoni, E., Vilkas, E., and Lederer, E. (1971a). *Science* **174**, 1240.
Bekierkunst, A., Yarkoni, E., Flechnes, I., Morecki, S., Vilkas, E., and Lederer, E. (1971b). *Infect. Immun.* **4**, 256.
Bergelson, L. D., Batrakov, S. G., and Pilipenko, T. V. (1970). *Chem. Phys. Lipids* **4**, 181.
Berger, B., Carty, C. E., and Ingram, L. O. (1980). *J. Bacteriol.* **142**, 1040.
Bessel, C. J., Fletcher, D. L., Mortimer, A. M., Anslow, W. A., Campbell, A. H., and Shaw, W. H. C. (1961). British Patent No. 884711.
Bloch, H., Defaye, J., Lederer, E., and Noll, H. (1957). *Biochim. Biophys. Acta* **23**, 312.
Bordet, C., and Michel, G. (1963). *Biochim. Biophys. Acta* **70**, 613.
Bordet, C., and Michel, G. (1969). *Bull. Soc. Chim. Biol.* **51**, 527.
Bouquet, A., and Negre, L. (1922). *C.R. Seances Soc. Biol. Fil.* **86**, 581.

Boyden, S. V. (1958). *Prog. Allergy* **5,** 149.
Brennan, P. J., and Ballou, C. E. (1967). *J. Biol. Chem.* **242,** 3046.
Brennan, P. J., and Ballou, C. E. (1968a). *J. Biol. Chem.* **243,** 2975.
Brennan, P. J., and Ballou, C. E. (1968b). *Biochem. Biophys. Res. Commun.* **30,** 69.
Brennan, P. J., and Lehane, D. P. (1971). *Lipids* **6,** 401.
Brennan, P. J., Lehane, D. P., and Thomas, D. W. (1970). *Eur. J. Biochem.* **13,** 117.
Brock, T. D. (1956). *Appl. Microbiol.* **4,** 131.
Buttke, T. M., and Ingram, L. O. (1978). *Biochemistry* **17,** 637.
Buttke, T. M., and Ingram, L. O. (1980). *Arch. Biochem. Biophys.* **203,** 565.
Campbell, I. M., and Naworal, J. (1969a). *J. Lipid Res.* **10,** 589.
Campbell, I. M., and Naworal, J. (1969b). *J. Lipid Res.* **10,** 593.
Canetti, G. (1955). "The Tubercle Bacillus in Pulmonary Lesions of Man," p. 94. Springer Publ., New York.
Cason, J., and Miller, W. T. (1963). *J. Biol. Chem.* **238,** 883.
Cason, J., Allen, C. F., Deacetis, W., and Fonken, G. J. (1956). *J. Biol. Chem.* **220,** 893.
Chandrashekar, S., De Monte, A. J. H., and Venkitasubramanian, T. A. (1958). *Indian J. Med. Res.* **46,** 643.
Chandramouli, V., and Venkitasubramanian, T. A. (1974). *Indian J. Chest Dis.* **16,** 199.
Cole, R., Benns, G., and Proulx, P. (1974). *Biochim. Biophys. Acta* **337,** 325.
Cossins, A. R. (1977). *Biochim. Biophys. Acta* **470,** 395.
Cronan, J. E., Jr. (1978). *Annu. Rev. Biochem.* **47,** 163.
Cronan, J. E., Jr., and Gelman, E. P. (1975). *Bacteriol. Rev.* **39,** 232.
Cronan, J. E., Jr., and Vagelos, P. R. (1972). *Biochim. Biophys. Acta* **265,** 25.
Crowe, J. L., and Urban, J. E. (1978). *Can. J. Microbiol.* **24,** 1277.
Crowle, A. J. (1958). *Bacteriol. Rev.* **22,** 183.
Dawes, F. A., and Ribbons, D. W. (1964). *Bacteriol. Rev.* **28,** 126.
Dawson, R. M. S. (1966). *Assays Biochem.* **2,** 69.
De Bruijn, J. H. (1966). *Br. J. Vener. Dis.* **42,** 125.
De Siervo, A. J., and Salton, M. R. J. (1971). *Biochim. Biophys. Acta* **239,** 280.
Dhariwal, K. R., and Venkitasubramanian, T. A. (1978). *Experientia* **34,** 303.
Dhariwal, K. R., Chander, A., and Venkitasubramanian, T. A. (1976). *Microbios* **16,** 169.
Dhariwal, K. R., Chander, A., and Venkitasubramanian, T. A. (1977). *Can. J. Microbiol.* **23,** 7.
Dhariwal, K. R., Chander, A., and Venkitasubramanian, T. A. (1978). *Arch. Microbiol.* **116,** 69.
Donetz, A. T., Gerasimova, N. M., Kotlev, V. V., and Bekhtereva, M. N. (1970). *Microbiology* **39,** 675.
Dubos, R. J. (1950). *J. Exp. Med.* **92,** 319.
Dunlap, K. R., and Perry, J. J. (1967). *J. Bacteriol.* **94,** 1919.
Efimova, T. P., and Tsyganov, V. A. (1969). *Microbiology* **38,** 571.
Ehrlich, J., Knudsen, M. P., and Bartz, Q. R. (1955). Canadian Patent No. 514894.
Etemadi, A. H. (1967a). *Bull. Soc. Chim. Biol.* **49,** 695.
Etemadi, A. H. (1967b). *Exp. Anal. Biochem. Med.* **28,** 77.
Etemadi, A. H., and Lederer, E. (1965). *Bull. Soc. Chim. Fr.* 2640.
Etemadi, A. H., and Pinte, F. (1966). *C.R. Acad. Sci. Ser. D* **262,** 1151.
Etemadi, A. H., Miquel, A. M., Lederer, E., and Barbier, M. (1964). *Bull. Soc. Chim. Fr.* p. 3274.
Etemadi, A. H., Gasche, J., and Sifferlen, J. (1965). *Bull. Soc. Chim. Biol.* **47,** 631.
Farrell, J., and Rose, A. H. (1967). *Annu. Rev. Microbiol.* **21,** 101.
Farshtchi, D., and McClung, N. M. (1970). *Can. J. Microbiol.* **16,** 243.

Faure, M., and Conlon-Morlec, M. J. (1963). *Ann. Inst. Pasteur (Paris)* **104**, 246.
Faure, M., and Marechal, J. (1962). *C.R. Acad. Sci.* **254**, 4518.
Fay, J. P., and Farias, R. N. (1975). *J. Gen. Microbiol.* **91**, 233.
Finnerty, W. R. (1978). *Adv. Microbiol. Physiol.* **18**, 177.
Finnerty, W. R., and Makula, R. A. (1975). *Crit. Rev. Microbiol.* **4**, 1.
Forssman, J. (1911). [Cited from Brady, R. O., and Trams, E. (1964). *Annu. Rev. Biochem.* **33**, 75.]
Foster, J. W. (1962). *Antonie van Leeuwenhoek* **28**, 241.
Frouin, A., and Guillaumie, M. (1928). *Ann. Inst. Pasteur (Paris)* **42**, 667.
Gangadharam, P. R. J., Cohn, M. L., and Middlebrook, G. (1963). *Tubercle* **44**, 452.
Gastambide-Odier, M. (1973). *Eur. J. Biochem.* **33**, 81.
Gastambide-Odier, M., and Sarda, P. (1970). *Pneumologie* **142**, 241.
Gastambide-Odier, M., Delaumeny, J. M., and Lederer, E. (1963). *Biochim. Biophys. Acta* **70**, 670.
Gastambide-Odier, M., Delaumeny, J. M., and Kuntzel, H. (1966). *Biochim. Biophys. Acta* **125**, 33.
Gatt, S., and Barenholz, Y. (1973). *Annu. Rev. Biochem.* **42**, 61.
Getz, G. F. (1970). *Lipids* **8**, 175.
Glaser, M., Simpkins, H., Singer, S. J., Sheetz, M., and Chan, S. J. (1970). *Proc. Natl. Acad. Sci. U.S.A.* **65**, 721.
Goldfine, H. (1972). *Adv. Microbiol. Physiol.* **8**, 1.
Goldfine, H. (1982). *Curr. Top. Membr. Transp.* **17**, 1.
Goldman, D. S. (1970). *Annu. Rev. Respir. Dis.* **102**, 543.
Goren, M. B. (1970a). *Biochim. Biophys. Acta* **210**, 116.
Goren, M. B. (1970b). *Biochim. Biophys. Acta* **210**, 127.
Goren, M. B. (1972). *Bacteriol. Rev.* **36**, 33.
Goren, M. B., and Brennan, P. J. (1979). *In* "Tuberculosis" (G. P. Youman, ed.), p. 63. Saunders, Philadelphia, Pennsylvania.
Goren, M. B., Brokl, O., Roller, P., Fales, H. M., and Das, B. C. (1976). *Biochemistry* **15**, 2728.
Granger, D. L., Yamamoto, K., and Ribi, E. (1976). *J. Immunol.* **116**, 482.
Guarnieri, M. (1974). *Lipids* **9**, 692.
Guinand, M., Michel, G., and Lederer, E. (1958). *C.R. Acad. Sci.* **246**, 848.
Hackett, J. A., and Brennan, P. J. (1975). *Biochem. J.* **148**, 253.
Hallas, L. E., and Vestal, J. R. (1978). *Can. J. Microbiol.* **24**, 1197.
Hamilton-Miller, J. M. T. (1973). *Bacteriol. Rev.* **37**, 166.
Hill, E. E., and Ballou, C. E. (1966). *J. Biol. Chem.* **241**, 895.
Hill, E. E., and Lands, W. E. M. (1970). *In* "Lipid Metabolism" (S. J. Wakil, ed.), p. 185. Academic Press, New York.
Hofheinz, W., and Grisebach, H. (1965). *Z. Naturforsch. B* **20**, 43.
Hung, J. G. C., and Walker, R. W. (1970). *Lipids* **5**, 720.
Ikawa, M. (1967). *Bacteriol. Rev.* **31**, 54.
Imamura, S., and Horiuti, Y. (1979). *J. Biochem.* **85**, 79.
Ingram, L. O. (1976). *J. Bacteriol.* **125**, 670.
Ingram, L. O. (1977a). *Appl. Environ. Microbiol.* **33**, 1233.
Ingram, L. O. (1977b). *Can. J. Microbiol.* **23**, 779.
Ingram, L. O., Dickens, B. F., and Buttke, T. M. (1980). *Adv. Exp. Med. Biol.* **126**, 299.
Inoue, K., and Nojima, S. (1967). *Biochim. Biophys. Acta* **144**, 409.
Inoue, K., and Nojima, S. (1969). *Chem. Phys. Lipids* **3**, 70.
Ioneda, T., Lederer, E., and Rozanis, J. (1970). *Chem. Phys. Lipids* **4**, 375.

Ito, F.; Coleman, C. M., and Middlebrook, G. (1961). *Kekku* **36**, 764.
Itoh, S., and Suzuki, T. (1974). *Agric. Biol. Chem.* **38**, 1443.
Jelsema, C. L., and Moore, J. (1978). *J. Biol. Chem.* **253**, 7960.
Johnson, M. J. (1964). *In* "Utilization of Hydrocarbons by Microorganisms," p. 1532. Chem. Ind., London.
Jolles, P., Bigler, F., Gendre, T., and Lederer, E. (1961). *Bull. Soc. Chim. Biol.* **43**, 177.
Jolles, P., Samour, D., and Lederer, E. (1962). *Arch. Biochem. Biophys.* **99**, 283.
Kaneda, T. (1967). *J. Bacteriol.* **93**, 894.
Kaneda, T. (1977). *Bacteriol. Rev.* **41**, 391.
Kanetsuna, F., and Bartoli, A. (1972). *J. Gen. Microbiol.* **70**, 209.
Kanetsuna, F., and San-Blas, G. (1970). *Biochim. Biophys. Acta* **208**, 434.
Kanfer, J., and Kennedy, E. P. (1964). *J. Biol. Chem.* **239**, 1720.
Kashiwabara, Y., Nakagawa, H., and Matsuki, G. (1980). *J. Biochem.* **88**, 1861.
Kataoka, T., and Nojima, S. (1967). *Biochim. Biophys. Acta* **144**, 683.
Kates, M. (1955). *Can. J. Biochem. Physiol.* **33**, 575.
Kates, M. (1964). *Adv. Lipid Res.* **2**, 17.
Kato, M. (1972). *Infect. Immun.* **5**, 203.
Kato, M. (1973a). *Infect. Immun.* **7**, 9.
Kato, M. (1973b). *Infect. Immun.* **7**, 14.
Kato, M., and Maeda, J. (1974). *Infect. Immun.* **9**, 8.
Kavanau, J. L. (1965). *In* "Structure and Function of Biological Membranes," Vol. 1, p. 122. Holden-Day, San Francisco, California.
Kawanami, J. (1971). *Chem. Phys. Lipids* **7**, 159.
Kawanami, J., Kimura, A., and Otsuka, H. (1968). *Biochim. Biophys. Acta* **152**, 808.
Kawasaki, N., and Saito, K. (1973). *Biochim. Biophys. Acta* **296**, 426.
Khuller, G. K. (1976). *Experientia* **32**, 1371.
Khuller, G. K. (1977). *Indian J. Med. Res.* **65**, 657.
Khuller, G. K., and Banerjee, B. (1978). *J. Chromatogr.* **150**, 518.
Khuller, G. K., and Brennan, P. J. (1972a). *Biochem. J.* **127**, 369.
Khuller, G. K., and Brennan, P. J. (1972b). *J. Gen. Microbiol.* **73**, 409.
Khuller, G. K., and Subrahmanyam, D. (1968). *Experientia* **24**, 851.
Khuller, G. K., and Subrahmanyam, D. (1970). *J. Bacteriol.* **2**, 654.
Khuller, G. K., and Subrahmanyam, D. (1971). *Immunochemistry* **8**, 251.
Khuller, G. K., and Subrahmanyam, D. (1974). *Indian J. Chest Dis.* **16**, 1.
Khuller, G. K., and Trana, A. K. (1977). *Experientia* **33**, 1422.
Khuller, G. K., Banerjee, B., Sharma, B. V. S., and Subrahmanyam, D. (1972). *Indian J. Biochem. Biophys.* **9**, 274.
Khuller, G. K., Malik, U., and Nalini, P. (1981a). *IRCS Med. Sci.* **9**, 492.
Khuller, G. K., Verma, J. N., and Grover, A. (1981b). *Curr. Microbiol.* **5**, 273.
Khuller, G. K., Kashinathan, C., Taneja, R., Nalini, P., and Bansal, V. S. (1982a). *Curr. Microbiol.* **7**, 49.
Khuller, G. K., Malik, U., and Kumar, B. (1982b). *Int. J. Lepr.* (in press).
Khuller, G. K., Malik, U., and Subrahmanyam, D. (1982c). *Tubercle* **63**, 111.
Khuller, G. K., Malik, U., and Verma, J. N. (1982d). *Actinomycetes* **16**, 145.
Khuller, G. K., Taneja, R., and Nath, N. (1983). *J. Appl. Bacteriol.* **54**, 63.
Khuller, G. K., Penumarti, N., Chakravarti, R. N., and Subrahmanyam, D. (1982e). *Austral. J. Exp. Biol. Med. Sci.* (in press).
Kimura, A., and Otsuka, H. (1969). *Agric. Biol. Chem.* **33**, 781.
Kimura, A., Kawanami, J., and Otsuka, H. (1967a). *Agric. Biol. Chem.* **31**, 1434.
Kimura, A., Kawanami, J., and Otsuka, H. (1967b). *J. Biochem.* **62**, 384.

King, D. H., and Perry, J. J. (1975). *Can. J. Microbiol.* **21**, 85.
King, M. E., and Spector, A. A. (1978). *J. Biol. Chem.* **253**, 6493.
Kovalchuk, L. P., Donetz, A. T., and Razumovsky, P. N. (1973). *Izv. Akad. Nauk Mold. SSR Ser. Biol. Khim. Nauk* **6**, 49.
Krzywy, T. (1963). *Arch. Immunol. Ther. Exp.* **11**, 521.
Kwapinsky, J. B. G. (1963). *J. Bacteriol.* **86**, 179.
Kwapinsky, J. B. G. (1972). *Can. J. Microbiol.* **18**, 1213.
Laneelle, G., and Asselineau, J. (1962). *Biochim. Biophys. Acta* **59**, 731.
Laneelle, G., and Asslineau, J. (1968). *Eur. J. Biochem.* **5**, 487.
Laneelle, M. A., Laneelle, G., and Asselineau, J. (1963). *Biochim. Biophys. Acta* **70**, 99.
Laneelle, G., Asselineau, J., and Chamoiseau, G. (1971). *FEBS Lett.* **19**, 109.
Landsteiner, K., and Simms, S. (1923). *J. Exp. Med.* **38**, 127.
Larsson, L., Mardh, P. A., and Odham, G. (1979). *J. Chromatogr.* **163**, 221.
Lechevalier, H. A., and Lechevalier, M. P. (1967). *Annu. Rev. Microbiol.* **21**, 27.
Lechevalier, M. P., Horan, A. C., and Lechevalier, H. (1971). *J. Bacteriol.* **105**, 313.
Lecrec, C., Lamensans, A., Chedid, L., Drapier, J. C., Petit, J. F., Wietzerbin, J., and Lederer, E. (1976). *Cancer Immunol. Immunother.* **1**, 227.
Lederer, E. (1967). *Chem. Phys. Lipids* **1**, 294.
Lederer, E. (1968). *Pure Appl. Chem.* **17**, 489.
Lederer, E. (1971). *Pure Appl. Chem.* **25**, 135.
Lederer, E. (1976). *Chem. Phys. Lipids* **16**, 91.
Lederer, E., Adam, A., Ciorbaru, R., Petit, J. F., and Wietzerbin, J. (1975). *Mol. Cell. Biochem.* **7**, 87.
Lee, A. G. (1976). *Biochemistry* **15**, 2448.
Lee, Y. C., and Ballou, C. E. (1965). *Biochemistry* **4**, 1395.
Lennarz, W. J., Scheuerbrandt, G., and Bloch, K., with a note by R. Ryhage (1962). *J. Biol. Chem.* **237**, 664.
Lettre, E., Inhoffen, H. H., and Tschesche, R. (1954). "Über Sterine, Gallensüaren und verwandte Naturstoffe" (2nd Ed.), Vol. 1, p. 138. Enke, Sttutgart.
Littleton, J. H., John, G. R., and Grieve, S. J. (1979). *Clin. Exp. Res.* **31**, 50.
Long, E. R. (1958). In "Chemistry and Chemotherapy of Tuberculosis." Balliere, London.
McCarthy, C. (1971). *Infect. Immun.* **4**, 199.
McCarthy, C. (1974). *Infect. Immun.* **9**, 363.
McCarthy, C. (1976). *Infect. Immun.* **14**, 1241.
McCarthy, F. J., Fisher, W. P., Charney, J., and Tytell, A. A. (1955). *Antibiot. Annu.* **1954**, 719.
McElhaney, R. N. (1974). *J. Mol. Biol.* **84**, 145.
McElhaney, R. N. (1976). In "Extreme Environments" (M. R. Heinrich, ed.), p. 255. Academic Press, New York.
Macheboenf, M., Levy, G., and Faure, M. (1935). *Bull. Soc. Chim. Biol.* **17**, 1210.
McLaughlin, C. A., Kelley, M. T., Milner, K. C., Parkes, R., Ribi, E., Smith, R., and Toubiana, R. (1976). *Proc. Joint Meet. Tuberc. Panel U.S.-Jpn. Coop. Med. Sci. Prog., 11th* p. 167.
Makula, R., and Finnerty, W. R. (1972). *J. Bacteriol.* **112**, 398.
Malik, U., and Khuller, G. K. (1982a). *IRCS Med. Sci.* **10**, 115.
Malik, U., and Khuller, G. K. (1982b). *Antonie van Leeuwenhoek* **48**, 285.
Malik, V. S. (1979). *Adv. Appl. Microbiol.* **25**, 75.
Marr, G. A., and Ingram, J. L. (1962). *J. Bacteriol.* **84**, 1260.
Mathur, A. K., Murthy, P. S., Saharia, G. S., and Venkitasubramanian, T. A. (1976). *Can. J. Microbiol.* **22**, 354.
Maurice, M. T., Vacheron, M. J., and Michel, G. (1971). *Chem. Phys. Lipids* **7**, 9.

Melchoir, D. L., and Steim, J. M. (1977). *Biochim. Biophys. Acta* **466**, 148.
Meyer, T. J., Ribi, E., Azuma, T., and Zbar, B. (1974). *J. Natl. Cancer Inst.* **52**, 103.
Michel, G. (1957). *C.R. Acad. Sci. (Paris)* **255**, 2429.
Michel, G., Bordet, C., and Lederer, E. (1960). *C.R. Acad. Sci.* **250**, 3518.
Michio, T., and Shoji, M. (1981). *Microbial. Immunol.* **25**, 75.
Middlebrook, G., Coleman, C. M., and Schaefer, W. B. (1959). *Proc. Natl. Acad. Sci. U.S.A.* **45**, 1801.
Migliore, D., and Jolles, P. (1968). *FEBS Lett.* **2**, 7.
Migliore, D., and Jolles, P. (1969). *C.R. Acad. Sci. Paris Ser. D* **269**, 2268.
Minnikin, D. E., and Goodfellow, M. (1976). *In* "The Biology of Nocardiae" (M. Goodfellow, G. H. Brownell, and J. A. Serrano, eds.), p. 160. Academic Press, New York.
Minnikin, D. E., and Polgar, N. (1967). *Chem. Commun.* **22**, 915.
Minnikin, D. E., Patel, P. V., Alshamaony, L., and Goodfellow, M. (1977). *Int. J. Syst. Bacteriol.* **27**, 104.
Mitra, R. S., and Maitra, P. K. (1966). *Indian J. Biochem.* **3**, 250.
Mohan, P. R., Piannotti, J. F., Martin, S., Ringel, M., Schwartz, B. S., Bailey, E. G., McDaniel, L. E., and Schaffner, C. P. (1964). *Antimicrob. Agents Chemother.* **1963**, 462.
Möllby, R. (1978). *In* "Bacterial Toxins and Cell Membranes" (J. Jelzaszewicz and T. Wadstrom, eds.), p. 367. Academic Press, New York.
Moore, V. L., Myrvik, Q. S., and Kato, M. (1972). *Infect. Immun.* **6**, 5.
Mordarska, H. (1966). *Arch. Immunol. Ther. Exp.* **14**, 436.
Moss, C. W., Dowel, V. R., Jr., Lewis, V. J., and Shckter, M. A. (1967). *J. Bacteriol.* **94**, 1300.
Motomiya, M., Fujimoto, M., Sato, H., and Oka, S. (1968a). *Sci. Rep. Res. Inst. Tohoku Univ. Ser. C* **15**, 10.
Motomiya, M., Mayama, A., Fujimoto, M., Sato, H., and Oka, S. (1968b). *Sci. Rep. Res. Inst. Tohoku Univ. Ser. C* **15**, 1.
Motomiya, M., Mayama, A., Fujimoto, M., Sato, H., and Oka, S. (1969). *Chem. Phys. Lipids* **3**, 159.
Nacash, C., and Vilkas, E. (1967). *C.R. Acad. Sci. (Paris) Ser. C* **265**, 413.
Nandedkar, A. K. N. (1975). *Biochem. Med.* **12**, 116.
Nantel, G., and Proulx, P. (1973). *Biochim. Biophys. Acta* **316**, 156.
Nesterenko, O. A., Panchemko, L. P., and Andreev, L. V. (1980). *Mikrobiol. Zh. (Kiev)* **42**, 556.
Nishijima, M., and Nojima, S. (1977). *J. Biochem.* **81**, 533.
Nishijima, M., Akamatsu, Y., and Nojima, S. (1974). *J. Biol. Chem.* **249**, 5658.
Ohara, T., Shimmyo, Y., and Sekikawa, E. (1957). *Jpn. J. Tuberc.* **5**, 128.
Okami, Y. (1975). *Jpn. J. Microbiol.* **19**, 411.
Okawa, Y., and Yamaguchi, T. (1975a). *J. Biochem.* **78**, 537.
Okawa, Y., and Yamaguchi, T. (1975b). *J. Biochem.* **78**, 363.
Okazaki, H., Ono, H., Yamada, K., Beppu, T., and Arima, K. (1973). *Agric. Biol. Chem.* **37**, 2319.
Okazaki, H., Beppu, T., and Arima, K. (1974). *Agric. Biol. Chem.* **38**, 1455.
Okuyama, H., Kankura, T., and Nojima, S. (1967). *J. Biochem.* **61**, 732.
Okuyama, H., Yamada, K., Kameyama, Y., Ikezawa, H., Akamatsu, Y., and Nojima, S. (1977). *Biochemistry* **16**, 2668.
O'Leary, W. M. (1967). *In* "Chemistry and Metabolism of Microbial Lipids." World, Cleveland, Ohio.
O'Neil, H. J., and Gershbein, L. L. (1976). *Trans. I. 11 State Acad. Sci.* **69**, 344.
Ono, Y., and Nojima, S. (1969a). *Biochim. Biophys. Acta* **176**, 111.

Ono, Y., and Nojima, S. (1969b). *J. Biochem.* **65**, 979.
Pangborn, M. C. (1968). *Ann. N.Y. Acad. Sci.* **154**, 133.
Pangborn, M. C., and McKinney, J. A. (1966). *J. Lipid Res.* **7**, 627.
Patterson, P. H., and Lennarz, W. J. (1971). *J. Biol. Chem.* **246**, 1092.
Paznokas, J. L., and Kaplan, A. (1977). *Biochim. Biophys. Acta* **487**, 405.
Penumarti, N., and Khuller, G. K. (1982a). *Arch. Microbiol.* **132**, 87.
Penumarti, N., and Khuller, G. K. (1982b). *Infect. Immun.* **37**, 884.
Penumarti, N., and Khuller, G. K. (1982c). Submitted.
Pigretti, M., Vilkas, E., Lederer, E., and Bloch, H. (1965). *Bull. Soc. Chem. Biol.* **47**, 2039.
Power, D. A., and Hanks, J. H. (1965). *Am. Rev. Respir. Dis.* **92**, 83.
Prabhudesai, A. V., Kaur, S., and Khuller, G. K. (1981a). *Indian J. Med. Res.* **73**, 181.
Prabhudesai, A. V., Malik, U., Subrahmanyam, D., and Khuller, G. K. (1981b). *Indian J. Biochem. Biophys.* **18**, 71.
Prevet, A. R. (1961). *In* "Traits de Systematique Bacterine," Vol. 2, p. 771. Dunoel, Paris.
Prottey, C., and Ballou, C. E. (1968). *J. Biol. Chem.* **243**, 6196.
Pudles, J., and Lederer, E. (1954). *Bull. Soc. Chim. Biol.* **36**, 759.
Qureshi, N., Takayama, K., and Schnoes, H. K. (1980). *J. Biol. Chem.* **255**, 182.
Raetz, C. R. H. (1978). *Microbiol. Rev.* **42**, 614.
Raffel, S. (1968). *Proc. Int. Soc. Blood Transfusion, Bibl. Haematol., 11th* **29**, 647.
Ramakrishnan, T., Murthy, P. S., and Gopinathan, K. P. (1972). *Bacteriol. Rev.* **36**, 65.
Rapport, M. M., Graf, L., and Scheider, M. (1964). *Arch. Biochem.* **105**, 431.
Rigomier, D., Bohin, J. P., and Lubochinsky, B. (1980). *J. Gen. Microbiol.* **121**, 139.
Rottem, S., Markowitz, O., and Razin, S. (1978). *Eur. J. Biochem.* **85**, 445.
Sabin, E. R., Doan, C. A., and Forknes, C. E. (1930). *J. Exp. Med. (Suppl.)* **3**, 1.
Saito, M., and Kanfer, J. (1975). *Arch. Biochem. Biophys.* **169**, 318.
Saito, R., Tanaka, A., Sugiyama, K., Azuma, I., Yamamura, Y., Kato, M., and Goren, M. B. (1976). *Infect. Immun.,* **13**, 776.
Sandermann, H., Jr. (1978). *Biochim. Biophys. Acta* **515**, 219.
Sarma, G. R., Chandramouli, V., and Venkitasubramanian, T. A. (1970). *Biochim. Biophys. Acta* **218**, 561.
Sasaki, A. (1974). *Iryo (Japan)* **28**, 703.
Sasaki, A. (1975). *J. Biochem.* **78**, 547.
Sasaki, A., and Takahashi, Y. (1974). *C.R. Soc. Biol.* **168**, 626.
Schaefer, W. B., and Lewis, C. W., Jr. (1965). *J. Bacteriol.* **90**, 1438.
Schubert, K., Rose, G., Wachtel, H., Horhold, C., and Ikekawa, N. (1968). *Eur. J. Biochem.* **5**, 246.
Schulz, J. C., and Elbein, A. D. (1974). *J. Bacteriol.* **117**, 107.
Sehgal, S. C., Trana, A. K., and Khuller, G. K. (1979). *IRCS Med. Sci.* **7**, 20.
Shaw, M. K., and Ingram, J. L. (1965). *J. Bacteriol.* **90**, 141.
Shaw, N. (1964). *Adv. Microbial Physiol.* **8**, 1.
Shaw, N. (1970). *Bacteriol. Rev.* **34**, 365.
Shaw, N. (1974). *Adv. Appl. Microbiol.* **17**, 63.
Shaw, N., and Dinglinger, F. (1968). *Biochem. J.* **109**, 700.
Shaw, N., and Dinglinger, F. (1969). *Biochem. J.* **112**, 769.
Short, S. A., and White, D. C. (1972). *J. Bacteriol.* **109**, 820.
Siewert, G., and Kieslich, K. (1971). *Appl. Microbiol.* **21**, 1007.
Silbert, D. F., Ulbright, T. M., and Honegger, J. L. (1973). *Biochemistry* **12**, 164.
Silvius, J. R., and McElhaney, R. N. (1978). *Can. J. Biochem.* **56**, 462.
Sinensky, M. (1971). *J. Bacteriol.* **106**, 449.
Sinensky, M. (1974). *Proc. Natl. Acad. Sci. U.S.A.* **71**, 522.

Smith, D. W., Randall, H. M., Gastambide-Odier, M. M., and Koevoet, A. L. (1957). *Ann. N.Y. Acad. Sci.* **69,** 145.
Smith, D. W., Randall, H. M., McLennan, A. P., and Lederer, E. (1960). *Nature (London)* **186,** 887.
Smith, H. (1977). *Bacteriol. Rev.* **41,** 475.
Sparling, P. F. (1971). *N. Engl. J. Med.* **824,** 642.
Stacey, M., and Kent, P. W. (1948). *Adv. Carbohydrate Chem.* **3,** 311.
Stanacev, N. Z., Chang, V. Y., and Kennedy, E. P. (1967). *J. Biol. Chem.* **242,** 3018.
Steck, P. A., Schwartz, B. A., Rosendahl, M. S., and Gray, G. R. (1978). *J. Biol. Chem.* **253,** 5625.
Stern, N., and Tiets, A. (1973a). *Biochim. Biophys. Acta* **296,** 130.
Stern, N., and Tiets, A. (1973b). *Biochim. Biophys. Acta* **296,** 136.
Stinson, M. W., and Solotrovsky, M. (1971). *Am. Rev. Respir. Dis.* **104,** 717.
Stjernholm, R. L., Noble, R. E., and Koch-Weser, D. (1962). *Biochim. Biophys. Acta* **70,** 670.
Stoss, B., and Herrmann, R. (1965). *Nature (London)* **208,** 1224.
Strain, S. M., Toubiana, R., Ribi, E., and Parker, R. (1977). *Biochem. Biophys. Res. Commun.* **77,** 449.
Subrahmanyam, D. (1964). *Life Sci.* **3,** 1267.
Subrahmanyam, D. (1965). *Indian J. Biochem.* **2,** 27.
Subrahmanyam, D., and Singhvi, D. R. (1965a). *Indian J. Biochem.* **2,** 112.
Subrahmanyam, D., and Singhvi, D. R. (1965b). *Proc. Soc. Exp. Biol. Med.* **120,** 102.
Subrahmanyam, D., Nandedkar, A. K. N., and Viswanathan, R. (1962). *Biochim. Biophys. Acta* **63,** 542.
Subrahmanyam, D., Singhvi, D. R., and Venkitasubramanian, T. A. (1964). *Indian J. Exp. Biol.* **2,** 56.
Subramoniam, A., and Subrahmanyam, D. (1981). *J. Gen. Microbiol.* **124,** 203.
Sullivan, K. H., Hegeman, G. D., and Cordes, A. H. (1979). *J. Bacteriol.* **138,** 133.
Suzuki, T., Tanaka, H., and Itoh, S. (1974). *Agric. Biol. Chem.* **38,** 557.
Takahashi, Y. (1962). *Am. Rev. Respir. Dis.* **85,** 708.
Takahashi, Y., Fujita, S., and Sasaki, A. (1961). *J. Exp. Med.* **113,** 1141.
Takeyama, K., and Armstrong, E. L. (1971). *FEBS Lett.* **18,** 67.
Takeyama, K., and Armstrong, E. L. (1976). *Infect. Immun.* **9,** 8.
Takeyama, K., and Goldman, D. S. (1969). *Biochim. Biophys. Acta* **176,** 196.
Talwar, P., and Khuller, G. K. (1977a). *Indian J. Biochem. Biophys.* **14,** 72.
Talwar, P., and Khuller, G. K. (1977b). *Indian J. Biochem. Biophys.* **14,** 85.
Taneja, R., and Khuller, G. K. (1980). *FEMS Microbiol. Lett.* **8,** 83.
Taneja, R., and Khuller, G. K. (1981). *Arch. Microbiol.* **129,** 81.
Taneja, R., and Khuller, G. K. (1982). *Indian J. Med. Res.* **75,** 796.
Taneja, R., Malik, U., and Khuller, G. K. (1979). *J. Gen. Microbiol.* **113,** 413.
Taneja, R., Malik, U., and Khuller, G. K. (1982). *Indian J. Med. Res.* **75,** 648.
Tepper, B. S. (1965). *Am. Rev. Respir. Dis.* **92,** 75.
Thirumalchar, M. P., Radhalkar, P. W., Sukapure, R. S., and Gopalkrishnan, K. S. (1964). *Hindustan Antibiot. Bull.* **6,** 108.
Thorpe, R. F., and Ratledge, C. (1972). *J. Gen. Microbiol.* **72,** 151.
Tornabene, T. G., Gelpi, E., and Oro, J. (1967). *J. Bacteriol.* **94,** 333.
Toubiana, R., Berlan, J., Sato, H., and Strain, M. (1979). *J. Bacteriol.* **139,** 205.
Trana, A. K., and Khuller, G. K. (1977). *Indian J. Biochem. Biophys.* **14,** 385.
Trana, A. K., and Khuller, G. K. (1978). *Indian J. Med. Res.* **67,** 734.
Trana, A. K., and Khuller, G. K. (1980). *Indian J. Biochem. Biophys.* **17,** 387.
Trana, A. K., and Khuller, G. K. (1981). *Indian J. Med. Res.* **74,** 534.

Trana, A. K., Khuller, G. K., and Subrahmanyam, D. (1980a). *Indian J. Med. Res.* **72,** 650.
Trana, A. K., Khuller, G. K., and Subrahmanyam, D. (1980b). *J. Gen. Microbiol.* **116,** 89.
Trana, A. K., Verma, J. N., and Khuller, G. K. (1982). *Arch. Microbiol.* **131,** 252.
Tuboly, S. (1968). *Acta Microbiol. Acad. Sci. (Hung.)* **15,** 207.
Uehara, S., Hasegawa, K., and Iwai, K. (1979). *Bull. Res. Inst. Food Sci. Kyoto Univ.* No. 42.
Uemura, K., and Kinsky, S. C. (1972). *Biochemistry* **11,** 4085.
van den Bosch, H. (1974). *Annu. Rev. Biochem.* **42,** 61.
Van den Bosch, H. (1981). *Biochim. Biophys. Acta* **604,** 191.
van den Bosch, H., and Vegelos, P. R. (1970). *Biochim. Biophys. Acta* **218,** 233.
Veerkamp, J. H. (1977). *Antonie van Leeuwenhoek* **43,** 101.
Verma, J. N., and Khuller, G. K. (1980). *FEMS Microbiol. Lett.* **9,** 73.
Verma, J. N., and Khuller, G. K. (1981a). *Antonie van Leeuwenhoek* **47,** 307.
Verma, J. N., and Khuller, G. K. (1981b). *FEMS Microbiol. Lett.* **11,** 55.
Verma, J. N., and Khuller, G. K. (1982). *Indian J. Biochem. Biophys.* **19,** 191.
Verma, J. N., Bansal, V. S., Khuller, G. K., and Subrahmanyam, D. (1980a). *Indian J. Med. Res.* **72,** 487.
Verma, J. N., Khera, A., Khuller, G. K., and Subrahmanyam, D. (1980b). *Curr. Microbiol.* **4,** 13.
Vestal, J. R., and Perry, J. J. (1971). *Can. J. Microbiol.* **17,** 445.
Vilkas, E. (1960). *Bull. Soc. Chim. Biol.* **42,** 1005.
Vilkas, E., and Lederer, E. (1960). *Bull. Soc. Chim. Biol.* **42,** 1013.
Vilkas, E., and Rojas, A. (1964). *Bull. Soc. Chim. Biol.* **46,** 689.
Vilkas, E., Adam, A., and Senn, M. (1968). *Chem. Phys. Lipids* **2,** 11.
Ville, C., and Gastambide-Odier, M. (1973). *Eur. J. Biochem.* **33,** 81.
Voiland, A., Bruneteau, M., and Michel, G. (1971). *Eur. J. Biochem.* **21,** 285.
Wadsworth, A., Mattener, F., and Mattener, E. (1925). *J. Immunol.* **10,** 241.
Walker, R. H., Barakat, H., and Hung, J. G. C. (1970). *Lipids* **5,** 684.
Watanabe, T., Fukushima, H., and Nojima, Y. (1979). *Biochim. Biophys. Acta* **575,** 365.
Weeks, G., and Wakil, S. J. (1970). *J. Biol. Chem.* **245,** 1913.
Weir, M. P., Langridge, H. R., III, and Walker, R. W. (1972). *Am. Rev. Respir. Dis.* **106,** 450.
Wilkinson, S. G., and Bell, M. G. (1971). *Biochim. Biophys. Acta* **248,** 293.
Witebsky, E., Klingenstein, R., and Kuhn, H. (1931). *Klin. Wochenschr.* **10,** 1068.
Wong, M. Y. H., Steck, P. A., and Gray, G. R. (1979). *J. Biol. Chem.* **254,** 5734.
Yano, I., and Kusunose, M. (1966). *Biochim. Biophys. Acta* **116,** 593.
Yano, I., Furakawa, Y., and Kusunose, M. (1969). *J. Bacteriol.* **98,** 124.
Yano, I., Furakawa, Y., and Kusunose, M. (1970). *Biochim. Biophys. Acta* **202,** 189.
Yano, I., Furakawa, Y., and Kusunose, M. (1971). *J. Gen. Appl. Microbiol.* **17,** 329.
Yano, I., Toriyama, S., Masui, M., Kusunose, M., Kusunose, E., and Akimori, N. (1978). *Koenshu-lyo Masu Kenkyukai* **3,** 169.
Yarkoni, E., Bekierkunst, A., Asselineau, J., Toubiana, R., and Lederer, E. (1973). *J. Natl. Cancer Inst.* **51,** 717.
Yarkoni, E., Meltzer, M. S., and Rapp, H. J. (1977). *Int. J. Cancer* **19,** 818.
Yarkoni, E., Goren, M. B., and Rapp, H. J. (1979). *Infect. Immun.* **24,** 586.
Youmans, A. D., and Youmans, G. P. (1954). *J. Bacteriol.* **67,** 731.
Youmans, G. P., Millman, I., and Youmans, A. S. (1955). *J. Bacteriol.* **70,** 557.
Yribarren, M., Vilkas, E., and Rozanis, J. (1974). *Chem. Phys. Lipids* **12,** 173.
Zbar, B., and Tanaka, T. (1971). *Science* **172,** 271.
Zbar, B., Rapp, H. J., and Ribi, E. E. (1972). *J. Natl. Cancer Inst.* **18,** 831.

The Role of Sterols in Sperm Capacitation

KATHRYN J. GO* AND DON P. WOLF†

Department of Obstetrics and Gynecology
University of Pennsylvania
Philadelphia, Pennsylvania

I.	Introduction	317
II.	Methodology for the Study of Capacitation	319
III.	Sterols of Mammalian Sperm	320
	A. Content and Composition	320
	B. Modulation of Sperm Lipids during Maturation	322
	C. Sterol Levels and Cold Shock	322
IV.	Modification of Sperm Lipid Levels during Capacitation	323
	A. *In Vitro* Studies	323
	B. *In Vivo* Studies	325
V.	Effects of Sterols on Sperm Fertility	325
VI.	Factors Which May Regulate Sperm Sterol Content	326
	A. Sterol Acceptors	326
	B. Sterol Sulfohydrolase Activity	327
VII.	Conclusions	328
	References	328

I. Introduction

In mammals, sperm emerging from the male reproductive tract are incapable of fertilizing eggs, but they acquire this ability during residence in the female reproductive tract (Austin, 1951; Chang, 1951). This requisite period of conditioning is called capacitation, and enables the spermatozoon to undergo irreversible changes involving a specialized lysosomal-like organelle at the anterior tip of the sperm head, the acrosome. This event, the acrosome reaction, is morphologically distinct at low resolution in some mammals, and arises from the fusion between the outer acrosomal membrane and the overlying plasma membrane, permitting the release of acrosomal contents (Fig. 1). Enzymes originating from the acrosome facil-

---

* Present address: Department of Physiology and Biochemistry, The Medical College of Pennsylvania, Philadelphia, Pennsylvania 19129.
† Present address: Department of Obstetrics and Gynecology, University of Texas Health Science Center at Houston Medical School, Houston, Texas 77030.

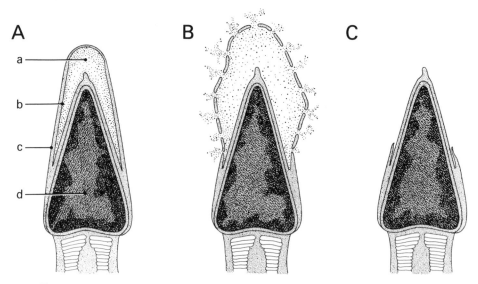

Fig. 1. Schematic representation of a "typical" mammalian spermatozoon: (a) acrosome; (b) outer acrosomal membrane; (c) plasma membrane; (d) nucleus. (A) Spermatozoon with intact membranes: this is the morphology of both capacitated and noncapacitated sperm. (B) Onset of the acrosome reaction: multiple-point fusions between the outer acrosomal membrane and adjacent plasma membrane have occurred, with release of the acrosomal contents. (C) Completion of the acrosome reaction: the hybrid membrane vesicles formed from outer acrosomal and plasma membranes have been sloughed, exposing the inner acrosomal membrane.

itate passage of the sperm through the investments of the egg, including the cumulus oophorus, corona radiata, and zona pellucida, and fusion with the egg plasma membrane.

Evidence that capacitation involves molecular modification at the level of the sperm plasma membrane has been gleaned from biochemical and ultrastructural studies. Sperm undergoing capacitation exhibit the following characteristics: (1) increased permeability to ions, particularly Ca^{2+} (Singh et al., 1978; Triana et al., 1980); (2) loss of adsorbed components originating from seminal plasma (Bedford and Chang, 1962; Oliphant and Brackett, 1973); (3) redistribution of surface components (Gordon et al., 1975; O'Rand, 1977; Kinsey and Koehler, 1978; Talbot and Franklin, 1978; Schwarz and Koehler, 1979); (4) changes in intramembranous particle distribution (Friend et al., 1977); (5) modification of lipid composition (Davis et al., 1979; Evans et al., 1980; Go and Wolf, 1984) and distribution in the plasma membrane (Bearer and Friend, 1982); and (6) changes in motility (Yanagimachi, 1981).

As major structural components of cellular plasma membranes, sterols are important regulators of membrane function. Cholesterol, the predominant sterol in mammalian cell membranes, has been studied most extensively in this regard. Through interaction with constituent phospholipids and proteins, membrane cholesterol can exert effects on the physicochemical characteristics of the bilayer and influence membrane processes such as transport, permeability, and mobility of intramembranous components, and activities of membrane-bound enzymes. This regulatory role of cholesterol has been the focus of studies employing several cell types and experimental approaches, and particular attention has been directed toward the modulation of membrane function by the level of cholesterol in the bilayer.

Studies in erythrocytes have demonstrated that above the phase transition temperature of the plasma membrane, cholesterol has a rigidifying and condensing effect on the lipid bilayer (Vanderkooi *et al.*, 1974). Thus, increased levels of cholesterol relative to phospholipid are associated with membranes of decreased fluidity. *In vitro* studies using erythrocytes have provided evidence that cholesterol enrichment of plasma membranes by incubation with sterol-rich liposomes alters cell shape and reduces permeability and transport processes (Cooper, 1978). The aggregation of intramembranous Band 3 protein (Muhlebach and Cherry, 1982) and exposure of blood group antigens (Shinitsky and Souroujon, 1979) are also modulated by the cholesterol content of red cell membranes.

Since capacitation involves modification of the sperm plasma membrane, the role that sterol content and its modulation may play in this process has been considered in studies using several mammalian species. A methodological challenge in these investigations is the isolation of sperm plasma membranes. Due to the close apposition of acrosomal and nuclear membranes, purified preparations of plasma membranes from sperm are difficult to obtain and have not been widely used in these studies. Thus, for the most part, assessment of sperm sterol levels in relation to capacitation has considered the total cellular content of a pool of sperm without drawing conclusions about its distribution within the cell. The findings of these investigations are discussed in this article.

II. Methodology for the Study of Capacitation

A functional assay for capacitation is measuring the ability of sperm to fuse with eggs, and the development of *in vitro* fertilization techniques using mammalian gametes has been seminal in advancing our knowledge concerning this process. The capacitation of ejaculated or epididymal sperm is achieved *in vitro* by incubation in media ranging from complex tissue culture media to simple balanced salt solutions supplemented with metabolic energy sources and female reproductive tract fluids, blood sera, egg components such as cumulus cells, or protein in the form of serum albumin (Gwatkin, 1977; Rogers, 1978; Wolf, 1979; Bavister, 1981). Sperm capacitated *in vivo* may be recovered from the reproductive tracts of appropriately stimulated females following an incubation period.

To assess the state of capacitation by *in vitro* techniques, sperm are incubated with eggs collected from the oviducts of females which are superovulated. The eggs may be intact (surrounded by their investments) or rendered cumulus or zona pellucida free by enzymatic or mechanical means. Following gamete interaction, eggs are collected and sperm–egg fusion is assessed by microscopic inspection. The criteria for penetration are the presence of the sperm head or male pronucleus and tail in the egg cytoplasm.

III. Sterols of Mammalian Sperm

A. Content and Composition

Evaluation of the sterol content of mammalian sperm has been undertaken (Table I). The considerable range in values may reflect interspecific variation, or differences arising from the source of the sperm (epididymal

Table I
CELLULAR STEROLS IN INTACT MAMMALIAN SPERM

Species	Micrograms of sterol per 10^7 sperm	Reference
Bull	3.45[a]	Darin-Bennett and White (1977)
Hamster	3.85[b]	Legault *et al.* (1979a)
Human	5.56[a]	Darin-Bennett and White (1977)
Mouse	12.9[c]	Go (1982)
Rabbit	5.45[a]	Darin-Bennett and White (1977)
Ram	2.8[a]	Quinn and White (1967)
Rat	5.85[a]	Davis *et al.* (1979)
	10.0[a]	Adams and Johnson (1977)

[a] Cholesterol.
[b] Desmosterol.
[c] Total sterol.

versus ejaculated), lipid extraction procedures, or methods of analysis. At present, there are insufficient data to differentiate among these possibilities.

Cholesterol is the major sterol in most mammalian cells, however, desmosterol (cholest-5,24-dien-3β-ol), a precursor of cholesterol, has been detected in hamster, rat, and human sperm (Bleau and VandenHeuvel, 1974). In fact, in epididymal hamster sperm, desmosterol represents the major sterol, present at four- to fivefold the concentration of cholesterol (Legault et al., 1979a). The occurrence of desmosterol is specific to sperm since it is not found in the free sterol fraction of hamster plasma and liver (Bleau and VandenHeuvel, 1974). In addition, another cholesterol precursor, identified as 5-cholesta-7,24-dien-3β-ol, is present in hamster sperm at a concentration threefold higher than cholesterol (Legault et al., 1978). The occurrence of cholesterol precursors in mammalian sperm is noteworthy since they are uncommon components of normal mammalian cells; however, their functional significance and influence on membrane properties are uncertain. Low levels of the Δ24-reductase or the presence of an inhibitor of this enzyme during spermatogenesis may account for their unique presence in sperm (Bleau and VandenHeuvel, 1974).

Of the total cellular cholesterol present in rat epididymal sperm, approximately 30% was detected in the esterified form (Davis et al., 1979); in addition, cholesteryl ester comprised about 37% of the total sterol in isolated rat sperm plasma membranes (Davis et al., 1980). Since cholesteryl ester is primarily an intracellular storage form of cholesterol and is found in only trace amounts in membranes, its presence as a major component of sperm plasma membranes probably arises from contamination.

Sulfo conjugated sterols have also been detected in mammalian sperm. While desmosteryl sulfate represents greater than 95% of the steryl sulfate associated with hamster epididymal sperm (Bleau and VandenHeuvel, 1974), cholesteryl sulfate comprised greater than 85% of the steryl sulfate fraction in human ejaculated sperm (Lalumiere et al., 1976). In both human (Lalumiere et al., 1976; Langlais et al., 1981) and hamster (Legault et al., 1979b), the uptake of isotopically labeled steryl sulfates by these cells has been described. Autoradiographic analysis of cholesteryl sulfate in human sperm and desmosteryl sulfate in hamster sperm localized these substances primarily to the head region. Based on the ability of cholesteryl sulfate to stabilize erythrocyte plasma membranes (Bleau et al., 1975a), the authors suggested that sulfated sterols may fulfill a comparable role in sperm during epididymal transit and regulate the fluidity of the sperm membrane during capacitation.

B. Modulation of Sperm Lipids during Maturation

Paralleling the acquisition of fertility by sperm as they migrate through and mature within the epididymis, changes in lipid content and composition occur. In general, a trend of decreasing phospholipid and sterol is observed in several species. A diminishing phospholipid content during traversal of the epididymis has been reported in bull (Poulos et al., 1973), ram (Grogan et al., 1966; Scott et al., 1967), boar (Evans and Setchell, 1979), and rat (Adams and Johnson, 1977) sperm. This depletion may reflect the use of phospholipids as endogenous energy substrates by sperm during epididymal transit and storage. The oxidation of phospholipids by sperm has been reported by Lardy and Phillips (1941) and Hartree and Mann (1961), although the relevance to sperm metabolism has been challenged by Darin-Bennett et al. (1973). In this regard, analysis of the energy requirements of sperm during residence in the epididymis suggests that endogenous lipids alone are insufficient to maintain sperm metabolism (Brooks, 1979).

A modulation of sperm sterol content during epididymal maturation has also been described. In the hamster, the cholesterol content of sperm decreases tenfold during passage from the caput to cauda epididymis, concomitant with an 18-fold increase in desmosteryl sulfate and unchanged free desmosterol levels (Legault et al., 1979a). A similar pattern for cholesterol diminution was described in rat and ram sperm. Caudal epididymal rat sperm contain lower levels of cholesterol than those from the caput (Adams and Johnson, 1977); ejaculated ram sperm have lower cholesterol levels than testicular sperm (Scott et al., 1967).

These alterations in sperm lipids during maturation may reflect changes in membrane structure and properties which could have important consequences for survival in and response to the capacitating environment of the female reproductive tract.

C. Sterol Levels and Cold Shock

A causal relationship between sperm membrane stability and cholesterol content is suggested by studies involving the phenomenon of cold shock. Sperm exposed to a cold environment for varying lengths of time show differences in the ability to recover motility upon rewarming. Ram and bull sperm, for example, which are sensitive to cold shock, have approximately half the amount of cholesterol of human and rabbit sperm, which are relatively resistant (Quinn and White, 1967; Darin-Bennett and White, 1977). Mouse sperm, which are sterol-rich cells, are also resistant to cold shock, as demonstrated by Heffner and Storey (1982). The validity of this correlation will be strengthened by more extensive comparisons and when resistance to cold shock can be related to sterol levels specifically in the plasma membrane.

IV. Modification of Sperm Lipid Levels during Capacitation

During passage through the epididymis, modifications in sperm sterol and phospholipid composition occur which may reflect changes in membrane lipids during maturation. As a locus of many biochemical changes during capacitation, the sperm plasma membrane may undergo changes in lipid composition as part of this process, and interest has centered on the role that sperm sterol content and its modulation may play in capacitation.

A. *In Vitro* STUDIES

Sperm from the human, rat, mouse, and hamster can be capacitated in chemically defined media containing serum albumin. This requirement for albumin is specific; a variety of other proteins including ovalbumin, cytochrome c, rabbit β- and γ-globulins, hemoglobin, ribonuclease, lysozyme, and fibrinogen, and the polymer, polyvinylpyrrolidone, did not substitute for albumin (Davis, 1976a). The role that albumin plays in capacitation remains to be elucidated. However, several mechanisms for its activity have been suggested: (1) removal or alteration of coating materials at the sperm surface (Yanagimachi, 1981); (2) promotion of plasma membrane protein hydrolysis (Davis and Gergely, 1979); and (3) alteration of sperm membrane cholesterol and phospholipid composition through lipid exchange (Davis *et al.*, 1979). The latter is consistent with the well-characterized abilities of albumin to bind phospholipids (Jonas, 1976) and mediate a sterol efflux from cells in culture (Bartholow and Geyer, 1981).

In an investigation into the role of albumin in the capacitation of rat epididymal sperm, Davis *et al.* (1979) reported a bidirectional transfer of lipid between sperm and protein. During incubation in a capacitating culture medium, rat sperm undergo an 11% decrease in cholesterol content concurrent with a 1.5-fold increase in phospholipid. Complementary changes in lipid content occur in albumin, with the majority of albumin-associated cholesterol recovered in the esterified form. Confirmation of capacitation by fertilization of eggs, however, was not undertaken. In an extension of this work, plasma membranes of rat sperm subjected to capacitating conditions were isolated. As in whole sperm, incubation with albumin resulted in a 46% increase in phospholipid content; however, the decrease in cholesterol levels was not statistically significant. Thus, the phospholipid uptake by sperm from albumin was considered to be the primary factor responsible for the lowered sterol-to-phospholipid ratio in the cell membranes (Davis et al., 1980).

Similarly, we have investigated the ability of albumin to modulate sperm lipid levels in the mouse with a secondary objective of correlating individual changes in lipid content with capacitation as assayed by an in vitro fertilization system (Go and Wolf, 1983). Two albumins, fraction V and fatty acid-free bovine serum albumins (BSA), were effective in supporting the capacitation of mouse epididymal sperm as evaluated by the fertilization in vitro of cumulus-free, zona pellucida-intact mouse eggs. Following incubation in medium supplemented with either of these albumin preparations for an interval sufficient to achieve capacitation, sperm undergo increases in phospholipid content and statistically significant decreases in total cellular sterol. The magnitudes of the lipid fluxes between sperm and albumin were dissimilar for the two protein preparations. While incubation in fraction V BSA-supplemented medium results in a 30% decrease in total sperm sterol concurrent with a twofold increase in phospholipid, sperm incubated with fatty acid-free BSA undergo a 50% decrease in sterol and a 70% increase in phospholipid. This difference in effect on sperm lipid levels may arise from the disparate amounts of cholesterol and phospholipid present in the albumin preparations as ligands of the protein or as components of contaminating serum lipoproteins (Fainaru et al., 1981). Albumin preparations containing greater amounts of phospholipid are more effective in releasing sterol from mouse (Chau and Geyer, 1978) and human (Bartholow and Geyer, 1981) fibroblasts. Consistent with these observations, fatty acid-free BSA, which contained tenfold more phospholipid than fraction V BSA, effected a greater depletion of sterol from mouse sperm.

To determine whether both sterol depletion and phospholipid enrichment of sperm by albumin are obligatory events in capacitation, a lipid-free albumin was prepared to eliminate phospholipid uptake (Go and Wolf, 1983). Fraction V BSA was rendered phospholipid free by delipidation and fractionation by molecular sieving under denaturing conditions. The ability to support sperm capacitation was conserved in the protein following these treatments, indicating that endogenous lipids are not required for albumin-mediated capacitation. The sterol-releasing activity of albumin was retained, as reflected in a 25% decrease in sperm sterol levels following incubation with this lipid-free BSA. These results contrast with the findings reported on the interaction of rat sperm and albumin in which phospholipid enrichment of sperm was suggested as a pivotal event in the transformation of the spermatozoal plasma membrane during capacitation (Davis et al., 1979, 1980). Further studies are required to determine if sterol depletion mediated by albumin or another acceptor is sufficient to support sperm capacitation in species other than the mouse.

B. *In Vivo* Studies

The lipid composition of ejaculated porcine sperm has been analyzed after incubation under *in vivo* capacitating conditions. Following incubation in the uterine horns of a gonadotrophin-stimulated gilt, no change in spermatozoal cholesterol levels occurred. In addition, although the phosphatidylcholine content increased significantly, the increase in total sperm phospholipid levels did not (Evans *et al.*, 1980).

In sperm isotopically labeled with cholesterol, evidence for sterol depletion during *in vivo* capacitation has been obtained. Transfer of labeled sterol from rabbit sperm to high-molecular-weight proteins has been described during incubation of these cells in the uteri of females induced to ovulate by gonadotrophin administration (Davis, 1982). Following incubation and sperm removal, uterine fluid contained significant amounts of labeled sterol. Chromatographic fractionation of the fluid revealed that most of the radioactivity was associated with a component tentatively identified as albumin. Labeled sterol also was eluted with an unidentified component which had a molecular weight of approximately 100,000, suggesting that other acceptors of sterol may be present in the uterus. Since this study was carried out with sperm only, it is not known whether exposure to the uterine milieu causes sterol depletion from other cell types.

V. Effects of Sterols on Sperm Fertility

In several species, exposure of sperm to sterol-enriched media has inhibitory effects on fertility. Incubation of uterine-capacitated rabbit sperm with cholesterol-containing phospholipid liposomes results in reduced levels of fertilization when treated sperm are inseminated into the oviducts of superovulated does (Davis, 1980). The inhibitory or decapacitating effect is concentration dependent and apparently specific for the free sterol since liposomes containing cholesteryl myristate are without adverse effects on fertilization. Sperm incubated with sterol-rich vesicles isolated from rabbit seminal plasma also exhibit reduced fertility when instilled into the uteri of ovulating does (Davis, 1974, 1980). The reversible nature of decapacitation mediated by either synthetic cholesterol–phospholipid vesicles or those from seminal plasma was demonstrated by the recovery of fertilizing ability following reincubation of sperm in a capacitating environment (Davis, 1974, 1976b).

Another approach to evaluate cellular sterol release by albumin on the ability of this protein to support *in vitro* capacitation and fertilization is to alter this activity by adding cholesterol. Presaturation of albumin with cholesterol results in reduced levels of fertilization in both the rat (Davis, 1976a) and mouse (Go and Wolf, 1983). In the latter, the inhibitory activity is cholesterol dependent.

Similarly, in studies measuring the ability of albumin to induce acrosome reactions, supplementation of the protein with cholesterol reduced the percentage of reacted sperm (Davis, 1980). In the guinea pig, a species in which albumin is not required for sperm capacitation *in vitro*, treatment of sperm with cholesterol–albumin complexes retarded the onset of the acrosome reaction (Fleming and Yanagimachi, 1981).

While the inhibitory effect of cholesterol on capacitation and the induction of the acrosome reaction seems real, the mechanism by which this inhibition is achieved is not known. Human and hamster sperm are capable of taking up labeled free and sulfo conjugated sterols from the medium, although it is not clear to what extent this represents an exchange process or a net accumulation of sterol in the cells. In the case of the latter, alterations in membrane properties arising from increased sterol levels may inhibit the development of functional changes associated with capacitation.

An inhibitory effect on capacitation by sulfated sterols apparently also exists. Inclusion of desmosteryl sulfate in an *in vitro* fertilization medium composed of Medium 199M2 and dispersed cumulus cells from hamster ova results in the inhibition of hamster sperm capacitation (Bleau *et al.*, 1975b). Free desmosterol and cholesterol had no effect. It was suggested that this suppression by desmosteryl sulfate of capacitation occurs through competitive inhibition of sterol sulfatase activity present in cumulus cells. A role for this enzymic activity in the modification of sperm steryl sulfate levels and subsequent membrane destabilization during capacitation has been postulated (Legault *et al.*, 1980).

VI. Factors Which May Regulate Sperm Sterol Content

A. Sterol Acceptors

Modulation of cellular sterol content may occur through interaction with proteins or protein–lipid complexes which can induce a depletion of sterol from the cell. These agents, sterol acceptors, may play a regulatory role in the sterol metabolism of some cells. The ability of albumin to function as a sterol acceptor *in vitro* has been demonstrated in a variety of cell types, including lymphoblasts (Bailey, 1967), fibroblasts (Burns and Rothblat, 1969; Chau and Geyer, 1978; Bartholow and Geyer, 1981), macrophages (Werb and Cohn, 1971a,b), and vesicular stomatitis virus (Pal *et al.*, 1981). Consistent with these reports, a diminution of total sterol levels in sperm in the presence of albumin has been demonstrated *in vitro* and *in vivo* as described in Section IV, suggesting a role for albumin as a modulator of sterol content in these cells.

In studies characterizing the composition of fluids from the female reproductive tract, albumin was identified as the major protein in the oviductal fluids from every species examined: humans (Moghissi, 1970), *Rhesus* monkeys (Mastroianni *et al.*, 1970), rabbits (Shapiro *et al.*, 1971), ewes (Roberts *et al.*, 1976), rats (Shalgi *et al.*, 1977), and bovines (Stanke *et al.*, 1974). Although it is tempting to assign albumin a role in capacitation *in vivo* on the basis of its activity as a sterol acceptor from sperm *in vitro*, it is important to be cognizant of the complexity of the fluid at the physiological site of capacitation and fertilization. Many other proteins, arising from transudation from the serum, follicular, and uterine fluids, as well as the low concentrations of secreted proteins by the oviductal epithelia, are present, and these may participate in sperm capacitation. Indeed, capacitation of guinea pig sperm *in vitro* can be achieved in the absence of albumin, and normal fertility has been reported in a strain of analbuminemic rats (Nagase and Shimamune, 1979). It is interesting to note, however, that these animals had elevated levels of cholesterol binding α-globulins in their sera.

B. STEROL SULFOHYDROLASE ACTIVITY

Sulfated sterols have been implicated in the stabilization of sperm membranes during transit in the epididymis (Legault *et al.*, 1979a). On the basis of the presence of hormonally dependent sterol sulfohydrolase activity in the human and hamster female reproductive tracts (Lalumiere *et al.*, 1976; Legault *et al.*, 1980), a hypothesis for the role of steryl sulfates and their modification in sperm capacitation has been presented by Langlais *et al.* (1981). The cleavage of sperm steryl sulfate moieties by sterol sulfatases is suggested to initiate a series of events leading to capacitation, including an alteration in sperm sterol-to-phospholipid ratio through cholesteryl ester and lysolecithin production. Thus, the following sequence of events is hypothesized: (1) free sterol is generated in the sperm plasma membrane following cleavage of the sulfate groups; (2) through the simulatory effect of albumin on their activities, phospholipases in the sperm membrane generate lysolecithin accompanied by the formation of cholesteryl esters; (3) the removal of esterified cholesterol from the membrane, in conjunction with the increased levels of lysolecithin in the sperm membrane, creates conditions conducive to cell fusion. While the presence of the key components in this hypothesis has been demonstrated, experimental evidence supporting this cascade of events leading to the acrosome reaction and fertilization is unavailable.

VII. Conclusions

The sperm plasma membrane is the site of many biochemical changes occurring during capacitation, including the removal of surface components, increased ion permeability, redistribution of intramembranous particles, alteration in surface charge, and modulation of lipid composition (Yanagimachi, 1981). Given the role of sterol as a determinant of membrane physicochemical properties, sperm sterol levels, particularly those in the plasma membrane, may influence these alterations by regulating the fluid characteristics of the bilayer. Evidence for a modulation of sperm sterol levels under capacitating conditions has been obtained in several species. These studies are limited, however, in that the measured changes in lipid content reflect an average value for a population of sperm under study and do not yield information on the cellular or subcellular sources of the depleted sterol, that is, whole-cell or plasma membrane and head, midpiece, or tail regions. Indeed, documentation of the presence of specialized plasma membrane domains with altered sterol content is a demanding undertaking. As an alternative, a morphologic approach can be taken: freeze fracture studies of filipin-stained membranes would allow identification of sterol in the bilayer through its interaction with this antibiotic (Montesano et al., 1979). Nevertheless, the requirement for membrane destabilization during capacitation and the recognized role of sterol as a regulatory agent in membrane fluidity and function encourage the idea that modulation of sperm sterol levels may be among the molecular mechanisms involved in capacitation.

ACKNOWLEDGMENT

The work of Kathryn J. Go was supported by the Andrew W. Mellon Foundation.

References

Adams, C. S., and Johnson, A. D. (1977). *Comp. Biochem. Physiol.* **58B**, 409.
Austin, C. R. (1951). *Aust. J. Biol. Sci. Ser.* **B4**, 581.

Bailey, J. M. (1967). *Am. J. Physiol.* **207**, 1221.
Bartholow, L. C., and Geyer, R. P. (1981). *Biochim. Biophys. Acta* **665**, 40.
Bavister, B. D. (1981). In "Fertilization and Embryonic Development *in Vitro*" (L. Mastroianni, Jr., and J. D. Biggers, eds.), pp. 42–60. Plenum, New York.
Bearer, E. L., and Friend, D. S. (1982). *J. Cell Biol.* **92**, 604.
Bedford, J. M., and Chang, M. C. (1962). *Am. J. Physiol.* **202**, 179.
Bleau, G., and VandenHeuvel, W. J. A. (1974). *Steroids* **24**, 549.
Bleau, G., Lalumiere, G., Chapdelaine, A., and Roberts, K. D. (1975a). *Biochim. Biophys. Acta* **375**, 220.
Bleau, G., VandenHeuvel, W. J. A., Andersen, O. F., and Gwatkin, R. B. L. (1975b). *J. Reprod. Fertil.* **43**, 175.
Brooks, D. E. (1979). In "The Spermatozoon" (D. W. Fawcett and J. M. Bedford, eds.), pp. 23–34. Urban and Schwarzenberg, Baltimore, Md.
Burns, C. H., and Rothblat, G. H. (1969). *Biochim. Biophys. Acta* **176**, 616.
Chang, M. C. (1951). *Nature (London)* **168**, 697.
Chau, I. Y., and Geyer, R. P. (1978). *Biochim. Biophys. Acta* **542**, 214.
Cooper, R. A. (1978). *J. Supramol. Struct.* **8**, 413.
Darin-Bennett, A., and White, I. G. (1977). *Cryobiology* **14**, 466.
Darin-Bennett, A., Poulos, A., and White, I. G. (1973). *J. Reprod. Fertil.* **34**, 543.
Davis, B. K. (1974). *J. Reprod. Fertil.* **41**, 241.
Davis, B. K. (1976a). *Proc. Soc. Exp. Biol. Med.* **151**, 240.
Davis, B. K. (1976b). *Proc. Soc. Exp. Biol. Med.* **152**, 257.
Davis, B. K. (1980). *Arch. Androl.* **5**, 249.
Davis, B. K. (1982). *Experientia* **38**, 1063.
Davis, B. K., and Gergely, A. F. (1979). *Biochem. Biophys. Res. Commun.* **88**, 613.
Davis, B. K., Byrne, R., and Hungund, B. (1979). *Biochim. Biophys. Acta* **558**, 257.
Davis, B. K., Byrne, R., and Bedigian, K. (1980). *Proc. Natl. Acad. Sci. U.S.A.* **77**, 1546.
Evans, R. W., and Setchell, B. P. (1979). *J. Reprod. Fertil.* **57**, 189.
Evans, R. W., Weaver, D. E., and Clegg, E. D. (1980). *J. Lipid Res.* **21**, 223.
Fainaru, M., Schaeffer, Z., and Decklebaum, R. J. (1981). *Prep. Biochem.* **11**, 273.
Fleming. A. D., and Yanagimachi, R. (1981). *Gamete Res.* **4**, 253.
Friend, D. S., Orci, L., Perrelet, A., and Yanagimachi, R. (1977). *J. Cell Biol.* **74**, 561.
Go, K. J. (1982). Ph.D. thesis, University of Pennsylvania.
Go, K. J., and Wolf, D. P. (1984). Manuscript in preparation.
Gordon, J., Dandekar, P. V., and Bartoszewicz, W. (1975). *J. Ultrastruct. Res.* **50**, 199.
Grogan, D. E., Mayer, D. T., and Sikes, J. D. (1966). *J. Reprod. Fertil.* **12**, 431.
Gwatkin, R. B. L. (1977). "Fertilization Mechanisms in Man and Mammals." Plenum, New York.
Hartree, E. F., and Mann, T. (1961). *Biochem. J.* **80**, 464.
Heffner, L. J., and Storey, B. T. (1982). *J. Exp. Zool.* **219**, 155.
Jonas, A. (1976). *Biochim. Biophys. Acta* **427**, 325.
Kinsey, W. H., and Koehler, J. K. (1978). *J. Ultrastruct. Res.* **64**, 1.
Lalumiere, G., Bleau, G., Chapdelaine, A., and Roberts, K. D. (1976). *Steroids* **27**, 247.
Langlais, J., Zollinger, M., Plante, L., Chapdelaine, A., Bleau, G., and Roberts, K. D. (1981). *Proc. Natl. Acad. Sci. U.S.A.* **78**, 7266.
Lardy, H. A., and Phillips, P. H. (1941). *Am. J. Physiol.* **133**, 602.
Legault, Y., VandenHeuvel, W. J. A., Arison, B. H., Bleau, G., Chapdelaine, A., and Roberts, K. D. (1978). *Steroids* **32**, 649.
Legault, Y., Bouthillier, M., Bleau, G., Chapdelaine, A., and Roberts, K. D. (1979a). *Biol. Reprod.* **20**, 1213.

Legault, Y., Bleau, G., Chapdelaine, A., and Roberts, K. D. (1979b). *Steroids* **34**, 89.
Legault, Y., Bleau, G., Chapdelaine, A., and Roberts, K. D. (1980). *Biol. Reprod.* **23**, 720.
Mastroianni, L., Jr., Urzua, M., and Stambaugh, R. (1970). *Fertil. Steril.* **21**, 817.
Moghissi, K. S. (1970). *Fertil. Steril.* **21**, 281.
Montesano, R., Perrelet, A., Vassalli, P., and Orci, L. (1979). *Proc. Natl. Acad. Sci. U.S.A.* **76**, 6391.
Muhlebach, T., and Cherry, R. J. (1982). *Biochem.* **21**, 4225.
Nagase, S., and Shimamune, K. (1979). *Science* **205**, 590.
Oliphant, G., and Brackett, B. G. (1973). *Biol. Reprod.* **9**, 404.
O'Rand, M. G. (1977). *Dev. Biol.* **55**, 260.
Pal, R., Barenholz, Y., and Wagner, R. R. (1981). *Biochemistry* **20**, 530.
Poulos, A., Voglmayr, J. K., and White, I. G. (1973). *Biochim. Biophys. Acta* **306**, 194.
Quinn, P. J., and White, I. G. (1967). *Aust. J. Biol. Sci.* **20**, 1205.
Roberts, G. P., Parker, J. M., and Symonds, H. W. (1976). *J. Reprod. Fertil.* **48**, 99.
Rogers, B. J. (1978). *Gamete Res.* **1**, 165.
Schwarz, M. A., and Koehler, J. K. (1979). *Biol. Reprod.* **21**, 295.
Scott, T. W., Voglmayr, J. K., and Setchell, B. P. (1967). *Biochem. J.* **102**, 465.
Shalgi, R., Kaplan, R., and Kraicer, P. F. (1977). *Biol. Reprod.* **17**, 333.
Shapiro, S. S., Jentsch, J. P., and Yard, A. S. (1971). *J. Reprod. Fertil.* **24**, 403.
Shinitsky, M., and Souroujon, M. (1979). *Proc. Natl. Acad. Sci. U.S.A.* **76**, 4438.
Singh, J. P., Babcock, D. F., and Lardy, H. A. (1978). *Biochem. J.* **152**, 549.
Stanke, D. F., Sikes, J. P., DeYoung, D. W., and Tumbleson, M. E. (1974). *J. Reprod. Fertil.* **38**, 493.
Talbot, P., and Franklin, L. E. (1978). *J. Exp. Zool.* **203**, 1.
Triana, L. R., Babcock, D. F., Lorton, S. P., First, N. L., and Lardy, H. A. (1980). *Biol. Reprod.* **21**, 883.
Vanderkooi, J., Fischkoff, S., Chance, B., and Cooper, R. A. (1974). *Biochem.* **13**, 1589.
Werb, Z., and Cohn, Z. A. (1971a). *J. Exp. Med.* **134**, 1545.
Werb, Z., and Cohn, Z. A. (1971b). *J. Exp. Med.* **134**, 1570.
Wolf, D. P. (1979). *In* "The Biology of the Fluids of the Female Genital Tract" (F. K. Beller and G. F. B. Schumacher, eds.), pp. 407–414. Elsevier, Amsterdam.
Yanagimachi, R. (1981). *In* "Fertilization and Embryonic Development *in Vitro*" (L. Mastroianni, Jr., and J. D. Biggers, eds.), pp. 82–182. Plenum, New York.

Author Index

A

Aaron, R., 230, *254*
Abano, D. A., 110, 111, 121, 122, 123, 124, 125, 141, *187*
Abbate, R., 236, *252*
Abdulah, Y. H., 208, *213*
Abe, M., 138, *185*
Åberg, G., 209, 210, 211, *213*
Abildgaard, U., 222, *254*
Able, K., 277, *307*
Abraham, S., 46, 51, *103*
Abrams, J. J., 182, *185*
Adam, A., 267, 276, *312, 316*
Adams, C. S., 320, 322, *328*
Adams, C. W. M., 236, 242, *246, 252*
Adams, G. H., 92, *104,* 152, *192*
Adams, P. W., 204, *215*
Addo, O., 236, 244, *246*
Addonizio, M. L., 221, 235, *247*
Adolphson, J. L., 120, 121, 122, 126, 128, 132, 142, 143, 153, 154, 155, 156, 157, 158, 166, *185*
Aftergood, L., 147, 148, *185*
Ahlgren, L. T., 228, *246*
Ahrens, E. H., Jr., 45, 46, 47, 48, 50, 51, 53, 55, 57, 58, 59, 61, 64, 65, 69, 73, 80, 88, *102, 103, 104,* 182, *189*
Akamatsu, Y., 259, 273, 276, 277, 284, 289, 293, 294, 298, *307, 313*
Akanuma, Y., 119, 121, 131, 141, 156, 157, 158, *185, 191, 194*
Akbar, H., 225, *246*
Akimori, N., 274, *316*
Akino, T., 138, *185*
Akman, N., 236, *246*
Aktories, K., 207, 208, 210, *213*
Alameda, S., 227, *246*
Alaupovic, P., 141, *187, 190*
Albers, J. J., 11, *41,* 83, 85, 88, 91, *102, 103,* 109, 116, 117, 120, 121, 122, 123, 124, 125, 126, 128, 130, 132, 133, 140, 141, 142, 143, 153, 154, 155, 156, 157, 158, 166, 173, *185, 186, 191, 193, 246*

Albers, J. S., 242, *246*
Albert, W., 3, *42*
Albertweil, J., 301, *307*
Albrechtsen, O. K., 233, *246*
Albright, F. R., 296, *307*
Alcindor, L. G., 85, *102,* 128, 129, 165, 183, *185, 187, 193*
Aldersberg, D., 82, 85, *103*
Alexander, M., 108, 138, 139, *192*
Alexandrova, N. P., 207, *217*
Alfin-Slater, R. B., 147, 148, *185*
Allan, G., 225, 226, *248*
Allen, C. F., 227, *246, 252,* 287, *309*
Allen, K. G. D., 148, *189*
Alshamaoni, L., 259, 261, 264, 274, 278, 279, *307, 313*
Altschul, R., 195, 203, 208, *213*
Alupovic, P., 1, 2, 4, 5, 10, 12, 14, 40, *41, 42, 43*
Amaducci, L., 119, *185*
Amar, C., 276, *307*
Ambron, R. T., 289, *307*
Ameryckx, J. P., 10, *42*
Ammon, H. P. T., 211, *213*
Amris, C. J., 240, *247*
Anderson, D. W., 10, *41,* 142, *185, 192*
Anderson, J. M., 68, 88, *101, 104*
Anderson, N. G., 222, *246*
Anderson, O. F., 326, *329*
Anderson, P., 235, *246*
Anderson, R. G., 209, 210, 211, *213*
Anderson, R. J., 258, 269, 274, 279, *307*
Anderson, W. H., 225, *252*
Andreev, L. V., 264, 273, *313*
Andrejew, A., 288, *308*
Anfinsen, C. B., 223, *249*
Angel, A., 74, *103*
Angelin, B., 92, *101*
Anslow, W. A., 306, *308*
Antonie, A. D., 271, 282, *308*
Antonio, M., 205, *216*
Antuono, P., 119, *185*
Applegate, K. R., 108, 109, 142, 165, *188, 191*
Ardlie, N. G., 225, 240, *246, 254*

Arima, K., 286, 306, *308, 313*
Arison, B. H., 321, *329*
Arky, R. A., 160, *194*
Armijo, M., 200, 203, 207, 208, *216*
Armstrong, E. L., 267, 291, *315*
Armstrong, M. L., 232, *246*
Aron, A., 136, 137, *186*
Aron, L., 111, 112, 113, 122, 123, 125, 133, *185*
Aronson, R., 238, *246*
Arroyave, G., 162, *190*
Ascenzi, J. M., 283, 284, *308*
Ashman, P., 176, *189*
Asplund, J., 229, *246*
Asselineau, C., 267, 273, 275, 277, *308*
Asselineau, J., 258, 269, 270, 271, 272, 275, 280, 304, *308, 312, 316*
Assmann, G., 2, 7, 9, 35, *41*, 111, 113, 120, 143, *185, 186, 192*, 205, *213*
Astrup, P., 242, 244, *246*
Astrup, T., 233, *246*
Atkins, E. D. T., 221, 243, *246*
Attie, A. D., 88, *104*
Aubert, P., 183, *193*
Aubrey, B. J., 233, *246*
Aune, K. C., 12, *43*
Austen, K. F., 226, 227, *250, 253, 254*
Austin, C. R., 317, *328*
Avigad, G., 292, 300, *308*
Avogaro, P., 12, 22, 25, 30, 33, *41, 42*, 156, 176, *186*, 197, 198, *213*
Azuma, I., 304, *314*
Azuma, T., 304, *313*

B

Babcock, D. F., 318, *330*
Babický, A., 147, *186*
Bagdade, J. D., 109, 116, 117, 140, 145, 172, 173, *186, 193*
Bailey, E. G., 306, *313*
Bailey, J. M., 327, *329*
Baird, E. E., 207, *213*
Baker, H. N., 109, 111, *189, 193*, 196, *214*
Baker, P. J., 227, *246*
Balart, L., 162, *190*
Balasse, E. O., 211, *213*
Baldoni, E., 198, 207, *214*
Ballio, A., 264, 272, 273, 274, *308*

Ballou, C. E., 260, 261, 279, 290, 291, *308, 309, 310, 312, 314*
Balmer, J., 99, *105*, 136, *194*
Baluda, M. V., 207, *215*
Baluda, V. P., 206, 207, *215*
Banerjee, B., 260, 261, 262, 281, 303, *308, 311*
Bansal, V. S., 286, 294, 303, *311, 316*
Barakat, H., 263, 264, 272, *316*
Barbend, J. C., 237, *251*
Barbier, M., 259, 275, *308, 309*
Barcellona, S., 264, 272, 273, 274, *308*
Barenholz, Y., 289, *310, 327, 330*
Bargoot, F. G., 139, *193*
Barksdale, L., 269, 272, 288, 303, *308*
Barnhart, R. L., 134, *189*
Barr, D. P., 204, 205, *213*, 238, *246*
Barter, P. J., 28, *41*, 85, 87, *101*, 131, 136, 145, 149, 154, *186, 194*
Bartholomé, M., 132, *186*
Bartholow, L. C., 323, 324, 327, *329*
Bartoli, A., 275, *311*
Bartolini, L., 119, *185*
Bartoszewicz, W., 318, *329*
Bartz, Q. R., 306, *309*
Basista, M., 206, *214*
Basset, D. R., 59, 60, 73, 74, 75, *103*
Batrakov, S. G., 259, 260, 266, 270, 271, *308*
Bavister, B. D., 320, *329*
Baydack, T., 229, *249*
Bayliss, O. B., 242, *246*
Bazasian, G. G., 229, *252*
Bearer, E. L., 318, *329*
Bearn, A. G., 134, *190*
Beaumont, J. E., 173, *186*
Becker, D. L., 235, *253*
Becker, M. A., 211, *213*
Bedford, J. M., 318, *329*
Bedigian, K., 321, 324, *329*
Behrends, C. L., 227, *246*
Bekhtereva, M. N., 280, *309*
Bekierkunst, A., 304, *308, 316*
Bell, F. B., 73, 74, 88, 90, *101, 103*
Bell, F. P., 134, 135, 149, 174, *186*
Bell, M. G., 266, *316*
Bellini, F., 182, *187*
Benaim, M. E., 198, 207, *213*
Bengtsson, G., 223, *246*
Benhamou, G., 128, 129, *185*

Author Index

Benito, M., 198, 207, *213*
Bennett, B., 244, *255*
Bennett-Clark, S., 119, 120, *186*
Benns, G., 295, *309*
Benzce, W. L., 197, *213*
Beppu, T., 286, 306, *208, 313*
Bereziat, G., 183, *193*
Berg, K., 6, *41*
Berg, T., 109, 118, 119, 147, 152, 158, *191*
Bergan, A., *192*
Bergelin, R. O., 154, *185*
Bergelson, L. D., 108, *194,* 259, 260, 266, 270, 271, *308*
Berger, B., 284, *308*
Bergham, L., 228, *251*
Bergheim, L. E., 228, *246*
Bergmann, K. V., 46, 49, *103*
Bergström, S., 209, 210, *217*
Berlan, J., 275, *315*
Berman, M., 74, *102,* 203, 204, 205, 209, *214*
Bermann, M., 204, 205, *213*
Bernstein, R. M., 156, 177, *186, 192*
Bernstein, V., 156, 177, *186*
Bersot, T. P., 8, *43*
Bertram, P. D., 142, *192*
Berwick, L., 242, *252*
Bessel, C. J., 306, *308*
Besterman, E. M. M., 225, *246*
Betzing, H., 134, *194*
Beubler, E., 29, *43*
Bezman, A., 47, *104*
Bhattacharyya, A. K., 52, 74, 78, 80, *101, 102*
Biagi, M., 197, *217*
Biale, Y., 208, *216*
Bíbr, B., 138, *187*
Bicker, S., 86, 91, *104*
Bierman, E. L., 88, 91, 99, *102,* 109, *186,* 239, 242, *246*
Biggs, M. W., 83, *101*
Biggs, R., 222, *246*
Bigler, F., 269, *311*
Bilheimer, D. W., 48, 50, 74, 77, 87, 91, 92, *102, 103,* 204, *214*
Bina, M., 231, *247*
Binak, K., 241, *253*
Bing, D. H., 227, *246*
Bing, R. J., 88, *102*

Bittolo Bon, G., 12, 22, 25, 30, 33, *41, 42,* 156, 176, *186,* 197, 198, *213*
Bjornson, L. K., 135, *186*
Blackshear, P. L., 241, *254*
Blackwelder, W. C., 142, *192*
Blackwell, J., 220, *246*
Blajckman, M. A., 222, *252*
Blanche, P. J., 12, *41,* 139, *191*
Blanchette-Mackie, E. J., 135, *186*
Blasko, G., 222, 228, 233, *251*
Bleau, G., 320, 321, 322, 326, 327, *329, 330*
Bleyl, H., 236, 244, *246*
Bloch, H., 261, 271, *308, 314*
Bloch, K., 262, 272, 273, 281, 282, *312*
Blomback, M., 236, *248*
Blomhoff, J. P., 131, 144, 149, 158, 163, 177, *186, 188, 190, 192,* 238, *254*
Blum, C. B., 11, 74, *43, 102,* 204, 205, *213*
Blümchen, G., 177, *189*
Blumel, G., 229, *254*
Blumenstock, F. A., 227, *246*
Boackle, R. J., 227, *247*
Boberg, J., 209, *214*
Bohin, J. P., 285, *314*
Bohlmann, H. G., 16, 17, 29, *42*
Bojensen, E., 134, *186*
Bolzano, K., 21, 22, *42*
Boncinelli, L., 182, *187*
Boniforti, L., 264, 272, 273, 274, *308*
Bordet, C., 267, 272, 273, 274, 278, *308, 313*
Borell, U., 229, *246*
Borer, J. S., 225, *254*
Borgstrom, B., 46, 56, 67, *102*
Bouquet, A., 301, *308*
Bourdillon, N. C., 205, *216*
Bouthillier, M., 320, 327, *329*
Bowyer, D. P., 239, *253*
Boyd, G. S., 149, *189*
Boyden, S. V., 301, *309*
Boyer, J. L., 117, *192*
Boyle, E., 223, 238, *246, 249*
Bracellona, S., 274, *308*
Brackett, B. G., 318, *330*
Braeuler, C., 203, *213*
Bragdon, J. H., 223, 224, 238, *246, 247*
Brattsand, R., 197, 198, 199, 200, 205, 206, 209, 210, 211, *213, 215*
Braunsteiner, H., 3, 11, 12, *42, 43,* 131, 154, *191*

Braunwald, E., 220, *247*
Brecker, P. I., 135, *187*
Breillot, J., 85, *102*
Brennan, P. J., 260, 261, 262, 265, 266, 267, 269, 272, 279, 288, 290, 291, *309, 310, 311*
Brenner, G., 197, *213*
Brenner, H., 197, *213*
Breslow, J. L., 7, 9, *43*
Bretherton, K. N., 167, *186*
Brewer, H. B., 2, 26, 29, *42*, 142, *192*
Bridges, J. M., 236, *252*
Bridoux, A. M., 142, *192*
Brock, T. D., 306, *309*
Brockner, J., 240, *247*
Brodan, V., 163, 164, 165, *187*
Brokl, O., 268, *310*
Bron, A. F., 165, *186*
Brooks, D. E., *329*
Brouty-Borge, H., 228, *254*
Brown, M. S., 23, 28, *41*, 69, 79, 87, 88, 91, 93, 97, 101, *102, 103*, 175, *189*
Brown, R. K., 223, 238, *246, 249*
Brown, R. R., 203, *213*
Brown, W. D., 223, *247*
Brown, W. V., 139, *192*
Bruckdorfer, K. R., 115, *187*
Brunelli, C., 236, *252*
Bruneteau, M., 269, *316*
Brunner, E., 8, *41*
Brunzell, J. D., 109, *186*
Buchanan, J. M., 233, *247*
Buchanan, M. R., 225, *247*
Buchwald, H., 241, *254*
Buck, K., 167, *188*
Buddecke, E., 206, *217*
Buja, M. L., 77, *102*
Bullock, B. C., 56, 58, 60, 65, 73, 78, 90, *102, 103*
Bunting, S., 210, *214*
Bütler, R., 8, *41*
Buonassi, V., 221, *247*
Burke, D. J., 11, *42*
Burns, C. H., 327, *329*
Burns, T. W., 207, 208, *213*
Burrows, B. A., 51, 53, 59, *102*
Busch, C., 226, 229, 234, *247, 249, 253, 254*
Butcher, R. W., 207, *213*
Buttke, T. M., 280, 285, 286, *309, 310*
Byrne, R., 318, 320, 321, 323, 324, *329*

C

Cabana, V. G., 121, 122, *185*
Caen, J. P., 225, *254*
Calatroni, A., 243, *247*
Calle, S., 225, *254*
Campanacci, L., 173, *189*
Campbell, A. H., 306, *308*
Campbell, I. M., 263, 272, 278, *309*
Canetti, G., 273, *309*
Cappelletti, R., 243, *254*
Capressi, E. S., 237, 245, *253*
Caprini, L. F., 237, 245, *253*
Capurso, A., 7, *41*
Caputi, A., 199, 200, *215*
Cardin, A. D., 134, *189*
Carlander, B., 238, *250*
Carlander, E., 232, *250*
Carlile, S. I., 87, *102*
Carlson, L. A., 196, 197, 198, 203, 204, 207, 208, 209, 210, 211, *213, 214, 215, 216*
Carpenter, C. B., 238, *252*
Carr, J., 231, *247*
Carroll, K. K., 53, 60, 67, 73, *102, 103*, 151, *186*
Carter, G. A., 52, 74, *102*
Carter, J., 225, *254*
Carty, C. E., 284, *308*
Caruso, V., 197, 200, *216*
Casaretto, A., 172, 173, *186*
Casey, C. R., 232, *249*
Cason, J., 263, 272, 287, *309*
Castaldi, P. A., 225, *247*
Castellanos, H., 237, 245, *253*
Castelli, W. P., 175, 176, *186, 189*, 205, *214*
Castellot, J. J., Jr., 221, 235, *247*
Cataldi, S., 199, 200, *215*
Caughman, G. B., 227, *247*
Cavallo, T., 221, 230, *250*
Cazenave, J. P., 225, 226, 233, *247, 248*
Cazzolato, G., 12, 22, 25, 30, 33, *41, 42*, 156, 176, *186*, 197, 198, *213*
Cella, G., 235, *247*
Chaikoff, I. L., 46, 51, 67, *103, 104*

Author Index

Chait, A., 88, 91, *102, 103,* 109, 163, 170, *186, 191*
Chajek, T., 99, *102,* 136, 137, *186,* 224, *247*
Chakrabarti, R., 236, *252*
Chakravarti, R. N., 303, *311*
Chamoiseau, G., 269, *312*
Chan, S. J., 300, *310*
Chance, B., 319, *330*
Chander, A., 264, 271, 280, 282, 283, 284, 288, 298, *309*
Chandler, A. B., 240, *252*
Chandramouli, V., 260, 280, 281, *309, 314*
Chandrashekar, S., 287, *309*
Chang, M. C., 317, 318, *329*
Chang, M. L. W., 200, 203, *213*
Chang, S. C. S., 121, *194*
Chang, V. Y., 289, *315*
Chapdelaine, A., 320, 321, 322, 326, 327, *329, 330*
Chapman, M. J., 11, *41*
Charles, A. F., 243, *247*
Charman, R. C., 203, *213*
Charney, J., 306, *312*
Charpentier, G., 183, *193*
Charra, E., 172, *190*
Chataing, B., 136, 137, 145, *189*
Chau, I. Y., 324, 327, *329*
Chaykin, S., 203, *213*
Chedid, L., 304, *312*
Chen, C.-H., 109, 116, 117, 120, 121, 122, 123, 124, 126, 128, 130, 132, 133, 140, 142, 143, 153, 154, 155, 156, 157, 158, 166, *185, 186, 193*
Chen, L. B., 233, *247*
Chen, Y. C., 225, *247*
Cheng, F. H., 232, *246*
Cherkes, A., 223, *249*
Cherry, R. J., 319, *330*
Cheung, M. C., 11, *41*
Chevallier, F., 56, 68, 71, 73, 88, *102, 103,* 145, *186*
Chevet, D., 165, *187*
Chiariello, C. M., 238, *252*
Chiarugi, V., 243, *254*
Chinello, M., 156, 176, *186*
Chinen, I., *193*
Chiostri, R., 182, *187*
Chisolm, G. M., 241, *247*
Chmelař, M., 208, *214*

Chmelařová, M., 208, *214*
Cho, E., 227, *246*
Chobanian, A. V., 51, 53, 59, *102,* 135, *187*
Choi, Y. C., 232, *247*
Chong, B. H., 225, *247*
Chong, K. S., 122, *187*
Chopra, P. S., 230, *254*
Christophe, J., 10, *42*
Chuang, H. Y. K., 225, *252*
Chung, B. H., 11, *41*
Chung, J. A., 10, *42,* 110, 111, 121, 122, 123, 124, 125, 138, 141, *187*
Cines, D. B., 231, *247*
Ciorbaru, R., 276, *312*
Cisternas, J. R., 144, 149, *187, 189*
Clarke, W. R., 232, *246*
Clarkson, T. B., 55, 56, 58, 60, 61, 73, 78, 88, 90, *102, 103, 104*
Classen, M., 233, *246*
Clegg, E. D., 318, 325, *329*
Clifton-Bligh, P., 87, *102,* 120, 181, *187*
Clowes, A. W., 234, *249*
Cochrane, C. G., 231, *247, 251*
Cohn, M. L., 268, *310*
Cohn, Z. A., 228, *247,* 327, *330*
Cole, R., 295, *309*
Coleman, C. M., 268, *311, 313*
Collen, D., 228, *247*
Colli, S., 197, 200, *216*
Colman, D., 83, *102*
Colombo, M. A., 231, 232, 237, *250*
Comp, P. C., 228, *247*
Conlon-Morlec, M. J., 302, *310*
Conner, W. E., 48, 50, 55, 80, *101, 103*
Connor, W. E., *251*
Constantinides, P., 244, *247*
Cooper, R. A., 319, *329, 330*
Cordes, A. H., 285, *315*
Corew, T. E., 88, *104*
Corkey, R., 231, *247*
Cornwell, D. G., 204, *214*
Corr, P. B., 108, *193*
Cossins, A. R., 285, *309*
Costa, J. C. M., 133, 134, 143, *191*
Costas, R., Jr., 237, *254*
Cotran, R. S., 230, 235, *252, 253*
Couchat, N. C., 205, *216*
Craddock, P. R., 226, 231, *249*

Creashaw, E. B., 228, *254*
Credner, K., 210, *214*
Crim, M., 59, 65, 76, 80, 89, *102*
Cronan, J. E., Jr., 280, 284, 285, *309*
Crowe, J. L., 284, *309*
Crowle, A. J., 301, *309*
Cullen, C. F., 240, 241, *247*
Cultrera, G., 198, 207, *214*
Curry, M. D., 141, *187*
Curtis, J. J., 173, *187*
Curwen, K. D., 231, *247*
Cuttin, S., 198, 207, *214*
Czervionka, R. L., 233, *248*
Czervionki, R. L., 226, 233, *247*

D

Da Col, P., 34, *41*
Daerr, W. H., 143, *187*
Dagani, M., 203, *213*
Dahl, R. M., *190*
Dahlem, G., 25, *41*
Dahlström, S., 142, 162, *190*
D'Alessandro, A., 182, *187*
Dandekar, P. V., 318, *329*
D'Angelo, V., 206, 207, *214*
Danilova, L. M., 207, *217*
Danishevsky, I., 243, *248*
Danner, R. N., 142, *192*
Danon, D., 230, *254*
Darby, W. J., 203, *214*
Darin-Bennett, A., 320, 322, *329*
Das, B. C., 268, *310*
David, A., 142, *187*
David, J. S. K., 144, 148, 149, *187*
Davidson, D., 241, *247*
Davidson, L. M., 122, *187*, 203, 209, *215*
Davies, J. A., 226, *247*
Davies, J. I., *215*
Davies, J. W. L., 87, 88, *104*
Davignon, J., 47, *103*, 158, *188*
Davis, B. K., 318, 320, 321, 323, 324, 325, 326, *329*
Davis, J. W., 235, 236, *247, 251*
Dawes, F. A., 282, *309*
Dawes, J., 229, 234, 243, *247*
Dawson, R. M. S., 300, *309*
Day, A. J., 167, *186*
Day, C. E., 232, *247*
Day, E. E., 110, *187*

Deacetis, W., 287, *309*
De Buijn, J. H., 302, *309*
Deckelbaum, R. J., 48, 50, *102*, 324, *329*
Defaye, J., 271, *308*
de Gaetano, G., 206, 207, *214*
Deganello, S., 142, *192*
De Jana, E., 225, *247*
de La Farge, F., 134, 174, *190*
De Lalla, O. F., 3, 39, *41*
De Laughter, M., 227, *246*
Delaumeny, J. M., 273, *310*
Dell, R. B., 45, 46, 51, 58, 59, 61, 69, 74, 76, *102, 104*
Del Rosso, M., 243, *254*
Demaree, R., 241, *249*
Dembinska-Kiec, A., 206, 207, 210, *214, 215*
De Medio, G. E., 119, *185*
Demel, R. A., 115, *187*
De Monte, A. J. H., 287, *309*
Desager, J. P., 119, 171, 181, 182, *189*
de Schmertzing, H., 277, *307*
Descovich, G. C., 199, 200, *216*
De Siervo, A. J., 290, *309*
de Sobugram, R. C., 238, *252*
Dewar, H. A., 198, 207, *213*
De Young, D. W., 327, *330*
Dhariwal, K. R., 264, 271, 280, 282, 283, 284, 288, 298, *309*
D'Holander, F., 145, *186*
Dickens, B. F., 280, *310*
Dieker, P., 165, *193*
Dieplinger, H., 132, *187*
Dietmann, K., 212, *213*
Dietrich, P., 228, *252*
Dietschy, J. M., 67, 68, 71, 88, *101, 102, 103, 104*
Di Ferrante, N., 243, *247*
Di Luzio, N. R., 227, *253*
Dinglinger, F., 261, 290, *314*
Djubeková, E., 163, 164, 165, *187*
Doan, C. A., 301, *314*
Dobiášová, M., 108, 129, 134, 135, 138, 139, 140, 144, 149, 152, 153, 154, 155, 156, 162, 163, 164, 165, 167, 169, 171, 178, 179, 182, 183, *187, 193*
Doi, Y., 109, 115, 116, 122, 123, 124, 125, *187, 190*
Dolowitz, D. A., 233, *247*

Donath, N., 113, *186*
Donati, M. B., 206, 207, *214*
Donatti, M. O., 225, *253*
Donetz, A. T., 259, 280, *309, 312*
Donnelly, P. V., 243, *247*
Doran, J. E., 228, *247*
Dougherty, T. F., 233, *247*
Douglas, A. S., 222, 244, *246, 255*
Douset, J. C., 134, *188*
Douset, N., 134, 149, *188*
Douste-Blazy, L., 134, 149, *188*
Dowel, V. R., Jr., 273, *313*
Downar, E., 74, *103*
Downs, D., 11, *42*
Doyle, E. A., 56, 71, 73, *102*
Doyle, J. T., 176, *186*
Drapier, J. C., 304, *312*
Drawber, W. R., 205, *214*
Drevon, C. A., 109, 147, 152, 158, *191*
Drummond, M. M., 237, *251*
Dubos, R. J., 288, *309*
Dudley, A., 242, 244, *248*
Duncan, D., 46, *104*
Duncan, L. E., Jr., 167, *188*
Dunlap, K. R., 283, *309*
Duron, F., 183, *193*
Dusser, A., 129, *185*
Dussert, A., 183, *193*
Dustan, H. P., 167, *188*
Dutter, R., 37, *43*
Dyck, P. J., 114, 121, 179, *194*
Dyke, P. W., 87, 88, *104*

E

Eckhardt, T., 237, *252*
Edelstein, C., 110, *188*
Eder, H. A., 134, *188*, 204, 205, *213*, 238, *246*
Edgill, M., 226, 236, *251*
Edmondson, H. T., 228, *247*
Edner, O. J., 139, *193*
Edson, J. R., 236, *247*
Edwards, K. D. G., 67, *103*
Edwards, P. A., 204, *213*
Efimova, T. P., 273, *309*
Egebert, O., 236, *248*
Egelrud, T., 224, *248*
Egge, H., 142, *190*

Eggen, D. A., 53, 54, 55, 56, 57, 59, 65, 66, 73, 78, 90, *101, 102*
Eggleton, P., 241, *247*
Ehrlich, J., 306, *309*
Eiber, H. B., 243, *248*
Eidinoff, M. K., 82, 85, *103*
Eika, C., 224, *248*
Eika, L., 225, *248*
Einarsson, K., 92, *101*, 209, *214*
Eisele, R., 199, 208, *215*
Eisenberg, S., 74, 87, 92, *102*, 135, 145, *188*, 204, 205, *213, 214*, 239, *246*
Ek, I., 236, *248*
Ekelund, L. G., 203, 207, 209, *213*
Eklund, B., 209, 210, 211, *214, 215*
Ekström-Jodal, B., 197, *214*
Elbeim, A. D., 225, 245, *253*, 266, 267, *314*
Eldor, A., 225, 226, 232, *248*
Ellsworth, J. L., 134, 136, 137, 145, *189, 194*
Emond, D., 138, *190*
Endo, A., 208, *214*
Engelberg, H., 219, 221, 223, 224, 230, 233, 239, 240, 241, 242, 244, 245, *248*
Enger, S. C., 177, *192*
Engvall, E., 228, *253*
Erbland, J. F., 117, *188*
Erdi, A., 237, *248*
Erickssen, J., 238, *254*
Ericsson, E., 209, 210, 211, *213*
Ericsson, M., 209, *214*
Erikson, I., 210, *213*
Erison, C., 25, *41*
Eskin, S. G., 232, *249*
Esmon, C. T., 222, *253*
Essien, E. M., 226, 233, *248*
Estler, C. J., 211, *213*
Etemadi, A. H., 274, 275, *309*
Etherington, M. D., 236, *252*
Evans, R. W., 318, 322, 325, *329*
Evelyn, K. A., 165, *188*
Evenson, S. A., 231, 232, 238, *248, 249, 250*
Ewert, M., 200, *216*

F

Fabien, H. D., 158, *188*
Fabricant, C. G., 231, *248*

Fabricant, C. J., 231, *252*
Fabricant, J., 231, *248, 252*
Fahmy, W. F., 144, 148, 149, *187*
Fain, J. E., 197, 208, *216*
Fain, J. N., 207, 211, *214*
Fainaru, M., 138, 139, 142, *188, 193,* 324, *329*
Faini, D., 200, *217*
Fales, H. M., 268, *310*
Falsone, D. J., 232, *248*
Faltová, E., 108, 135, 138, 139, 140, 149, *187*
Fansal, N. O., 46, 51, *103*
Farias, R. N., 286, *310*
Farrell, J., 280, 284, 285, *309*
Farshtchi, D., 282, *309*
Faure, M., 259, 301, 302, *310, 312*
Favreux, L. V., 225, 226, 235, *247, 253*
Fawcett, B., 222, *246*
Fay, J. P., 286, *310*
Fearon, D. T., 227, *250, 254*
Fekete, L. L., 244, *248, 251*
Feldman, E. B., 56, 71, 73, *102*
Feldner, M. A., 65, 73, *102*
Felker, T. E., 142, *188*
Felts, J. M., 46, 51, *103*
Feres, A. C., 144, *187*
Ferrans, V. J., 205, *216*
Ferri, L., 173, *189*
Fibbi, G., 243, *254*
Fidge, N. H., 86, *104*
Fiedel, B. A., 227, *253*
Fielding, C. J., 28, *41,* 99, *102,* 109, 111, 112, 113, 118, 119, 120, 121, 122, 123, 125, 133, 135, 136, 137, 139, 141, 142, *185, 186, 188, 189,* 205, *214,* 223, *248*
Fielding, P. E., 28, *41,* 99, *102,* 109, 119, 120, 121, 135, 137, 139, 141, *188,* 205, *214,* 223, *248*
Fillius, J. P., 227, *253*
Finnerty, W. R., 259, 270, 284, 289, 292, 295, *310, 312*
First, N. L., 318, *330*
Fischkoff, S., 319, *330*
Fishbein, M. C., 238, *252*
Fisher, J., 225, *254*
Fisher, W. P., 306, *312*
Flechnes, I., 304, *308*

Fleming, A. D., 326, *329*
Fless, G. M., 110, 111, 121, 122, 123, 124, 125, 141, *187*
Fletcher, D. L., 306, *308*
Florey, H. W., 229, *248*
Flower, R. J., 210, *214*
Foggie, P., 241, *247*
Foldes, I., 228, *250*
Fonken, G. J., 287, *309*
Foolis, S., 236, *252*
Forbes, C. D., 224, 237, *251, 254*
Forester, G. P., 11, *43*
Forknes, C. E., 301, *314*
Forssman, J., 301, *310*
Forte, T. M., 10, 12, *41,* 109, 139, 142, *185, 188, 191*
Fosbrook, A. S., 165, *186*
Foster, D. W., 88, 91, *103*
Foster, J. W., 283, *310*
Fourcans, B., 85, *102*
Fox, I. H., 211, *214*
Franceschini, G., 7, *41*
Francey, I., 233, *253*
Frandoli, G., 206, 207, 210, *216, 217*
Frank, O., 196, *214*
Franklin, L. E., 318, *330*
Franzblau, C., 231, *250*
Fredrickson, D. S., 40, *41,* 111, 175, *185, 188,* 205, *216*
Freeman, L., 242, 244, *248*
Freeman, N. K., 223, *252*
French, J. E., 223, *253*
Frey, G. L., 226, 233, *247*
Frey, H., 197, 198, *216*
Friedman, H. M., 231, *247, 248*
Friedman, R. J., 232, *248*
Friend, D. S., 318, *329*
Frigas, E., 231, *248*
Fritz, R. W., 232, *250*
Fröberg, S. O., 209, *214*
Frohlich, J., 126, 150, 165, *188, 189, 194*
Fromme, H. G., 221, *249*
Froschauer, J., 199, 207, *217*
Frouin, A., 280, *310*
Fry, G. L., 233, *248*
Fugitsugu, M., 229, *252*
Fujimoto, M., 261, 302, *313*
Fujimoto, Y., 133, *191*
Fujita, S., 301, *315*

Jain, S. K., 260, 262, *308*
Janeway, C. A., 204, *214*
Janiak, M. L., 136, *189*
Jansen, W., 177, *189*
Janssen, E. T., 205, 206, *214*
Jaques, L. B., 219, 220, 229, 230, 233, 243, 244, *250, 251, 254*
Jaramillo, J., 91, *103*
Jelsema, C. L., 281, *311*
Jennings, P. B., 240, *252*
Jensen, L. C., 11, *42*
Jentsch, J. P., 327, *330*
Jeske, D. J., 68, 71, 88, *103*
John, G. R., 285, *312*
Johnson, A. D., 320, 322, *328*
Johnson, B. C., 200, 203, *213*
Johnson, J. D., 134, *189*
Johnson, L., 203, *213*
Johnson, M. J., 283, *311*
Johnson, U., 227, *251*
Jokay, I., 228, *250*
Jolles, P., 269, 277, *311, 313*
Jolley, R. L., 201, 203, *216*
Jonas, A., 115, *189, 190,* 323, *329*
Jones, A. L., 203, *216*
Jones, D. B., 232, *249*
Jones, H. B., 83, *102,* 223, 224, *249, 250*
Jones, M. E., 28, *41,* 136, *186*
Jones, R. L., 237, *250*
Jones, S., 111, 112, 113, 122, 123, 125, 133, *185*
Jordan, R., 236, *250*
Jorgensen, L., 226, *249*
Joyner, C. R., Jr., 241, *250, 251*
Julian, P., 74, *103*
Juliano, J., 142, 158, 162, *192*
Jung, K., 173, *189*
Jürgens, G., 21, *41*
Juves, M. W., 153, *192*

K

Kadish, J. L., 231, *250*
Kafalides, N. A., 231, *248*
Kagan, A., 176, *186*
Kaijser, B., 209, 210, *215*
Kaijser, L., 209, 210, 211, *214*
Kaiser, E. T., 109, 110, 117, 141, *194*
Kakkar, V. J., 224, 225, 237, *248, 252*
Kakki, M., 238, *252*
Kako, M., 165, *189*
Kalinsky, H., 197, 200, *216*
Kambayashi, J., 225, *254*
Kameyama, Y., 284, *313*
Kamio, Y., 115, 116, 122, 123, 124, *190*
Kane, J. P., 8, 9, *41,* 163, 176, *189*
Kaneda, T., 264, 273, 276, 282, *311*
Kanetsuna, F., 275, *311*
Kanfer, J., 289, 295, *311, 314*
Kankura, T., 263, 264, *313*
Kannel, W. B., 173, 175, 180, *189*
Kaplan, A., 296, *314*
Kaplan, J. E., 228, 245, *250*
Kaplan, K. L., 226, 234, *255*
Kaplan, R., 327, *330*
Karczag, F., 228, *250*
Karen, P., 129, 155, 162, *187*
Kariya, T., 165, *189*
Karnofsky, M. J., 234, 235, *247, 249, 250*
Kase, M., 225, *254*
Kashinathan, C., 286, 303, *311*
Kashiwabara, Y., 294, *311*
Kashub, E., 117, *190*
Kashyap, M. L., 163, 176, *189,* 238, *249, 254*
Kataoka, T., 258, 264, 273, 278, *311*
Kates, M., 258, 271, 277, 281, 296, *311*
Kather, H., 208, 210, *215*
Kato, H., 165, *189*
Kato, M., 267, 277, 304, *311, 313, 314*
Katsua, M., 134, *191*
Katz, A. J., 7, *43*
Kaur, S., 269, *314*
Kavanau, J. L., 283, *311*
Kawanami, J., 270, 271, 272, *311*
Kawasaki, N., 293, *311*
Kayden, H. L., 135, *186*
Kazatchkine, M. D., 227, *250*
Keen, H., 179, *190*
Kefalides, N. A., 231, *247*
Kekki, M., 52, *103*
Kelley, M. T., 304, *312*
Kelley, R. E., 148, 152, *192*
Kelley, V. E., 221, 230, *250*
Kellner, A., 244, *250*
Kellogg, T. F., 68, *103*
Kennedy, D. W., 228, *255*
Kennedy, E. P., 289, *311, 315*

Kennedy, R., 205, 206, *214*
Kent, P. W., 301, *315*
Kessel, W. S. M. G., 115, *187*
Keyden, H. J., 10, *42*
Kézdy, F. J., 109, 110, 117, 141, *188, 194*
Khera, A., 198, 299, *316*
Khuller, G. K., 260, 261, 262, 263, 264, 265, 266, 268, 269, 270, 271, 272, 273, 274, 276, 277, 278, 279, 280, 281, 282, 284, 285, 286, 290, 291, 294, 295, 296, 297, 298, 299, 300, 302, 303, 304, 305, *311, 312, 314, 315, 316*
Kieslich, K., 306, *314*
Kilemenics, K., 228, *250*
Kim, D. M., 240, *251*
Kim, K. S., 269, 272, 288, 303, *308*
Kimura, A., 270, 271, 272, *311*
Kimura, S., 151, 162, *194*
King, D. H., 283, *312*
King, M. E., 286, *312*
King, W. C., 133, 142, *188, 191*
Kingdon, H. S., 222, 233, *249*
Kinlough, R. L., 240, *253*
Kinlough-Rathbone, R. L., 226, *249*
Kinnunen, P. J., 238, *254*
Kinsey, W. H., 318, *329*
Kinsky, S. C., 302, *316*
Kirchman, D. E. H., 48, 50, 55, 57, 58, 59, 61, 65, 73, *104*
Kirk, J. E., 206, *215*
Kirkpatrick, B., 231, 232, 237, *250*
Kirstein, A., 210, *215*
Kissebah, A. H., 204, *215*
Kitabatake, K., 115, 116, 122, 123, 124, *190*
Kitagawa, M., 203, 209, *215*
Klauda, H. C., 83, 85, 92, 95, 96, *103*
Klein, E., 244, *248, 251*
Klein, K., 222, *251*
Klein, M. S., 108, *193*
Klein, P. D., *190*
Klevay, L., 148, *190*
Klingenstein, R., 301, *316*
Klocking, H. P., 228, *252*
Klopfenstein, W. E., 139, *190*
Knicker, W. T., 231, *251*
Knisely, M. H., 240, *255*
Knopp, R. H., 160, *194*
Knortman, J. D., 238, *252*

Knudsen, M. P., 306, *309*
Kobayashi, H., 134, *191*
Koch-Weser, D., 273, *315*
Koehler, J. K., 318, *329, 330*
Koevoet, A. L., 269, *315*
Kojima, S., 133, 134, *191*
Kolář, J., 147, *186*
Komata, K., 228, *255*
Konova, I. V., 259, *308*
Kopecká, J., 129, 149, 171, 183, *187*
Korn, E. D., 223, 224, *251*
Kostka-Trabka, E., 206, *214*
Kostner, G. M., 2, 3, 4, 5, 8, 10, 11, 12, 13, 14, 15, 16, 17, 18, 19, 21, 22, 23, 24, 25, 26, 28, 29, 30, 33, 34, 37, *41, 42, 43,* 132, 141, 142, 162, 163, *187, 190*
Kotite, L., 9, *41*
Kotlev, V. V., 280, *309*
Kováč, J., 178, *193*
Kovacey, V. P., 211, *214*
Kovalchuk, L. P., 259, *312*
Kovanen, P. T., 77, 87, 88, 91, 93, 97, 101, *102, 103*
Kraemer, P. N., 221, *251*
Kraicer, P. F., 327, *330*
Kramer, W., 237, *252*
Kramsch, D. M., 231, *250*
Kratky, O., 18, *42*
Krauss, R. M., 11, *42*
Krempler, F., 21, 22, 24, *42*
Kresse, H., 221, *249*
Krishna, G., 208, *215*
Kristl, J., 144, *187*
Kritchevsky, D., 83, *102,* 203, 209, *215*
Krivit, W., 236, *247*
Krueski, A. W., 225, 245, *253*
Kruglyak, E. D., 271, *308*
Kruse, W., 200, 204, *215*
Krzywy, T., 301, *312*
Kuczyńska, K., 158, *190*
Kudchodkar, B. J., 46, 47, 51, 52, 57, 64, 74, 80, 82, 83, 85, 87, 91, 92, 99, 100, *103, 104,* 108, 109, 129, 131, 152, 154, 155, 158, *190, 193,* 203, 208, 209, *215, 217*
Kuhn, E., 163, 164, 165, *187, 190*
Kuhn, H., 301, *316*
Kuhn, R., 245, *248*
Kuksis, A., 113, 134, 140, *193*

Kuller, L., 176, *189*
Kumar, B., 304, 305, *311*
Kummer, M., 200, *215*
Kunkel, H. G., 134, *190*
Kuntzel, H., 273, *310*
Kuo, P. T., 241, *251*
Kupferberg, J. P., 109, 110, 117, 141, *194*
Kuroda, M., 208, *214*
Kuron, G. W., 181, *193*
Kushwaha, R. S., 83, 85, *103*
Kusunose, E., 274, *316*
Kusunose, M., 259, 260, 264, 267, 273, 274, *316*
Kuusi, T., 238, *254*
Kuzuya, N., 156, 157, 158, *185*
Kuzuya, T., 156, 157, 158, *185*
Kvist, M., 142, 162, *190*
Kwaan, H. C., 237, *251*
Kwapinsky, J. B. G., 258, 301, *312*
Kymla, J., 135, 138, 139, 140, *187*

L

Lacko, A. G., 87, *102*, 121, 122, 127, 129, 133, 141, 142, 144, 150, 158, 162, 171, 177, 179, *187, 189, 190, 192, 193*
Laffin, R., 227, *246*
Laga, H., 227, *252*
Lageder, H., 182, *190*
Lagegren, H., 226, 228, *251*
Lagente, M., 134, 174, *190*
Lages, B. A., 226, 234, *255*
Laggner, P., 3, 11, 13, 14, 15, 17, 18, *42*
Lagrange, D., 11, *41*
Lahnborg, G., 228, *251*
Lahwborg, E., 228, *246*
Lakart, C. J., 232, *249*
Lakin, K. M., 206, 207, *215*
Lally, J. I., 136, 145, *186*
Lalumiere, G., 321, 327, *329*
Lamensans, A., 304, *312*
Lands, W. E. M., 296, *310*
Landsteiner, K., 301, *312*
Lane, D. A., 224, 225, 245, *251, 252*
Lane, D. M., 160, *190*
Laneelle, G., 269, 270, 271, *312*
Laneelle, M. A., 271, *312*
Langer, K. H., 142, 165, *193*
Langer, T., 74, *103*

Langer, W., 138, *190*
Langlais, J., 321, 327, *329*
Langley, P. E., 207, 208, *213*
Langridge, H. R., III, 272, *316*
Lani, D. A., 237, *248*
Laplaud, P. M., 11, *41*
Lardy, H. A., 318, 322, *329, 330*
Larking, P. V., 148, *190*
Larsen, V., 240, *247*
Larson, R., 226, *251*
Larsson, L., 272, *312*
Larsson, R., 222, 226, *251*
Lasser, N., 176, *189*
Lau, B. W. C., 148, *190*
Laudat, M. H., 142, 144, 150, *192*
Laurell, S. M., 227, 243, 245, *251*
Lavelli, S. M., 236, *251*
Law, J. H., 273, *307*
Lawrie, T. D. V., 86, 91, *104*, 205, *216*
Lazzarini-Robertson, A., Jr., 229, 232, *251*
LeBlanc, P., 225, *250*
Lechevalier, H. A., 262, 278, 279, 301, *312*
Lechevalier, M. P., 262, 278, 279, 301. *312*
Lecrec, C., 304, *312*
Lederer, E., 259, 260, 261, 265, 266, 267, 269, 271, 273, 274, 275, 276, 279, 304, *308, 309, 310, 311, 312, 313, 314, 315, 316*
Lee, A. G., 260, 285, *312*
Lee, D. M., 10, 11, *42*
Lee, K. T., 240, *251*
Lee, Y. C., 201, *215*, 260, 290, *312*
Lees, R. S., 40, *41*, 48, 50, *102*
Legault, Y., 320, 321, 322, 326, 327, *329, 330*
Lehane, D. P., 262, 265, 272, *309*
Lehner, N. D. M., 53, 54, 65, 73, 78, 88, 90, *102, 103, 104*
Lei, I. M., 71, *103*
Lei, K. Y., 71, *103*
Leijd, B., 92, *101*, 209, *214*
Leiss, O., 142, *190*
Leitchner, C., 222, *251*
Lekim, D., 113, *186*
Lengsfeld, H., 209, *215*
Lennarz, W. J., 262, 272, 273, 281, 282, 289, 296, *307, 312, 314*
Lenzi, S., 199, 200, *216*
LePogam, 165, *187*
Lettre, E., 275, *312*

Leventhal, M., 230, *252*
Lever, W. F., 244, *248, 251*
Levhuck, T. P., 229, *252*
Levi, R. I., 205, *216*
Levij, I. S., 304, *308*
Levin, E., 229, *251*
Levy, G., 301, *312*
Levy, R. I., 40, *41*, 74, 87, 92, *102, 103*, 110, 142, 175, *187, 188, 192,* 204, 205, *213*
Levy, R. J., 204, *214*
Levy, R. S., 232, *247*
Lewis, B., 86, *104,* 109, 120, 163, 170, *191,* 238, *253*
Lewis, C. W., Jr., 288, *314*
Lewis, L. A., 166, *190*
Lewis, V. J., 273, *313*
Lichtenstein, A. H., 144, 150, *190*
Lieberman, S., 46, 51, 58, 64, 74, *104*
Lilljeqvist, A. C., 121, *191*
Lim, C. T., 10, *42*
Lin, C. I., 235, *251*
Lin, D. S., 48, 50, 52, 55, 74, *102, 103*
Lin, J. T., 109, 124, 125, *185*
Lindahl, U., 222, 223, 226, 243, *250, 251*
Linde, H., 200, *215*
Linder, A., 172, *190*
Lindgren, F. T., 10, 11, *41, 42,* 83, *102,* 142, *185,* 204, *214*
Lindon, J. N., 225, 226, *251, 253*
Linhart, J., 134, 135, 138, *187*
Lint, T. F., 227, *246*
Lisch, H. J., *191*
Lithell, H., 23, *43*
Littleton, J. H., 285, *312*
Littrento, M. M., 231, *252*
Liu, S. F., 167, *194*
Ljungman, C., 226, *247*
Lloyd, J. K., 165, *186*
Loegering, D. A., 231, *248*
Lofland, H. B., 53, 54, 56, 58, 59, 60, 61, 65, 73, 78, 88, 90, *102, 103, 104*
Logue, G. L., 227, *251*
Loire, R., 205, *216*
Lollar, P., 221, 233, *251*
Long, E. R., 271, *312*
Long, R. A., 111, 115, *189*
Lopez, S. A., 162, *190*
Lorenzen, I., 242, *250*
Lorimer, A. R., 237, *251*

Lorton, S. P., 318, *330*
Loskutoff, D. J., 229, *251*
Lough, J., 231, *251*
Lowe, D. G. O., 237, *251*
Lubochinsky, B., 285, *314*
Lucas, T., 230, *254*
Lucy, J. A., 108, 138, *190*
Luke, R. G., 173, *186, 187*
Lundblad, R. J., 222, 223, 225, *249, 255*
Lundholm, L., 197, 206, 209, 210, 211, *213, 215, 217*
Lundkvist, L., 25, *41*
Lutton, C., 68, *103*
Lyford, C. I., *251*
Lykoyanowa, I. I., 206, 207, *215*
Lynch, A., 167, *188*
Lyon, T. P., 83, *102,* 223, *249*
Lyss, A. P., 231, *247*

M

Macarak, E. J., 231, *248*
McCarthy, C., 268, 272, 306, *312*
McClung, N. M., 282, *309*
McConathy, W., 160, *190*
McConnathy, W. J., 12, *43*
McDaniel, L. E., 306, *313*
McDonald, L., 226, 236, *251*
McElhaney, R. N., 280, 284, 286, *312, 314*
MacFarlane, R. G., 222, *246*
McGee, D. L., 173, 180, *189*
McGill, H. C., Jr., 205, 215
McGovern, V. J., 226, 230, *251*
MacGregor, I. R., 231, 245, *248, 251*
Macheboenf, M., 301, *312*
Machovich, R., 222, 228, 233, 244, *251*
MacIntyre, D. E., 225, *251*
MacIomhair, M., 236, *251*
McKay, E. J., 227, 243, 245, *251*
MacKenzie, M. R., 224, 235, *249*
McKinney, J. A., 260, 261, 279, 303, *314*
Mackinnon, A. M., 149, *194*
McLain, S., 227, *252*
McLaughlin, C. A., 304, *312*
McLean, M. R., 224, *251*
McLennan, A. P., 269, *315*
McLeod, B. C., 227, *246*
McNamara, D. J., 67, 71, *103*
McTaggart, F., 244, *255*
Mäder, Ch., 206, *216*

Author Index

Maeda, J., 267, 277, *311*
Maggi, F., 197, 200, *216*
Magide, A. A., 204, *215*
Magill, P. J., 238, *253*
Magnani, H. N., *190*
Magnusson, O., 206, *215*
Mahadoo, J., 233, *251*
Mahley, R. W., 7, 8, 29, *41, 42, 43,* 79, 91, *103*
Maitra, P. K., 260, *313*
Major, F., 211, *213*
Majuro, G., 230, *252*
Makula, R. A. 284, 289, 292, 295, *310, 312*
Malakkova, E. A., 229, *252*
Malik, U., 268, 269, 277, 284, 285, 286, 304, 305, *311, 312, 314, 315*
Mälkönen, M., 144, 149, *190*
Malmberg, R., 197, *214*
Malmemdier, C. L., 10, *42*
Man, E. B., 223, *254*
Mann, T., 322, *329*
Mannien, V., 144, 149, *190*
Mansberger, A. R., 228, *247*
Marcel, Y. L., 122, 131, 136, 138, 142, 154, 158, *188, 190, 193*
Marchioro, T. L., 172, 173, *186*
Mardh, P. A., 272, *312*
Marechal, J., 259, *310*
Marek, J., 108, 138, 139, *187*
Marin, H. M., 221, *252*
Marinetti, G. V., 117, *188*
Markowitz, O., 284, *314*
Markwardt, F., 228, *252*
Marmo, E., 199, 200, *215*
Marniemi, J., 142, 162, *190*
Maroko, P. R., 238, *252*
Maroto, M. L., 200, 203, 208, *216*
Marqueta, S., 203, *213*
Marr, G. A., 284, *312*
Marsh, J. B., 117, 118, *190,* 204, *215*
Marth, E., 21, 22, 25, *41, 42*
Martin, B. M., 235, *252*
Martin, G. J., 240, *252*
Martin, J., *192*
Martin, S., 306, *313*
Martyakov, A. T., 271, *308*
Masana, L., 238, *252*
Masden, S., 238, *252*
Mason, D. T., 46, 52, 74, *104,* 109, 152, *193*

Mason, R. G., 225, *252*
Massaro, E. R., 53, 55, 56, 60, 73, 80, *103*
Mastroianni, L., Jr. 327, *329*
Masui, M., 274, *316*
Mathe, D., 56, 68, 71, 73, 88, *103*
Mathur, A. K., 289, *312*
Matoušek, V., 144, 171, 182, *187*
Matsuki, G., 294, *311*
Matsuo, T., 241, *248*
Mattener, E., 301, *316*
Mattener, F., 301, *316*
Matthews, L. B., 203, *213*
Mattock, M., 179, *190*
Matz, C. E., 115, *189, 190*
Maurawski, U., 142, *190*
Maurice, M. T., 279, *312*
Mayama, A., 261, 302, *313*
Mayer, D. T., 322, *329*
Mayne, E. E., 236, *252*
Mazzone, T., 88, 91, *103*
Meade, T. W., 236, *252*
Medina, J. M., 211, *213*
Meek, G. A., 229, *248*
Meetinger, K. L., 236, *252*
Megan, M. B., 232, *246*
Melchoir, D. L., 286, *313*
Melin, B., 128, 129, *185*
Meliv, B., 85, *102*
Melnik, B., 143, *192*
Meltzer, M. S., 304, *316*
Mengs, V. C., 226, *247*
Menzel, H. J., 142, *193*
Merrill, E., 225, *251*
Mertz, D. P., 199, 208, *215*
Metcalfe, D. D., 227, *250*
Metzger, A. L., 46, *103*
Meyer, L. J., 211, *213*
Meyer, T. J., 304, *313*
Michalski, R., 224, 225, *252*
Michel, G., 267, 269, 271, 272, 273, 274, 278, *308, 310, 312, 313, 316*
Michio, T., 268, *313*
Middlebrook, G., 268, *310, 311, 313*
Middelhoff, G., 139, *192*
Miettinen, T. A., 48, 50, 52, 71, *103, 104,* 203, 209, *215*
Migliore, D., 277, *313*
Miles, B. C., 56, 71, 73, *102*
Miller, C. J., 175, *190*

Miller, J. P., 118, 144, 147, 149, 163, 170, 171, 180, 181, *190*
Miller, M., 238, *249*
Miller, N. E., 87, *102*, 109, 120, 175, 181, *187, 190, 191*, 205, *215*, 238, *253*
Miller, O. N., 209, *215*
Miller, W. T., 263, 272, *309*
Millman, I., 301, *316*
Milne, R. W., 138, *190*
Milner, K. C., 304, *312*
Milstein Kuschnaroff, T., 144, *187*
Mims, C. A., 229, *252*
Minick, C. R., 227, 231, 232, *248, 252*
Minnikin, D. E., 259, 261, 264, 274, 275, 278, 279, *307, 313*
Minno, G. D., 225, *253*
Miquel, A. M., 275, *309*
Mitchell, C. D., 108, 109, 142, 165, *188, 191*
Mitra, R. S., 260, *313*
Mo, A., 121, *191*
Moberg, B., 171, 178, *194*
Modan, M., 236, *249*
Moghissi, K. S., 327, *330*
Mohammed, S. F., 225, *252*
Mohan, P. R., 306, *313*
Mok, H. Y. I., 46, 49, *103*, 203, 204, 205, 209, *214*
Mollby, R., 292, 293, 295, 300, *313*
Molnar, J., 227, *246, 249, 252*
Moncada, S., 210, *214, 217*
Monger, E. A., 82, 83, 87, 91, 92, *104*, 152, 154, *191*
Montes, A., 136, *191*
Montesano, R., 328, *330*
Montrozier, H. L., 267, 273, 275, 277, *308*
Moolten, S. E., 240, *252*
Moore, J. H., 144, *191*, 281, *311*
Moore, S., 226, 231, 232, 233, *248, 249, 251*
Moore, V. L., 304, *313*
Mora, P. T., 219, *252*
Moracchiello, M., 173, *189*
Mordarska, H., 279, 301, *307, 313*
Morecki, S., 304, *308*
Moreno, F. J., 197, 208, 211, *213, 216*
Moretti-Rojas, I., 56, 71, 73, *102*
Morgan, H. G., 86, 91, *104*, 205, *216*
Morganti, G., 8, *41*
Morin, R. J., 128, 130, 173, *191, 192*

Morrice, J. J., *251*
Morris, M. D., 46, 51, 60, *103*
Morrisett, J. D., 12, *43*, 109, 110, *189, 190, 192*
Morrison, A. D., 242, *252*
Mortensson, J., 227, *251*
Mortimer, A. M., 306, *308*
Moses, J. W., 225, *254*
Mosher, L. R., 211, 212, *216*
Moss, C. W., 273, *313*
Motojima, S., 133, *191*
Motomiya, M., 261, 302, *313*
Mott, G. E., 60, *103*
Moutafis, C. D., 91, *103*, 203, *216*
Mrochek, I. E., 201, 203, *216*
Mugnaini, C., 236, *252*
Muhlebach, T., 319, *330*
Muir, J. R., 223, *252*
Mulholland, S. G., 232, *250*
Müller, M., 18, *41, 42*
Muller-Berghauf, G., 237, *252*
Munder, P. G., 138, *190*
Murase, T., *194*
Murata, K., 133, *191*
Murawski, U., 177, *189*
Murmann, W., 197, *217*
Murthy, P. S., 288, 289, *312, 314*
Mustard, J. F., 225, 226, 233, 234, 240, *247, 248, 249, 253*
Myant, N. B., 91, 92, *103, 104*, 120, *192, 193*, 203, 204, *215, 216*
Myrvik, Q. S., 304, *313*
Mysliewiec, M., 206, 207, *214*

N

Nacash, C., 269, *313*
Nachev, P., 200, *216*
Nader, H. B., 228, *252*
Nagasaki, T., 131, *191*
Nagase, S., 327, *330*
Nahoo, K., 211, *215*
Naito, C., 165, *189*
Naito, H. K., 166, *190*
Nakagawa, H., 294, *311*
Nakagawa, M., 133, 134, 139, *191*, 225, *254*
Nakamura, H., 241, *254*
Nakaya, Y., 29, *42*
Nalini, P., 286, 303, 305, *311*

Author Index

Nanai, I., 232, *249*
Nandedkar, A. K. N., 259, 289, 296, *313, 315*
Nantel, G., 296, *313*
Nath, N., 286, *311*
Navarro, L. T., 232, *249*
Naworal, J., 263, 272, 278, *309*
Neef, A., 211, *213*
Negre, L., 301, *308*
Neiduszynski, I. A., 221, 243, *246*
Nelson, C. A., 11, *43*
Neri-Serneri, G. G., 236, *252*
Nervi, F. O., 68, *103, 104*
Nestel, P. J., 47, 51, 82, 83, 86, 87, 91, 92, 99, 100, *102, 104*, 109, 120, 129, 152, 154, 181, 182, *187, 189, 191*, 209, 210, *217*
Nesterenko, O. A., 264, 273, *313*
Neumann, R., 173, *189*
Newland, H., 240, *252*
Nichols, A. V., 10, 12, *41*, 109, 111, 113, 137, 139, 142, *185, 188, 191*, 223, *252*
Nicola, P., 206, 207, *216*
Nicoll, A., 86, *104*, 109, 120, *191*
Nicolosi, R. J., 144, 150, *190*
Niedmann, D., 132, *186*
Niewiarowski, S., 235, *253*
Nikasato, D., 204, *214*
Nikkari, T., 80, *104*
Nikkilä, E. A., 162, *191*, 204, 205, *216*, 223, 238, *252*
Nilsson, A., 80, *104*
Nilsson, I. M., 242, *252*
Nilsson, M., 118, 119, *191*
Ninikowski, J., 240, *250, 252*
Nisankowski, R., 210, *217*
Nishida, T., 109, 111, 112, 114, 115, 116, 122, 123, 124, 125, 133, 134, 139, *187, 188, 190, 191, 192*
Nishijima, M., 293, 294, *313*
Nobel, R. P., 45, 46, 51, 58, 59, 61, 69, 74, 76, 88, *102, 104*
Noble, R. C., 144, *191*
Noble, R. E., 273, *315*
Noll, H., 271, *308*
Nojima, S., 258, 259, 263, 264, 273, 276, 277, 278, 284, 289, 293, 294, 298, *307, 310, 311, 313, 314*
Nojima, Y., 284, *316*

Norbeck, H. E., 172, *191*
Norbing, B., 234, *254*
Nordby, A., 163, *188*
Nordby, G., 118, 119, 147, 149, *191*
Nordoy, A., 226, 232, 240, *252*
Normann, E. R., 121, *191*
Norstrand, E., 153, *192*
North, W. R. S., 236, *252*
Norum, K. R., 16, *43*, 84, 85, 92, *102, 104*, 108, 109, 118, 119, 120, 121, 127, 128, 131, 133, 142, 147, 149, 150, 152, 158, 165, 166, *186, 188, 189, 191*
Nossel, H. L., 226, 234, *255*
Nowak, J., 209, 210, 211, *214*
Nowotny, A. H., 126, *194*
Nubiola, A. R., 238, *252*
Nugel, E., 173, *189*
Nye, E. R., 154, 158, *193*, 197, *213*
Nyhrman, R., 227, *253*

O

Obenberger, J., 108, 138, 139, *187*
O'Brien, B. C., 67, *104*
O'Brien, J. R., 225, 235, 236, *252*
O'Brink, B., 226, *247*
Ochlich, P., 209, *216*
Oda, T., 165, *189*
Odenthal, J., 134, *194*
Odham, G., 272, *312*
Oelschläger, H., 200, *215, 216*
Ofosu, F., 222, *252*
Ohara, T., 303, *313*
Ohta, G., 229, *252*
Oka, H., 165, *189*
Oka, S., 261, 302, *313*
Okami, Y., 277, *313*
Okamoto, R., 241, *248*
Okawa, Y., 295, 296, *313*
Okazaki, H., 286, 306, *308, 313*
O'Kelly, J. C., 144, *191*
Okuyama, H., 263, 264, 284, *313*
Oldberg, A., 234, *253*
O'Leary, W. M., 271, *313*
Oliphant, G., 318, *330*
Olivecrona, T., 223, *246*
Olson, A. G., 198, 210, 212, *215, 216*
Olsson, P., 222, 226, *251*
Oncley, J. L., 204, *214*

O'Neil, H. J., 264, 272, *313*
Ono, H., 286, 306, *308, 313*
Ono, Y., 259, 276, 277, 289, 293, 294, 298, *307, 313, 314*
Oosta, G. M., 235, *253*
O'Rand, M. G., 318, *330*
Orci, L., 242, *252*, 318, 328, *329, 330*
Oro, J., 273, *315*
Orö, L., 196, 197, 198, 203, 204, 207, 212, *213, 216*
Ortega, M. P., 200, 207, *216*
Osborne, J. C., 2, 26, *42*
Ošťádalová, I., 147, *186*
Ostman, J., 196, 204, *213*
Östrem, T., 144, *186*
Ostwald, R., 59, 65, 76, 80, 89, *102*
Osuga, T., 117, 119, *191*
Otsuka, H., 270, 271, 272, *311*
Ott, W. H., 181, *193*
Otway, S., 205, *216*
Owen, J. S., 133, 134, 143, *191*
Owen, W. G., 221, 222, 233, *246, 251, 253*
Owens, G. K., 231, *253*
Ozdamar, E., 243, 244, *253*

P

Packard, C. J., 86, 91, *104*, 205, *216*
Packham, M. A., 234, *253*
Paddock, J., 231, 232, 237, *250*
Padovan, D., 237, *253*
Pagella, P. G., 200, *217*
Pais, M., 197, *213*
Pal, R., 327, *330*
Palade, G. E., 221, 230, *252, 254*
Palos, A., 222, 228, 233, *251*
Palumbo, P. J., 179, *194*
Pan, Y. T., 225, 245, *253*
Panchemko, L. P., 264, 273, *313*
Pangborn, M. C., 260, 261, 279, 303, *314*
Pangburn, S., 88, *104*
Panosyan, A. G., 259, *308*
Pao, Q., 130, *192*
Paoletti, R., 197, 200, 206, 207, *216*
Papadopoulos, N. M., 3, *42*
Pařízek, J., 147, *186*
Parker, J. M., *330*
Parker, R., 275, *315*

Parkes, R., 304, *312*
Parks, E., 228, *247*
Parks, J. S., 53, 54, 78, *104*
Parodi, O., 236, *252*
Parsons, C. L., 232, *253*
Parsons, W. B., Jr., 203, 205, 211, 212, *216*
Parwaresch, M. R., 206, *216*
Pasquali, F., 243, *254*
Pasternack, B. S., 236, *255*
Patch, J. R., 205, *216*
Patel, P. V., 259, 261, 264, 274, 278, *313*
Patsch, J. R., 3, 11, 12, *42, 43*
Patsch, W., 3, 11, *43*, 131, 154, *191*
Patten, R. L., 224, *253*
Patterson, P. H., 289, *314*
Pattnaik, N. M., 136, *191*
Paul, D., 235, *253*
Paulus, H. E., 8, *41*
Paznokas, J. L., 296, *314*
Peck, S. D., 237, *253*
Peeters, H., 139, *192*
Pelikan, D., 230, *253*
Penner, J. A., 224, 225, *254*
Penumarti, N., 276, 281, 291, 299, 303, *311, 314*
Pepin, D., 183, *193*
Pepper, D. S., 229, 234, 243, *247*
Perisutti, E., 238, *254*
Perrelet, A., 242, *252*, 318, 328, *329, 330*
Perrin, A., 205, *216*
Perry, J. J., 283, *309, 312, 316*
Pertsemlidis, D., 48, 50, 55, 57, 58, 59, 61, 65, 73, *104*
Peters, J. P., 223, *254*
Peterson, C. M., 237, *250*
Peterson, J. I., 277, *307*
Peterson, R. E., 232, *246*
Petit, J. F., 276, 304, *312*
Petrack, B., 197, 200, *216*
Petrassi, G. M., 237, *253*
Pfeiffer, K. P., 37, *43*
Pflug, J. J., 120, *192*
Phillips, G. B., 139, *191*
Phillips, P. E., 235, *251*
Phillips, P. H., 322, *329*
Philp, R. B., 233, *253*
Piannotti, J. F., 306, *313*
Pierce, F. T., 224, *250*

Author Index

Pieringer, R. A., 289, *307*
Pigretti, M., 261, *314*
Pilger, E., 37, *43*
Pilgeram, L. O., 244, *253*
Pilipenko, T. V., 266, 270, 271, *308*
Piña, M., 200, 203, 208, *216*
Pinney, S., 179, *190*
Pinon, J. C., 142, 144, 150, *192*
Pinte, F., 275, *309*
Piot, M. C., 85, *102*, 128, 129, *185*
Piran, U., 112, 114, 115, 116, 122, 123, 124, 128, 130, 133, 173, *190, 191, 192*
Pitney, W. R., 225, *247*
Pittman, R. C., 88, *104*
Plante, L., 321, 327, *329*
Polgar, N., 275, *313*
Pollett, R. W., 238, *249*
Polonovski, J., 85, *102*, 129, *185*
Polz, E., 10, 29, 33, *43*
Pomerance, A., 230, *254*
Pomerantz, M. W., *253*
Poole, J. C. F., 229, *248*
Poon, R. W. M., *192*
Porcellati, G., 119, *185*
Portman, O. W., 108, 114, 117, 119, 138, 139, 145, *191, 192, 193*
Poulos, A., 322, *329, 330*
Powell, J. R., 232, *247*
Power, D. A., 288, *314*
Pownall, H. J., 122, 130, *192, 193*
Prabhudesai, A. V., 268, 269, *314*
Pratt, R. M., 228, *255*
Precht, K., 173, *189*
Preiss, J., 195, *216*
Prentice, C. R. M., 224, 237, *251, 254*
Prevet, A. R., 301, *314*
Prexl, H. H., 17, *42*
Price, J. M., 203, *213*
Pridachina, N. N., 271, *308*
Priddle, W. W., 167, *194*
Priego, J. G., 200, 203, 207, 208, *216*
Pristautz, H., 37, *43*
Privett, O. S., 143, 148, 149, *193*
Proia, A., 67, 71, *103*
Prokasova, N. V., 108, *194*
Prome, J. C., 267, 273, 275, 277, *308*
Prottey, C., 261, 290, *314*
Proudfit, W. L., 167, *192*

Proulx, P., 295, 296, *309, 313*
Pudles, J., 274, *314*
Puglisi, L., 197, 200, 206, 207, *216*
Pulver, R., 243, *253*

Q

Quarfordt, S. H., 118, *192*
Quigley, T. W., Jr., 224, *251*
Quinci, G. B., 25, 30, *41, 42*, 156, 176, *186*
Quinn, P. J., 320, 322, *330*
Quintao, E., 46, *104*
Qureshi, N., 275, *314*

R

Rachmilewitz, D., 135, 145, *188*
Radergran, K., 226, *251*
Radhalkar, P. W., 306, *315*
Radway, P., 238, *252*
Raetz, C. R. H., 289, 298, 300, *314*
Raetzer, H., 200, 204, *215*
Raffel, S., 270, *314*
Ragland, J. B., 142, *192*
Rainwater, J., 240, *254*
Raivio, K. O., 211, *213*
Ramakrishnan, R., 61, *102*
Ramakrishnan, T., 288, *314*
Ramalho, V., 133, 134, 143, *191*
Ramée, M. P., 165, *187*
Rand, R. W., 226, *249*
Randall, H. M., 269, *315*
Rao, S. N., 238, *253*
Rapp, H. J., 304, 305, *316*
Rapport, M. M., 301, *314*
Rasmussen, J., 233, *246*
Ratledge, C., 283, *315*
Rautureau, J., 68, *103*
Raymond, T. L., 56, 59, 61, 73, 90, *104*
Razin, S., 284, *314*
Razumovsky, P. N., 259, *312*
Reardon, M., 86, *104*
Reaven, G. M., 156, *192*
Redgrave, T. G., 11, *43,* 239, *253*
Redish, W., 211, *217*
Rees, E. D., 173, *186, 187*
Reese, A. C., 228, *247*
Reeve, C. E., 165, *188*
Regan, T. J., 241, *253*

Rehunen, S., 162, *191*
Reichl, D. 120, *192*, 204, *215*
Reichle, R. M., 144, 148, 149, *187*
Reidy, M. A., 239, *253*
Reiser, R., 67, *104*
Reman, F., 138, *187*
Renaud, G., 79, *104*
Renaud, S., 236, 240, *249, 253*
Rent, R., 227, *253*
Rhode, M., 130, *192*
Ribbons, D. W., 282, *309*
Ribi, E., 275, 304, *310, 312, 313, 315, 316*
Richards, K. C., 203, *216*
Richardson, M., 226, *249*
Richle, F. A., 144, 148, 149, *187*
Ridgon, R. H., 229, *253*
Rigomier, D., 285, *314*
Rimondi, S., 199, 200, *216*
Ringel, M., 306, *313*
Ritland, S., 131, 158, 177, *186, 192*
Robbins, J., 205, 206, *214*
Robert, A., 183, *193*
Roberts, C. K., 11, *43*
Roberts, G. P., 109, 124, 125, *185, 330*
Roberts, H. R., 231, *253*
Roberts, K. D., 320, 321, 322, 326, 327, *329, 330*
Roberts, W. C., 205, *216*
Robertson, A. L., Jr., 242, *253*
Robinson, A. K., 108, *193*
Robinson, D. S., 205, *216,* 223, *253*
Robinson, G. A., 207, 208, *213*
Robinson, K., 238, *254*
Roccano, E., 227, *246*
Rodionov, S. V., 207, *217*
Roelcke, D., 14, *43*
Rogers, B. J., 320, 327, *330*
Rogers, M. P., 205, *216*
Rogmunt, J., 238, *246*
Rojas, A., 267, *316*
Roka, L., 236, 244, *246*
Roller, P., 268, *310*
Romanovskaya, V. N., 206, 207, *215*
Ronsaglione, M. C., 225, *253*
Roscher, A., 24, *42*
Rose, A. H., 280, 284, 285, *309*
Rose, H. G., 109, 133, 142, 158, 162, *192,* 275, *314*
Rose, J., *192*

Rosenberg, R. D., 221, 222, 225, 226, 227, 234, 235, 236, *246, 247, 249, 250, 251, 253*
Rosendahl, M. S., 274, 278, *315*
Rosenfeld, R. S., 82, 85, 88, *103*
Rosenhamer, G., 207, *216*
Ross, R., 220, 234, 235, *249, 253*
Rosseneu, M., 139, *192*
Rössner, S., 198, 212, *216*
Rosynov, B. V., 271, *308*
Rothblat, G. H., 327, *329*
Rothe, O., 209, *216*
Rothley, D., 200, *216*
Rottem, S., 284, *314*
Rowen, R., *192*
Rozanis, J., 259, 267, *310, 316*
Rubenstein, B., 134, *192*
Rubies-Prat, J., 238, *252*
Rubin, L., 223, *252*
Rubinstein, D., 134, *192*
Rucker, S., 235, *253*
Ruddy, S., 226, *253*
Rudermann, N. B., 203, *216*
Ruggiergo, H. A., 237, 245, *253*
Ruoslahti, E., 228, *253*
Russ, E. M., 204, 205, *213,* 238, *246*
Russel, B. S., 56, 71, 73, *102*
Russell, R. L., 119, 147, 149, *192, 193*
Russo, R., 235, *247*
Rutenberg, H. L., 121, 133, 144, 150, 158, 162, 171, 177, 179, *190, 192, 193*
Ruuskanen, O., 80, *104*
Ryan, R. J., 121, *194*

S

Saarni, H., 80, *104*
Saba, H. I., 225, 231, *253*
Saba, S. R., 225, *253*
Saba, T. M., 227, 228, 245, *246, 250, 253*
Sabesin, S. M., 142, *192*
Sabin, E. R., 301, *314*
Saharia, G. S., 289, *312*
Saier, E. L., 153, *192*
Sailer, S., 3, 11, 12, *42, 43,* 131, 154, *191*
Saito, K., 293, *311*
Saito, M., 295, 304, *314*
Sakma, M., 211, *215*
Salen, G., 65, *103,* 182, *189*

Author Index

Salmi, H. A., 197, 198, *216*
Salton, M. R. J., 290, *309*
Salvatori, T., 273, 274, *308*
Salzman, E. W., 225, 226, *251, 253*
Samour, D., 269, *311*
Samuel, P., 46, 51, 58, 64, 74, *104*
Samuels, P. B., 226, 229, *253*
San-Blas, G., 275, *311*
Sandermann, H., Jr., 286, *314*
Sandhofer, F., 11, 21, 22, 24, *42, 43*
Santa Rosa, C. A., 144, *187*
Sarda, P., 269, *310*
Sarma, G. R., 260, *314*
Sarma, J. S. M., 88, *102*
Sasaki, A., 261, 262, 301, 303, *314, 315*
Sato, H., 261, 275, 302, *313, 315*
Sauar, J., 238, *254*
Savagnac, A. M., 277, *308*
Sawyer, P. N., 230, *254*
Scanu, A. M., 10, *42,* 110, 111, 121, 122, 123, 124, 125, 138, 141, 142, *187, 188, 192*
Scardsudd, K., 25, *41*
Schaefer, E. J., 29, *42,* 142, *192*
Schaefer, W. B., 268, 288, *313, 314*
Schaeffer, Z., 324, *329*
Schaffner, C. P., 306, *313*
Schatton, W., 199, 200, 208, *215, 216*
Scheider, M., 301, *314*
Scheig, R., 166, *193*
Scheuerbrandt, G., 262, 272, 273, 281, 282, *312*
Schildt, B. E., 228, *246, 251*
Schlierf, G., 204, *216*
Schmidt, J. D., 232, *253*
Schmitz, G., 113, 120, 143, *186, 192*
Schnoes, H. K., 275, *314*
Schnare, F. H., 56, 71, 73, *102*
Schodt, K. P., 220, *246*
Schoenborn, W., 165, *193*
Scholz, D., 173, *189*
Schreibman, P. H., 182, *189*
Schroeder, S., 228, *254*
Schubert, K., 275, *314*
Schultz, J. S., 19, *41*
Schulz, J. C., 266, 267, *314*
Schumaker, V. N., 92, *104,* 152, *192*
Schwandt, P., 30, *43*
Schwartz, B. S., 274, 278, 306, *313, 315*

Schwartz, C. G., 205, *216*
Schwarz, M. A., 318, *330*
Sclarof, H., 226, *249*
Scott, D. A., 243, *247*
Scott, P. J., 74, 75, 87, 88, *104*
Scott, R. F., 240, *251*
Scott, T. W., 322, *330*
Scow, R. O., 135, *186*
Scribner, B. H., 172, *190*
Segrest, J. P., 11, *41,* 110, *192*
Sehgal, S. C., 302, *314*
Seidel, D., 17, *43,* 132, *186*
Sekikawa, E., 303, *313*
Selbekk, B., 121, *191*
Semeraro, N., 228, *247*
Senn, M., 267, *316*
Seplowitz, A. H., 61, *102*
Seppänen, A., 142, 162, *190*
Serenthà, P., 198, 207, *214*
Setchell, B. P., 322, *329, 330*
Sgoutas, D. S., 113, *192*
Shafir, E. 208, *216*
Shalgi, R., 327, *330*
Shanberge, J. N., 225, *254*
Shapiro, S. S., 327, *330*
Sharma, B. V. S., 281, *311*
Sharrant, E. P., 232, *249*
Shaw, M. K., 261, 265, 266, 285, 290, *314*
Shaw, N., 258, 259, 275, 277, 279, *313*
Shaw, W. H. C., 306, *308*
Shckter, M. A., 273, *313*
Shea, G. M., 230, *252*
Sheetz, M., 300, *310*
Shen, B. W., 110, *188*
Shepherd, J., 86, 91, *104,* 205, *216*
Shepherd, R. E., 197, 208, *216*
Sheppard, H., 67, *104*
Sherrard, D. J., 172, *190*
Sherwin, R., 176, *189*
Shestakov, V. A., 207, *217*
Shimamune, K., 327, *330*
Shimizu, T., 165, *189*
Shimmyo, Y., 303, *313*
Shinitsky, M., 319, *330*
Shipley, G. G., 136, *189*
Shirey, E. K., 167, *192*
Shoji, M., 268, *313*
Shore, B., 8, *43,* 223, *252*

Shore, V. G., 8, *43,* 109, 121, 135, 139, 141, *188,* 205, *214,* 223, *248*
Short, S. A., 259, *314*
Shub, M. M., 271, *308*
Shultz, G., 207, 208, 210, *213*
Siegel, H., 181, *193*
Siegel, R. R., 173, *186*
Siewert, G., 306, *314*
Sifferlen, J., 274, *309*
Sigler, G. F., 109, *192, 193*
Sigroth, K., 197, 199, *215*
Sigurdsson, G., 86, *104*
Sikes, J. P., 322, 327, *329, 330*
Silberbauer, K., 222, *251*
Silbert, D. F., 286, *314*
Silbert, J. E., 221, 227, *250, 254*
Silva, T. B., 154, 158, 159, 170, *188*
Silvius, J. R., 280, 286, *314*
Simionescu, M., 221, *254*
Simionescu, N., 221, *254*
Simko, V., 148, 152, *192*
Simms, S., 301, *312*
Simon, B., 208, 210, *215*
Simon, J. B., 109, 117, 166, *192, 193*
Simons, K., 139, *189*
Simons, L. A., 120, *192*
Simonton, J., 223, *249*
Simpkins, H., 300, *310*
Sinensky, M., 280, 284, *314*
Singal, D. P., 232, *248*
Singer, S. J., 300, *310*
Singh, J. P., 318, *330*
Singhvi, D. R., 302, 303, *315*
Sinzinger, H., 222, *251*
Siperstein, M. D., 67, *102*
Sirtori, C. R., 7, *41*
Sjoholm, A. G., 227, *251*
Skawinski, S., 210, *217*
Skinner, S. L., 167, *186*
Skizume, K., 211, *215*
Skrabal, P., 18, *41*
Skrede, S., 131, 158, 177, *186, 192,* 238, *254*
Skutelsky, E., 230, *254*
Slabý, P., 160, *193*
Small, D. M., 48, 50, *102,* 108, 109, 136, *189, 193*
Smith, D. W., 269, *315*
Smith, F. R., 61, 76, *102, 104*
Smith, G. F., 222, 233, 244, *254*
Smith, H., 270, *315*
Smith, J. B., 225, 233, *248, 252*
Smith, L. C., 109, 122, 137, *191, 192, 193,* 232, *249*
Smith, M. H., 225, 226, *253*
Smith, N. B., 113, 134, 140, *193*
Smith, R., 304, *312*
Sniderman, A., 136, *193*
Sobel, B. E., 108, *193*
Soboth, H., 196, *214*
Soden, A., 240, *252*
Sodhi, H. S., 46, 47, 51, 52, 53, 57, 64, 74, 80, 82, 83, 87, 91, 92, *103, 104,* 108, 109, 129, 131, 152, 154, 155, 158, *190, 193,* 209, *217*
Sodki, S., 203, 208, 209, *215*
Soetewey, F., 139, *192*
Solera, M., L., 134, *193*
Soler-Argilaga, C., 119, 147, 149, *192, 193*
Soloff, J. S. K., 142, *187*
Soloff, L. A., 121, 125, 126, 133, 142, 144, 150, 158, 162, 171, 176, 177, 179, *189, 190, 192, 193, 194*
Solomon, H. M., 196, 211, *214*
Solotrovsky, M., 288, *315*
Sones, F. M., 167, *192*
Sorenson, H., 237, 238, *249*
Sortie, P. D., 237, *254*
Souroujon, M., 319, *330*
Soula, G., 134, *188*
Soutar, A. K., 91, *104,* 109, 111, 122, *192, 193*
Spady, D. K., 68, 71, *104*
Sparling, P. F., 302, *315*
Sparrow, J. T., 109, 111, *189, 192, 193*
Spector, A. A., 80, *101,* 286, *312*
Spengel, F. A., 91, *104*
Spitzer, J. J., 224, *254*
Spöttl, F., 199, 207, *217*
Spreafico, P. L., 210, *217*
Srinivasan, S., 230, *254*
Srivastava, L. S., 238, *254*
Stabinger, H., 3, 11, *42*
Stacey, M., 301, *315*
Stachelberger, H., 222, *251*
Stahl, Y. D. B., 121, 122, *185*
Stambaugh, R., 327, *330*
Stanacev, N. Z., 289, *315*

Stanczewski, B., 230, *254*
Stanke, D. F., 327, *330*
Starr, P., 182, *193*
St. Clair, R. W., 53, 54, 55, 73, 78, 88, 90, *103, 104*
Steck, P. A., 274, 275, 278, *315, 316*
Steele, J. M., 211, *217*
Steele, P., 240, *254*
Steim, J. M., 139, *193*, 286, *313*
Stein, O., 99, *102*, 138, 139, 175, *193*, 224, 239, *246, 247*
Stein, Y., 99, *101*, 138, 139, 175, *193*, 224, 239, *246, 247*
Steinbach, J. H., 241, *254*
Steinberg, D., 88, *104*, 134, 136, *188, 193*, 209, 210, *217*
Steiner, P. M., 238, *249*
Steinman, M., 245, *248*
Stemberger, A., 229, *254*
Stemerman, M. B., 234, *249*
Stephen, J. D., 195, 203, 208, *213*
Stephenson, R. C., 236, *249*
Stern, N., 266, *315*
Stevens, G. R., 11, *42*
Steward, J. M., Third, 205, *216*
Stinson, M. W., 288, *315*
Stirling, Y., 236, *252*
Stjernholm, R. L., 273, *315*
Stokke, K. T., 16, *43*, 120, 131, 132, 133, 144, 145, 176, *193*
Stone, A. L., 220, *254*
Stone, N. J., 48, 50, 74, 77, 87, 92, *102*
Stoner, G. E., 241, *247*
Storey, B. T., 322, *329*
Stork, H., 212, *214*
Stoss, B., 301, *315*
Stoudemire, J. B., 79, *104*
Stover, G., 230, *254*
Stožický, F., 160, *193*
Strain, S. M., 275, *315*
Strand, O., 209, 210, *217*
Strauss, A. H., 228, *252*
Stříbrná, J., 144, 167, 169, 178, *187, 193*
Stringer, K., 179, *190*
Strober, W., 74, *103*
Stuffer, C., 232, *253*
Subbaiah, P. V., 109, 116, 117, 140, 145, *193*
Subissi, A., 197, *217*

Subrahmanyam, D., 259, 260, 261, 262, 263, 264, 268, 270, 271, 279, 281, 282, 284, 285, 286, 294, 298, 299, 300, 302, 303, 304, *308, 311, 314, 315, 316*
Subramoniam, A., 282, *315*
Sue, T. K., 244, *254*
Suenram, C. A., 12, *43*, 141, *187*
Suermann, I., 199, 208, *215*
Sugano, M., 114, 144, 145, *192, 193, 194*
Sugiyama, K., 304, *314*
Sukapure, R. S., 306, *315*
Sullivan, A. C., 209, *215*
Sullivan, K. H., 285, *315*
Sundboom, J. L., 222, 244, *254*
Sunkel, C., 200, 203, 207, 208, *216*
Sutherland, E. W., 207, *213*
Sutherland, W. H. F., 148, 154, 158, *190, 193*
Suzaki, H., 229, *252*
Suzue, G., 122, 138, *190, 193*
Suzuki, T., 265, 303, *311, 315*
Svedmyr, N., 196, 197, 211, *213, 217*
Swack, N. S., 232, *247*
Swank, R. L., 240, 241, *247, 254*
Swedenborg, J., 226, *251*
Swell, L. C., 205, *216*
Swenson, B., 226, 232, *252*
Symonds, H. W., *330*
Szczeklik, A., 210, *217*
Szekely, J., 232, *249*
Sznajderman, M., 158, *190*
Szondy, E., 232, *249*

T

Taal, A. R., 135, *187*
Tacquet, A., 288, *308*
Taek-Goldman, K., 225, *254*
Takada, N., 134, *191*
Takahashi, H. K., 228, *252*
Takahashi, Y., 301, 303, *314, 315*
Takajama, M., *194*
Takamura, M., 133, *191*
Takatori, T., 143, 148, 149, *193*
Takayama, K., 275, *314*
Takeyama, K., 267, 290, 291, *315*
Takinen, M. R., 238, *254*
Talbot, P., 318, *330*

Tall, A. R., 11, *43*, 108, 109, *193*
Talwar, P., 260, 261, 262, 265, 280, 281, *315*
Tamaka, A., 304, *314*
Tanaka, H., 265, 303, *315*
Tanaka, K., 231, *250*
Tanaka, T., 304, *316*
Tanaka, Y., 114, 144, *194*
Taneja, R., 284, 285, 286, 298, 299, *311, 315*
Tanishima, K., 229, *252*
Tanzawa, K., 208, *214*
Taroni, G. C., 197, *213*
Tarpila, S., 48, 50, *104*
Taskinen, M.-R., 162, *191*, 238, *252*
Tateson, J. E., 210, *217*
Taunton, O. D., 109, *189*
Taylor, F. B., Jr., 228, *247*
Teien, A. N., 222, *254*
Telesforo, P., 228, *247*
Telford, A. M., 245, *254*
Temple, W. A., 154, 158, *193*
Teng, B., 136, *193*
Tennent, D. M., 181, *193*
Tepper, B. S., 271, 280, 282, *308, 315*
Tepper, S. A., 203, 209, *215*
Teramoto, T., 165, *189*
Terrana, M., 238, *249*
Tesi, M., 210, *217*
Thanabalasingham, S., *193*
Thide, H., 17, *42*
Thirumalchar, M. P., 306, *315*
Thomas, D. P., 237, *248*
Thomas, D. W., 265, 272, *309*
Thomas, J., 67, *104*
Thomas, M., 183, *193*
Thomas, R., 165, *187*
Thomas, W. A., 240, *251*
Thomis, J., 23, *43*
Thompson, G. R., 91, *104*, 171, 180, *190, 193*
Thompson, S. G., 236, *252*
Thomson, C., 224, *254*
Thonnard-Neumann, E., 233, *254*
Thormälen, D., 209, *216*
Thorpe, R. F., 283, *315*
Thunell, S., 236, *248*
Tiets, A., 266, *315*
Tiffany, M. L., 224, 225, *254*
Timmis, G., 241, *253*

Tocanne, G., 275, *308*
Tocanne, J. F., 275, *308*
Todd, A. S., 228, *254*
Todhunter, T., 238, *249*
Tomkins, G. M., 67, *104*
Toriyama, S., 274, *316*
Tornabene, T. G., 273, *315*
Toubiana, R., 275, 304, *312, 315, 316*
Toulernonde, F., 228, *254*
Toustar, O., 203, *214*
Traber, M., 59, 65, 76, 80, 89, *102*
Trana, A. K., 261, 264, 280, 281, 282, 284, 285, 286, 298, 299, 300, 302, 303, *311, 314, 315*
Trayner, I., *193*
Triana, L. R., 318, *330*
Tripathi, R. C., 165, *186*
Tschesche, R., 275, *312*
Tseyita, Y., 208, *214*
Tsyganov, V. A., 273, *309*
Tsuskima, T., 211, *215*
Tu, A. T., 244, *253*
Tuboly, S., 264, *316*
Tumbleson, M. E., 327, *330*
Turazza, G., 210, *217*
Turba, C., 200, *217*
Turley, S. D., 68, 71, 88, *101, 104*
Turner, W. G., 201, 203, *216*
Turpaev, T. M., 108, *194*
Tytell, A. A., 306, *312*

U

Uchiyama, M., 134, *191*
Uehara, S., 295, *316*
Uemura, K., 302, *316*
Ulbright, T. M., 286, *314*
Ulutin, O. M., 236, *246*
Unanne, E. R., 235, *252*
Unna, K., 211, *217*
Urban, J. E., 284, *309*
Ursini, F., 173, *189*
Urzua, M., 327, *329*
Utermann, G., 9, *43*, 142, 165, *193*

V

Vacheron, M. J., 279, *312*
Vagelos, P. R., 284, *309*
Vaheri, A., 232, *254*

Author Index

Vairel, E. G., 228, *254*
Valdiguie, P., 134, 174, *190*
Válek, J., 144, 171, 182, *187*
Valente, M., 173, *189*
Valles de Bourges, V., 203, *216*
van Deenen, L. L. M., 115, *187*
Van den Bosch, H., 289, 296, 300, *316*
Vanden Heuvel, W. J. A., 321, 326, *329*
Vanderkooi, J., 319, *330*
Van de Water, L., 228, *254*
Vane, J. R., 210, *214*, *217*
Van Eck, W. F., 223, *254*
van Eyes, J., 203, *214*
van Itallie, T. B., 181, *189*
Vannucchi, S., 243, *254*
van Tol, V., 139, *189*
Van Winkle, W. B., 130, *192*
Varco, R. L., 241, *254*
Varma, K. G., 122, 125, 126, 133, 142, 176, *189, 190, 193, 194*, 235, *253*
Varughese, P., 46, *104*
Vassalli, P., 328, *330*
Vaughan, M., 145, *186*, 209, 210, *217*
Vaurik, M., 167, *194*
Vaver, V. A., 108, *194*
Veerkamp, J. H., 284, *316*
Vegelos, P. R., 296, *316*
Vejayagopal, P., 240, *254*
Velican, C., 221, 234, *254*
Velican, D., 234, *254*
Venkilasubramanian, T. A., 260, 264, 271, 280, 281, 282, 283, 284, 287, 288, 289, 298, 302, *309, 312, 314, 315*
Verdery, R. B., 127, 133, 138, 139, *190, 191, 194*
Verma, J. N., 264, 268, 269, 272, 273, 274, 276, 277, 278, 281, 282, 294, 295, 296, 297, 298, 299, 300, *311, 316*
Verstraete, M., 228, *247*
Vesely, J., 227, *247*
Vessby, B., 23, *43*
Vesselinovich, D., 240, 242, *249*
Vestal, J. R., 283, 284, *308, 310, 316*
Vettor, R., 237, *253*
Vezina, C., 122, 131, 136, 138, 142, 154, *190, 193*
Vial, R., 162, *190*
Viikari, J., 80, *104*
Vikrot, O., 129, 131, 142, 156, 157, 158, 163, *194*
Vilkas, E., 259, 261, 267, 269, 276, 279, 304, *307, 308, 313, 314, 316*
Villa, S., 206, 207, *214*
Ville, C., 269, *316*
Vinazzer, H., 229, *254*
Vincent, J. E., 196, 207, 208, *217*
Viswanathan, R., 259, *315*
Vítek, V., 149, 171, 183, *187*
Vitello, L., 142, *192*
Vlahcewics, Z. R., 205, *216*
Vlček, J., 178, *193*
Voglmayr, J. K., 322, *330*
Voiland, A., 269, *316*
Voleníková, L., 160, *193*
Vondra, K., 129, 144, 152, 153, 154, 155, 156, 162, 163, 164, 165, 167, 169, 171, 178, 179, 182, *187*
Vuoristo, M., 48, 50, *104*

W

Wachtel, H., 275, *314*
Wada, M., *193*
Wadsworth, A., 301, *316*
Wagner, R. R., 327, *330*
Wagner, W. D., 65, 73, *102*
Wahl, P. W., 154, *185*, 206, *217*
Wahlstrom, B., 52, *103*
Wakil, S. J., 284, *316*
Walker, R. W., 272, *310*
Walker, R. H., 263, 264, 272, *316*
Wallentin, L., 92, *101*, 129, 131, 142, 154, 155, 156, 157, 158, 163, 170, 171, 176, 178, 181, 182, *194*
Wallis, J., 225, *254*
Walsh, P. N., 225, *254*
Walton, K. W., 87, 88, *104*
Wan, L. S., 236, *255*
Wang, C. I., 82, 85, *103*
Warner, E. D., *251*
Warren, B. A., 233, *253*
Warth, M. R., 160, *194*
Wasteson, A., 234, *247, 254*
Wastson, E., 226, *253*
Watanabe, S., 229, *252*
Watanabe, T., 284, *316*
Wautier, J. L., 225, *254*
Wayland, J. H., 231, *249*
Weaver, D. E., 318, 325, *329*
Weaver, J. A., 236, *252*

Weber, P., 227, *246*
Webster, D. R., 226, 229, *253*
Webster, G. R., 145, *194*
Weeks, G., 284, *316*
Weiber, J. M., 227, *254*
Weicker, H., 14, *43*
Wein, A. J., 232, *250*
Weiner, M., 210, 211, *217*
Weinstein, D., 88, *104*
Weir, M. P., 272, *316*
Weis, H. J., 68, *104*
Weisgraber, K. H., 7, 8, *41, 43,* 79, 91, *103*
Weiss, B., *215*
Weiss, H. J., 226, 234, *255*
Weithmann, K. U., 210, *217*
Weitzel, G., 206, *217*
Weksler, B. B., 225, 226, 232, *248, 254*
Welby, M., 277, *308*
Weld, C. B., 222, *255*
Weltzien, H. U., 138, *190*
Wenckert, A., 242, *252*
Wennmalm, A., 209, 210, 211, *214, 215*
Werb, Z., 327, *330*
Wessler, S., 236, *249, 255*
West, C. E., 11, *43*
Westermark, B., 234, *254*
Westphal, O., 138, *190*
White, D. C., 259, 296, *307, 314*
White, G. C., 225, 233, *255*
White, H. S., 221, *252*
White, I. G., 320, 322, *329, 330*
White, J. G., 236, *247*
White, S., 223, *249*
Whyte, H. M., 51, *104*
Wiebe, D., 226, 232, *252*
Wieland, H., 132, *186*
Wiencke, I., 163, *188*
Wietzerbin, J., 276, 304, *312*
Wiklund, O., 177, *194*
Wilcox, H. G., 11, *43*
Wilkinson, S. G., 266, *316*
Wilkinson, T., 11, *41*
Wille, L. E., 139, *191, 194*
Williams, A. V., 240, *255*
Williams, D. R., 149, *194*
Williams, M. C., 118, 120, 142, *189*
Williams, P. G., 241, *250*
Williams, S., 235, *255*
Wills, R. D., 11, *42*

Wilson, C., 245, *254*
Wilson, D. B., 134, *194*
Wilson, H., 229, *253*
Wilson, J. D., 51, 59, 62, 63, 71, 73, 74, 76, *105*
Winder, A. F., 165, *186*
Winegrad, A. I., 242, *252*
Wing, D. R., 205, *216*
Winter, J. H., 244, *255*
Wissler, R. W., 240, 242, *249*
Witebsky, E., 301, *316*
Witkowski, F. K., 108, *193*
Witte, L. D., 226, 234, *255*
Wolf, D. P., 318, 320, 324, 326, *329, 330*
Wolf, F. J., 181, *193*
Wolf, P., 235, *255*
Wolfe, J., 231, *248*
Wolfram, G., 199, 200, *217*
Wong, M. Y. H., 275, *316*
Wood, L. L., 55, *104*
Woodford, S. Y., 173, *187*
Workman, E. F., Jr., 225, 233, *255*
Wright, C. J., 233, *251*
Wright, J. L., 121, 128, *188*
Wright, L. D., 203, *214*
Wurm, H., 10, 29, *43*
Wu, K. K., 225, *247*
Wynn, V., 204, *215*

Y

Yakovlev, V. A., 229, *252*
Yamada, K., 228, *255,* 284, 286, 306, *308, 313*
Yamada, O., 226, *249*
Yamaguchi, T., 295, 296, *313*
Yamamoto, K., 304, *310*
Yamamoto, M., 114, 144, *194*
Yamamura, Y., 304, *314*
Yamazaki, I., 138, *185*
Yanagimachi, R., 318, 323, 326, *329, 330*
Yankley, A., 223, *249*
Yano, I., 259, 260, 264, 267, 273, 274, *316*
Yao, J. K., 114, 121, 179, *194*
Yard, A. S., 327, *330*
Yarkoni, E., 304, 305, *308, 316*
Yashiro, M., 151, 162, *194*
Yokoyama, S., 109, 110, 117, 141, *185, 194*
Youmans, A. D., 288, *316*
Youmans, A. S., 301, *316*

Youmans, G. P., 288, 301, *316*
Young, D. S., 201, 203, *216*
Young, G. B., 219, *252*
Yribarren, M., 259, *316*
Yue, K. T. N., 235, *251*
Yurt, R. W., 227, *254*
Yu-Yan-Yeh, 207, *217*

Z

Zanetti, M. E., 181, *193*
Zannis, V. I., 7, 9, *43*
Zbar, B., 304, *313*
Zbruzkova, V., 87, *103*
Zech, L., 203, 204, 205, 209, *214*
Zemplinyi, T., 242, *255*
Zierenberg, O., 134, *194*
Zijlstra, F. J., 196, 207, 208, 210, *217*
Zilversmit, D. B., 46, 53, 55, 56, 59, 60, 73, 80, 83, 85, 87, 92, 95, 96, 99, *103, 104, 105,* 136, *191, 194,* 239, *255*
Živný, K., 129, *187*
Zmuda, A., 206, 207, 210, *214, 215*
Zollinger, M., 321, 327, *329*
Zöllner, N., 199, 200, 211, *217*
Zucconi, A., 182, *187*
Zugibe, F. T., 221, *255*
Zukel, W. J., 176, *186*
Zvezdina, N. D., 108, *194*

Subject Index

A

Abetalipoproteinemia, lipoproteins in, 16–17
Actinomycetes
  immunological properties of lipids, 300–301
    glycolipids, 303–305
    phospholipids, 301–303
  influence of environmental factors on lipid composition
    age, 280–282
    growth temperature, 284–285
    variation of carbon and nitrogen source, 282–284
    various growth-medium additives, 285–288
  lipid composition
    fatty acids, 272–274
    glycerides, 271–272
    glycolipids, 265–270
    lipoamino acids, 270–271
    mycolic acids, 274–275
    phospholipids, 258–264
    sterols, 275
  lipids and production of antibiotics, 305–306
  metabolism of lipids
    biosynthesis, 288–292
    catabolism and lipolytic enzymes, 292–298
    lipid turnover, physiological significance of, 298–300
  subcellular distribution of lipids, 275–276
    cell wall, 276–277
    cytosolic globular lipids, 277
    membrane, 276
  taxonomic significance of lipids
    fatty acids, 277–279
    mannophosphoinositides, 279–280
Antibiotics, production, actinomycete lipids and, 305–306
Apolipoproteins
  of plasma and lymph
    isoforms and genetic variants, 6–10

Apolipoproteins, of plasma and lymph (*contd.*)
    major apolipoproteins, 5–6
    proteins associated with lipids and lipoproteins, 10
    quantification in clinical chemical laboratory, 29–30
      Laurell electrophoresis, 30–31
      nephelometric quantitation, 33–34
      radial immunodiffusion, 31–33
      radioimmunoassay, enzyme-linked immunoabsorbent assay and fluoroimmunoassay, 34–35
      standards and control samples, 35
      values of normal plasma, 35
    as risk indicators for myocardial infarction and atherosclerosis, 35–39
    role in lipid metabolism, 26–29
Atherosclerosis
  actions of heparin in systems involved in
    coagulation, 222
    complement, 226–227
    endothelium, 221–222
    fibrinolysis, 228–229
    macrophages, 227–228
    platelets, 224–226
    triglyceride transport, 222–224
  apolipoproteins as risk indicators for, 35–39
  etiology of, 220–221

B

Bloodstream, heparin effects in, 235–245

C

Cell wall, actinomycete lipids and, 276–277
Cholesterol, *see also* Sterol
  body turnover of, 47–50
  individual pool sizes, 69–80
  measurement of true body turnover rate by tracer kinetic methods
    compartmental analysis, 51–57
    input–output analysis, 58

Subject Index

Cholesterol, measurement of true body turnover rate by tracer kinetic methods (*contd.*)
  isotopic steady-state method, 57
  plasma, exchange with tissue, 80–85
  synthesis, absorption, esterification and catabolism
    quantitative relationship with metabolism of plasma lipoproteins, especially LDL, 92–99
  total body mass, 58–69
Cholesteryl ester, plasma turnover, LDL and apoB turnover, 86–91
Coagulation, heparin and, 222
Cold shock, sperm sterol levels and, 322–323
Complement, heparin and, 226–227
Cytosol, actinomycete lipids and, 277

E

Endothelial surface, heparin effects at, 230–235
Endothelium, heparin and, 221–222

F

Fatty acids of actinomycetes, 272–274
  taxonomic significance of, 277–279
Fibrinolysis, heparin and, 228–229

G

Glycerides of actinomycetes, 271–272
Glycolipids of actinomycetes, 265–270
  immunological properties of, 303–305

H

Heparin
  actions in systems involved in atherosclerosis
    coagulation, 222
    complement, 226–227
    endothelium, 221–222
    fibrinolysis, 228–229
    macrophages, 227–228
    platelets, 224–226
    triglyceride transport, 222–224

Heparin (*contd.*)
  effects at different sites, 229
    bloodstream, 235–245
    endothelial surface, 230–235
  nature of activity, 219–220
  results of therapy, 245

L

Lecithin:cholesterol acyltransferase (LCAT)
  clinical studies, 151–153
  in disease, 165–180
    effect of hypolipemic drugs, 180–183
  in health, 154–165
  isolation and properties of, 121–126
  methods and, 127–132
  reaction pattern of
    activation, 109–112
    hydrolytic activity of, 112–115
    reverse reaction, 116–117
    transfer of acylgroup to acyl receptor, 115–116
  regulation of activity, 132–143
  sources of, 117–121
  studies in experimental animals, 143–145
    rat as model, 145–150
      some comments on animal models, 151
      some other animals, 150–151
    substrate and conditions of reaction, 126–127
Lipid
  metabolism, role of apolipoproteins, 26–29
  modification of levels during sperm capacitation, 323–325
  sperm, modulation during maturation, 322
Lipoamino acids of actinomycetes, 270–271
Lipoproteins
  classification and nomenclature, 2–3
    density classes, 3–4
    electrophoretic fractions, 3
    families, 4–5
  not commonly found in normal plasma, 15–16
    in abetalipoproteinemia, 16–17

Lipoproteins, not commonly found in normal plasma (*contd.*)
  lipoprotein-a, 19–26
  in obstructive jaundice, 17–19
  subfractions
    other preparative methods, 12–15
    ultracentrifugation, 10–12
Low-density lipoproteins (LDL)
  apoB turnover, relationship with plasma cholesteryl ester turnover, 86–91
  metabolism, quantitative relationship to cholesterol metabolism, 92–99

M

Macrophages, heparin and, 227–228
Mannophosphoinositide, actinomycete, taxonomic significance of, 279–280
Membrane, actinomycete lipids and, 276
Mycolic acids of actinomycetes, 274–275
Myocardial infarction, apolipoproteins as risk indicators for, 35–39

N

Nicotinic acid
  mechanism of action, 207–210
  pharmacology
    pharmacodynamics, 203–207
    pharmacokinetics, 196–203
    toxicology and side effects, 210–212

O

Obstructive jaundice, lipoproteins in, 17–19

P

Phospholipids of actinomycetes, 258–264
  immunological properties of, 301–303
Platelets, heparin and, 224–226
Proteins associated with lipids and lipoproteins, 10

S

Sperm
  factors regulating sterol content
    acceptors, 326–327
    sulfohydrolase activity, 327–328
  fertility, effect of sterols on, 325–326
Sperm capacitation
  methodology per study of, 319–320
  modification of lipid levels during
    in vitro studies, 323–325
    in vivo studies, 325
Sterols, *see also* Cholesterol
  of actinomycetes, 275
  effects on sperm fertility, 325–326
  factors regulating sperm content of
    acceptors, 326–327
    sulfohydrolase activity, 327–328
  of mammalian sperm
    cold shock and, 322–323
    content and composition, 320–321
    modulation of lipids during maturation, 322
Sterol sulfohydrolase activity, sperm sterol content and, 327–328

T

Triglyceride transport, heparin and, 222–224

CONTENTS OF PREVIOUS VOLUMES

Volume 1
The Structural Investigation of Natural Fats
  M. H. Coleman
Physical Structure and Behavior of Lipids and Lipid Enzymes
  A. D. Bangham
Recent Developments in the Mechanism of Fat Absorption
  John M. Johnston
The Clearing Factor Lipase and Its Action in the Transport of Fatty Acids between the Blood and Tissues
  D. S Robinson
Vitamin E and Lipid Metabolism
  Roslyn B. Alfin-Slater and Rosemary Shull Morris
Atherosclerosis—Spontaneous and Induced
  Thomas B. Clarkson
Chromatographic Investigations in Fatty Acid Biosynthesis
  M. Pascaud
Carnitine and Its Role in Fatty Acid Metabolism
  Irving B. Fritz
Present Status of Research on Catabolism and Excretion of Cholesterol
  Henry Danielsson
The Plant Sulfolipid
  A. A. Benson
AUTHOR INDEX—SUBJECT INDEX

Volume 2
Triglyceride Structure
  R. J. VanderWal
Bacterial Lipids
  M. Kates
Phosphatidylglycerols and Lipoamino Acids
  Marjorie G. Macfarlane
The Brain Phosphoinositides
  J. N. Hawthorne, and P. Kemp
The Synthesis of Phosphoglycerides and Some Biochemical Applications
  L. L. M. van Deenen and G. H. DeHaas

The Lipolytic and Esterolytic Activity of Blood and Tissues and Problems of Atherosclerosis
  T. Zemplényi
Evaluation of Drugs Active against Experimental Atherosclerosis
  Robert Hess
Comparative Evaluation of Lipid Biosynthesis in Vitro and in Vivo
  P. Favarger
AUTHOR INDEX—SUBJECT INDEX

Volume 3
The Metabolism of Polyenoic Fatty Acids
  E. Klenk
The Analysis of Human Serum Lipoprotein Distributions
  Alicia M. Ewing, Norman K. Freeman, and Frank T. Lindgren
Factors Affecting Lipoprotein Metabolism
  Angelo M. Scanu
The Action of Drugs on Phospholipid Metabolism
  G. B. Ansell
Brain Sterol Metabolism
  A. N. Davison
Lipases
  E. D. Wills
AUTHOR INDEX—SUBJECT INDEX

Volume 4
The Role of Lipids in Blood Coagulation
  Aaron J. Murcus
Lipid Responses to Dietary Carbohydrates
  I. Macdonald
Effects of Catecholamines on Lipid Mobilization
  Max Wenke
The Polyunsaturated Fatty Acids of Microorganisms
  Robert Shaw
Lipid Metabolism in the Bacteria
  W. J. Lennarz

Quantitative Methods for the Study of Vitamin D
  Padmanabhan P. Nair
Labeling and Radiopurity of Lipids
  Fred Snyder and Claude Piantadosi
AUTHOR INDEX—SUBJECT INDEX

Volume 5
Fatty Acid Biosynthesis and the Role of the Acyl Carrier Protein
  Philip W. Majerus and P. Roy Vagelos
Comparative Studies on the Physiology of Adipose Tissue
  Daniel Rudman and Mario Di Girolamo
Ethionine Fatty Liver
  Emmanuel Farber
Lipid Metabolism by Macrophages and Its Relationship to Atherosclerosis
  Allan J. Day
Dynamics of Cholesterol in Rats, Studied by the Isotopic Equilibrium Methods
  F. Chevallier
The Metabolism of Myelin Lipids
  Marion Edmonds Smith
Brain Cholesterol: The Effect of Chemical and Physical Agents
  Jon J. Kabara
The Analysis of Individual Molecular Species of Polar Lipids
  Ossi Renkonen
Phase Diagrams of Triglyceride Systems
  J. B. Rossell
AUTHOR INDEX—SUBJECT INDEX

Volume 6
Practical Methods for Plasma Lipoprotein Analysis
  Frederick T. Hatch and Robert S. Lees
The Lipids of *Mycoplasma*
  Paul F. Smith
Lipid Quinones
  T. Ramasarma
Comparative Pathogenetic Patterns in Atherosclerosis
  Robert W. Wissler and Dragoslava Vesselinovitch
Chemistry and Metabolism of Bile Alcohols and Higher Bile Acids
  Takahiko Hoshita and Taro Kazuno
Hydroxy Fatty Acid Metabolism in Brain
  David M. Bowen and Norman S. Radin
Gas Phase Analytical Methods for the Study of Steroids
  E. C. Horning, C. J. W. Brooks, and W. J. A. Vanden Heuvel
AUTHOR INDEX—SUBJECT INDEX

Volume 7
Lipid Histochemistry
  C. W. M. Adams
Control of Plasma and Liver Triglyceride Kinetics by Carbohydrate Metabolism and Insulin
  Esko A. Nikkilä
Lipid Metabolism in Tissue Culture Cells
  George H. Rothblat
Carcinogenic Effects of Steroids
  Fritz Bischoff
The Chemical and Biologic Properties of Heated and Oxidized Fats
  Neil R. Artman
AUTHOR INDEX—SUBJECT INDEX

Volume 8
Cholesterol Turnover in Man
  Paul J. Nestel
Arterial Composition and Metabolism: Esterified Fatty Acids and Cholesterol
  Oscar W. Portman
The Essential Fatty Acids
  Michael Guarnieri and Ralph M. Johnson
Lipids in Membrane Development
  Godfrey S. Getz
Plant Phospholipids and Glycolipids
  M. Kates
Metabolism of Long-Chain Fatty Acids in the Rumen
  Romano Viviani
Surface Chemistry of Lipids
  Dinesh O. Shah
AUTHOR INDEX—SUBJECT INDEX

Volume 9
Light and Electron Microscopic Radioautography of Lipids: Techniques and Biological Applications
  O. Stein and Y. Stein
The Origin of Hydrogen in Fatty Synthesis
  Simonne Rous
Fatty Acid Biosynthesis in Aorta and Heart
  Arthur F. Whereat
Structure of Membranes and Role of Lipids Therein
  Frantz A. Vandenheuvel
Glycosphingolipids
  Herbert Weigandt
Biosynthesis of Pregnane Derivatives
  Shlomo Burstein and Marcel Gut
Lipid Composition of Vegetable Oils
  Enzo Fedeli and Giovanni Jacini
AUTHOR INDEX—SUBJECT INDEX

Contents of Previous Volumes

Volume 10

Application of Electron Microscopy to the Study of Plasma Lipoprotein Structure
Trudy Forte and Alex V. Nichols

Employment of Lipids in the Measurement and Modification of Cellular, Humoral, and Immune Responses
Nicholas R. Di Luzio

Microsomal Enzymes of Sterol Biosynthesis
James L. Gaylor

Brain Lipids
Robert B. Ramsey and Harold J. Nicholas

Enzymatic Systems That Synthesize and Degrade Glycerolipids Possessing Ether Bonds
Fred Snyder

Lipids in the Nervous System of Different Species as a Function of Age: Brain, Spinal Cord, Peripheral Nerve, Purified Whole Cell Preparations, and Subcellular Particulates: Regulatory Mechanisms and Membrane Structure
George Rouser, Gene Kritchevsky, Akira Yamamoto, and Claude F. Baxter
AUTHOR INDEX—SUBJECT INDEX

Volume 11

The Metabolic Role of Lecithin : Cholesterol Acyltransferase: Perspectives from Pathology
John A. Glomset and Kaare B. Norum

Lipoprotein-Polyanion-Metal Interactions
M. Burstein and H. R. Scholnick

Uptake and Transport of Fatty Acids into the Brain and the Role of the Blood Brain Barrier System
Govind A. Dhopeshwarkar and James F. Mead

Microbiological Transformation of Bile Acids
Shohei Hayakawa

Phytosterols
George A. Bean

Metabolism of Steroids in Insects
M. J. Thompson, J. N. Kaplanis, W. E. Robbins, and J. A. Svoboda

Lipids in Viruses
Herbert A. Blugh and John M. Tiffany
AUTHOR INDEX—SUBJECT INDEX

Volume 12

The Relationship Between Plasma and Tissue Lipids in Human Atherosclerosis
Elspeth B. Smith

Lipid Metabolism in Cultured Cells
Barbara V. Howard and William J. Howard

Effect of Diet on Activity of Enzymes Involved in Fatty Acid and Cholesterol Synthesis
Dale R. Romsos and Gilbert A. Leveille

Role of Phospholipids in Transport and Enzymatic Reactions
Beatrix Fourcans and Mahendra Kumar Jain

The Composition and Possible Physiologic Role of the Thyroid Lipids
Leon A. Lipshaw and Piero P. Foà

Glycosyl Glycerides
P. S. Sastry

Inhibition of Fatty Acid Oxidation by Biguanides: Implications for Metabolic Physiopathology
Sergio Muntoni
AUTHOR INDEX—SUBJECT INDEX

Volume 13

Lipoprotein Metabolism
Shlomo Eisenberg and Robert I. Levy

Diabetes and Lipid Metabolism in Nonhuman Primates
Charles F. Howard, Jr.

Biliary Lipids and Cholesterol Gallstone Formation
Oscar W. Portman, Toshiaki Osuga, and Naomi Tanaka

The Composition and Biosynthesis of Milk Fat
Stuart Smith and S. Abraham
AUTHOR INDEX—SUBJECT INDEX

Volume 14

Regulation of HMG-CoA Reductase
Victor W. Rodwell, Jeffrey L. Nordstrom, and Jonathan J. Mitschelen

Fatty Acid Activation: Specificity, Localization, and Function
P. H. E. Groot, H. R. Scholte, and W. C. Hülsmann

Polyene Antibiotic–Sterol Interaction
Anthony W. Norman, Ann M. Spielvogel, and Richard G. Wong

The Lipids of Plant Tissue Cultures
S. S. Radwan and H. K. Mangold

Synthesis of Some Acylglycerols and Phosphoglycerides
Robert G. Jensen and Robert E. Pitas
AUTHOR INDEX—SUBJECT INDEX

Volume 15
Long-Range Order in Biomembranes
Mahendra K. Jain and Harold B. White III
The Pharmacodynamics and Toxicology of Steroids and Related Compounds
Fritz Bischoff and George Bryson
Fungal Lipids
Momtaz K. Wassef
The Biochemistry of Plant Sterols
William R. Nes
AUTHOR INDEX—SUBJECT INDEX

Volume 16
Metabolism of Molecular Species of Diacylglycerophospholipids
B. J. Holub and A. Kuksis
Fatty Acids and Immunity
Christopher J. Meade and Jurgen Mertin
Marginal Vitamin C Deficiency, Lipid Metabolism, and Atherogenesis
Emil Ginter
Arterial Enzymes of Cholesteryl Ester Metabolism
David Kritchevsky and H. V. Kothari
Phospholipase D
Michael Heller
Screening for Inhibitors of Prostaglandin and Thromboxane Biosynthesis
Ryszard J. Gryglewski
Atherosclerosis, Hypothyroidism, and Thyroid Hormone Therapy
Paul Starr
AUTHOR INDEX—SUBJECT INDEX

Volume 17
Body Cholesterol Removal: Role of Plasma High-Density Lipoproteins
Alan R. Tall and Donald M. Small
High-Density Lipoprotein Metabolism
A. Nicoll, N. E. Miller, and B. Lewis
Cholesterol Metabolism in Clinical Hyperlipidemias
Harbhajan S. Sodhi, Bhalchandra J. Kudchodkar, and Dean T. Mason
On the Mechanism of Hypocholesterolemic Effects of Polyunsaturated Lipids
Ranajit Paul, C. S. Ramesha, and J. Ganguly
Lipid Peroxidation in Mitochondrial Membrane
Yu. A. Vladimirov, V. I. Olenev, T. B. Suslova, and Z. P. Cheremisina
Membrane Cooperative Enzymes as a Tool for the Investigation of Membrane Structure and Related Phenomena
Ricardo Norberto Farías
AUTHOR INDEX—SUBJECT INDEX

Volume 18
Techniques in Pathology in Atherosclerosis Research
Dragoslava Vesselinovitch and Katti Fischer-Dzoga
Effects of Hypolipidemic Drugs on Bile Acid Metabolism in Man
Tatu A. Miettinen
Cholesterol Metabolism by Ovarian Tissue
Jerome F. Straus III, Linda A. Schuler, Mindy F. Rosenblum, and Toshinobu Tanaka
Metabolism of Sulfolipids in Mammalian Tissues
Akhlaq A. Farooqui
Influence of Dietary Linoleic Acid Content on Blood Pressure Regulation in Salt-Loaded Rats (with Special Reference to the Prostaglandin System)
Peter Hoffmann and Werner Förster
The Role of Dietary Fiber in Lipid Metabolism
Jon A. Story
Current Techniques of Extraction, Purification, and Preliminary Fractionation of Polar Lipids of Natural Origin
A. V. Zhukov and A. G. Vereshchagin
AUTHOR INDEX—SUBJECT INDEX

Volume 19
The Interaction of Lipids and Lipoproteins with the Intercellular Matrix of Arterial Tissue: Its Possible Role in Atherogenesis
Germán Camejo
Apoprotein C Metabolism in Man
Paul J. Nestel and Noel H. Fidge
Lecithin: Cholesterol Acyltransferase and Intravascular Cholesterol Transport
Yves L. Marcel
Development of Bile Acid Biogenesis and Its Significance in Cholesterol Homeostasis

M. T. R. Subbiah and A. S. Hassan
Biosynthesis and Transport of
Phosphatidylserine in the Cell
   Jolanta Barańska
Analysis of Prostanoids by GC/MS
Measurement

   Christine Fischer and Jürgen C. Frölich
Morphological Aspects of Dietary Fibers
in the Intestine
   *Marie M. Cassidy, Fred G. Lightfoot, and
   George V. Vahouny*
AUTHOR INDEX—SUBJECT INDEX

202837